VOLUME EIGHTY NINE

CURRENT TOPICS IN
DEVELOPMENTAL BIOLOGY

Tissue Remodeling and
Epithelial Morphogenesis

## Series Editor

Paul M. Wassarman
Department of Developmental and Regenerative Biology
Mount Sinai School of Medicine
New York, NY 10029-6574
USA

Olivier Pourquié
Howard Hughes Medical Institute
Stowers Institute for Medical Research
Kansas City, MO 64110-2262
USA

## Editorial Board

Blanche Capel
Duke University Medical Center
Durham, NC, USA

B. Denis Duboule
Department of Zoology and Animal Biology
NCCR 'Frontiers in Genetics'
Geneva, Switzerland

Anne Ephrussi
European Molecular Biology Laboratory
Heidelberg, Germany

Janet Heasman
Cincinnati Children's Hospital Medical Center
Department of Pediatrics
Cincinnati, OH, USA

Julian Lewis
Vertebrate Development Laboratory
Cancer Research UK London Research Institute
London WC2A 3PX, UK

Yoshiki Sasai
Director of the Neurogenesis and Organogenesis Group
RIKEN Center for Developmental Biology
Chuo, Japan

Cliff Tabin
Harvard Medical School
Department of Genetics
Boston, MA, USA

## Founding Editors

**A. A. Moscona**
**Alberto Monroy**

VOLUME EIGHTY NINE

CURRENT TOPICS IN
DEVELOPMENTAL BIOLOGY

# Tissue Remodeling and Epithelial Morphogenesis

Edited by

THOMAS LECUIT
Institut de Biologie du Développement de Marseille-Luminy
(IBDML)UMR6216 CNRS-Université de la Méditerranée
Campus de Luminy
Marseille, France

ELSEVIER

AMSTERDAM • BOSTON • HEIDELBERG • LONDON
NEW YORK • OXFORD • PARIS • SAN DIEGO
SAN FRANCISCO • SINGAPORE • SYDNEY • TOKYO
Academic Press is an imprint of Elsevier

Academic Press is an imprint of Elsevier
525 B Street, Suite 1900, San Diego, CA 92101-4495, USA
30 Corporate Drive, Suite 400, Burlington, MA 01803, USA
32, Jamestown Road, London NW1 7BY, UK
Linacre House, Jordan Hill, Oxford OX2 8DP, UK

First edition 2009

Copyright © 2009 Elsevier Inc. All rights reserved.

No part of this publication may be reproduced, stored in a retrieval system
or transmitted in any form or by any means electronic, mechanical, photocopying,
recording or otherwise without the prior written permission of the publisher

Permissions may be sought directly from Elsevier's Science & Technology Rights
Department in Oxford, UK: phone (+44) (0) 1865 843830; fax (+44) (0) 1865 853333;
email: permissions@elsevier.com. Alternatively you can submit your request online by
visiting the Elsevier web site at http: //elsevier.com/locate/permissions, and selecting
*Obtaining permission to use Elsevier material*

Notice

No responsibility is assumed by the publisher for any injury and/or damage to persons
or property as a matter of products liability, negligence or otherwise, or from any use
or operation of any methods, products, instructions or ideas contained in the material
herein. Because of rapid advances in the medical sciences, in particular, independent
verification of diagnoses and drug dosages should be made

ISBN: 978-0-12-374902-4
ISSN: 0070-2153

For information on all Academic Press publications
visit our website at elsevierdirect.com

Printed and bound in USA
09  10  11  12  13      9  8  7  6  5  4  3  2  1

Working together to grow
libraries in developing countries

www.elsevier.com | www.bookaid.org | www.sabre.org

ELSEVIER    BOOK AID International    Sabre Foundation

# Contents

| | |
|---|---|
| *Contributors* | *ix* |
| *Preface* | *xi* |

**1. Intercellular Adhesion in Morphogenesis: Molecular and Biophysical Considerations**    **1**

Nicolas Borghi and W. James Nelson

| | | |
|---|---|---|
| 1. | Introduction | 2 |
| 2. | Molecular Basis of Intercellular Adhesion | 4 |
| 3. | Modulation of Intercellular Adhesion in Development | 10 |
| 4. | Intercellular Adhesion Energy and Morphogenesis | 13 |
| 5. | Conclusions | 24 |
| | Acknowledgments | 25 |
| | References | 26 |

**2. Remodeling of the Adherens Junctions During Morphogenesis**    **33**

Tamako Nishimura and Masatoshi Takeichi

| | | |
|---|---|---|
| 1. | Introduction | 34 |
| 2. | Basic Machinery of the Adherens Junction | 36 |
| 3. | Remodeling by Small GTPase | 37 |
| 4. | Remodeling by Cadherin Turnover and Endocytosis | 39 |
| 5. | Remodeling by Nonclassic Cadherins and Nectins | 41 |
| 6. | Junctional Remodeling During Morphogenesis | 44 |
| 7. | Perspectives | 48 |
| | Acknowledgments | 49 |
| | References | 49 |

**3. How the Cytoskeleton Helps Build the Embryonic Body Plan: Models of Morphogenesis from *Drosophila***    **55**

Tony J. C. Harris, Jessica K. Sawyer, and Mark Peifer

| | | |
|---|---|---|
| 1. | Introduction | 56 |
| 2. | Establishing Epithelial Structure | 57 |
| 3. | Epithelial Morphogenesis | 63 |

|  |  |  |
|---|---|---|
| 4. | Coordinating Cytoskeletal Machinery in Epithelial Cells | 78 |
| 5. | Concluding Statement | 79 |
|  | References | 80 |

## 4. Cell Topology, Geometry, and Morphogenesis in Proliferating Epithelia     87

William T. Gibson and Matthew C. Gibson

|  |  |  |
|---|---|---|
| 1. | Introduction | 88 |
| 2. | Conservation of Epithelial Architecture | 89 |
| 3. | Introduction to Cellular Topology | 92 |
| 4. | Conservation of Topological Structure in Proliferating Epithelia | 92 |
| 5. | Topological Inference in Epithelia: Maximum Entropy Methods | 94 |
| 6. | Topological Models: The Simplest Models of Epithelia | 95 |
| 7. | The Smallest Geometrical Model: Cleavage Plane Orientation in a Single Cell | 97 |
| 8. | Nongeometric Mechanisms of Division Orientation: A Larger Morphogenetic Space | 99 |
| 9. | Scaling Up: Geometrical Models and Cellular Mechanics in Proliferating Epithelia | 100 |
| 10. | Putting It All Together: Genetics, Geometry, and Biophysics | 104 |
| 11. | Future Directions | 108 |
| 12. | Conclusion | 109 |
|  | References | 110 |

## 5. Principles of *Drosophila* Eye Differentiation     115

Ross Cagan

|  |  |  |
|---|---|---|
| 1. | Introduction | 116 |
| 2. | The Developing *Drosophila* Eye | 116 |
| 3. | Patterning the *Drosophila* Eye: Morphogenesis and Cell Movements | 124 |
| 4. | Conclusion | 131 |
|  | Acknowledgments | 132 |
|  | References | 132 |

## 6. Cellular and Molecular Mechanisms Underlying the Formation of Biological Tubes     137

Magdalena M. Baer, Helene Chanut-Delalande, and Markus Affolter

|  |  |  |
|---|---|---|
| 1. | Introduction | 137 |
| 2. | Architecture and Mechanisms of Formation of Tubular Structures | 138 |
| 3. | Molecular Basis of Cellular Aspects of Tube Formation | 139 |

|     | 4. Summary and Perspectives | 156 |
| --- | --- | --- |
|     | 5. Open Questions | 158 |
|     | Acknowledgments | 159 |
|     | References | 159 |

## 7. Convergence and Extension Movements During Vertebrate Gastrulation — 163

Chunyue Yin, Brian Ciruna, and Lilianna Solnica-Krezel

|     |     |     |
| --- | --- | --- |
| 1.  | Introduction | 164 |
| 2.  | The Regional and Temporal Pattern of C&E Movements in the Mesoderm | 165 |
| 3.  | Diverse Cellular Behaviors Underlie the Regional Differences of C&E Movements | 166 |
| 4.  | Molecular Regulation of C&E Movements | 172 |
| 5.  | Cell Polarization During C&E Movements | 183 |
| 6.  | Conclusion | 185 |
|     | Acknowledgments | 186 |
|     | References | 186 |

| *Index* | *193* |
| --- | --- |
| *Contents of Previous Volumes* | *201* |

# Contributors

**Markus Affolter**
Biozentrum der Universität Basel, Klingelbergstrasse, Basel, Switzerland

**Magdalena M. Baer**
Biozentrum der Universität Basel, Klingelbergstrasse, Basel, Switzerland

**Nicolas Borghi**
Department of Biology, Stanford University, Stanford, California, USA

**Ross Cagan**
Department of Developmental and Regenerative Biology, Mount Sinai School of Medicine, New York, NY, USA

**Helene Chanut-Delalande**
Biozentrum der Universität Basel, Klingelbergstrasse, Basel, Switzerland, and Centre de Biologie du Développement, CNRS UMR5547, Toulouse, France

**Brian Ciruna**
Program in Developmental and Stem Cell Biology, The Hospital for Sick Children, Toronto, Ontario, Canada, and the Department of Molecular Genetics, University of Toronto, Ontario, Canada

**Matthew C. Gibson**
The Stowers Institute for Medical Research, Kansas City, Missouri, USA

**William T. Gibson**
Program in Biophysics, Harvard University, Cambridge, Massachusetts, USA

**Tony J. C. Harris**
Department of Cell and Systems Biology, University of Toronto, Toronto, Ontario, Canada

**W. James Nelson**
Department of Biology, and Department of Molecular and Cellular Physiology, Stanford University, Stanford, California, USA

**Tamako Nishimura**
RIKEN Center for Developmental Biology, Chuo ku, Kobe, Japan

**Mark Peifer**
Department of Biology, and Lineberger Comprehensive Cancer Center, University of North Carolina at Chapel Hill, Chapel Hill, North Carolina, USA

**Jessica K. Sawyer**
Department of Biology, University of North Carolina at Chapel Hill, Chapel Hill, North Carolina, USA

**Lilianna Solnica-Krezel**
Department of Biological Sciences, Vanderbilt University, Nashville, Tennessee, USA

**Masatoshi Takeichi**
RIKEN Center for Developmental Biology, Chuo-ku, Kobe, Japan

**Chunyue Yin**
Department of Biochemistry and Biophysics, Program in Developmental Biology, Genetics, and Human Genetics, University of California at San Francisco, California, USA

# Preface

Understanding how biological tissues such as epithelia change shape, grow, and ultimately acquire defined proportions is one of the most fundamental and fascinating open questions in biology. Embryos and organs acquire characteristic structures and size. In a number of model organisms, tissue remodeling is driven by a limited set of cell morphogenetic behaviors, such as cell constriction for tissue invagination, and cell intercalation for tissue elongation. Distinct signaling pathways regulate in space and time these behaviors through the control of the actin cytoskeleton and of myosin-based tension required for cell shape changes. Yet, the physical properties of cells and tissues are in general poorly understood, especially in living, developing animals. These properties are expected to constrain how signaling pathways can control morphogenesis. An important challenge is thus to characterize cell and tissue physical properties, to dissect their molecular basis, in order to ultimately understand the interplay between tissue mechanics and genetic regulation of cell behavior.

Epithelial tissues form robust barriers to chemical diffusion between different physiological compartments and protect the organisms against microorganisms. This robustness is key to maintaining a polarized organization and ensuring tissue architecture both during embryonic and organ development and in the adult. However, epithelia are not static structures and display a remarkable plasticity, that is, the capacity to adapt to intrinsic or extrinsic signals or perturbations. For instance, they are extensively remodeled as they acquire their three-dimensional shape during morphogenesis. Understanding how epithelia maintain a proper balance between robustness and plasticity is a major challenge. Disruptions in this equilibrium mark key steps in solid tumor progression and other diseases.

The correct balance between robustness and plasticity relies on unique mechanical properties of cell contacts. For a long time, these properties have been largely, if not exclusively, considered from the perspective of cell–cell adhesion by cadherin molecules which engage in homophilic or heterophilic complexes. The strength of cadherin association controls the extent of cell aggregation and, thus, the robustness of epithelial tissues. In the context of the *D*ifferential *A*dhesion *H*ypothesis (DAH) proposed by Steinberg, differences of affinities between cadherins expressed on different cells can explain certain dynamic processes and remodeling events such as cell sorting, drawing a simple analogy between tissues and viscous fluids on long time scales. The DAH provides a powerful framework to explain tissue envelopment behaviors and certain developmental processes.

However, the central role of adhesion has largely overshadowed other important contributions to the mechanical properties of cell contacts, in particular force generation by acto-myosin networks and force transmission at the cell cortex by the same molecules that ensure adhesion, namely cadherins. Force generation emerges from the interaction and movement of actin filaments (F-actin) by Myosin-II (Myo-II) motors. Robust interactions between tensile acto-myosin networks and E-cadherin (E-cad) unsure that tension is transmitted properly at the cell cortex and drives cell shape changes, such as cell constriction or polarized junction remodeling. The reversible nature of actin and cadherin bonds on the one hand, and of cadherin–cadherin interactions within homophilic complexes on the other, must also be invoked to account for the complete remodeling of cell contacts and tissue shape changes.

This volume provides a broad coverage of key findings and concepts on the mechanisms of tissue morphogenesis. The chapters emphasize the fact that, today more than ever before, tissue morphogenesis is tackled from different perspectives and is a particularly active and multidisciplinary field of investigation, at the crossroad of cell biology, developmental biology in model organisms, physics, and even mathematics. Since the core molecular mechanisms underlying adhesion and cytoskeletal regulation are very strongly conserved, we learn much from studies in both invertebrates (e.g., worms, flies) and vertebrates (zebrafish, mouse, humans). One of the major challenges of morphogenesis is to connect different scales of description and understanding of biological processes, from multimolecular complexes and machines to tissue level dynamics. This volume will also reflect this diversity of scales of description of tissue morphogenesis.

The first three chapters cover basic cell biological mechanisms of epithelial organization and dynamics in a range of different organisms.

In Chapter 1, "Intercellular Adhesion in Morphogenesis: Molecular and Biophysical Considerations," Borghi and Nelson present a detailed historical perspective on intercellular adhesion, reviewing the molecular mechanisms of adhesion *per se*, and how adhesion energy can be used to describe morphogenetic processes.

Nishimura and Takeichi address how adherens junctions are remodeled during development in Chapter 2, "Remodeling of Adherens Junctions During Morphogenesis." The role and regulation of different kinds of cadherins and junctional proteins are reviewed.

In Chapter 3, Harris *et al.* review extensively how *Drosophila melanogaster* can be used to reveal core mechanisms underlying epithelial formation, maintenance, and dynamics during morphogenesis: "How the Cytoskeleton Helps Build the Embryonic Body Plan: Models of Morphogenesis from *Drosophila*."

Gibson and Gibson then cover in Chapter 4 a so far little appreciated aspect of epithelial organization, namely the influence of cell division Mathematical descriptions of multicellular packing are reviewed: "Cell Topology, Geometry, and Morphogenesis in Proliferating Epithelia."

Chapters 5–7 address how multicellular interactions drive organ formation and embryogenesis. The formation of complex three-dimensional organs relies on hierarchical levels of cell regulation. Cagan reviews in Chapter 5 how the fly eye is formed: "Principles of *Drosophila* Eye Differentiation."

In Chapter 6, Baer *et al.* review extensively the mechanisms of tube formation in different organisms: "Cellular and Molecular Mechanisms Underlying the Formation of Biological Tubes."

Finally, Yin *et al.* present a broad review of "Convergence and Extension Movements During Vertebrate Gastrulation," with a special focus on cell motility.

It is hoped that this will prove useful to those discovering this active field of research as well as to others who wish to read recent advance in the mechanisms of tissue morphogenesis.

THOMAS LECUIT, PH.D.
Marseille

CHAPTER ONE

# INTERCELLULAR ADHESION IN MORPHOGENESIS: MOLECULAR AND BIOPHYSICAL CONSIDERATIONS

Nicolas Borghi* and W. James Nelson*,†

## Contents

| | |
|---|---|
| 1. Introduction | 2 |
| 2. Molecular Basis of Intercellular Adhesion | 4 |
|    2.1. Formation of *trans*-binding interactions by classical cadherins | 4 |
|    2.2. Intracellular signaling by cadherins | 6 |
|    2.3. Cytoskeletal responses | 7 |
|    2.4. Cell interfacial tension response and the rise of adhesion energy | 8 |
| 3. Modulation of Intercellular Adhesion in Development | 10 |
|    3.1. Modulation of cadherin–catenin membrane levels | 10 |
|    3.2. Modulation of cortical cytoskeleton organization and response | 11 |
|    3.3. Differential expression of cadherin subtypes | 12 |
| 4. Intercellular Adhesion Energy and Morphogenesis | 13 |
|    4.1. Intercellular adhesion energy and tissue surface tension | 13 |
|    4.2. Cell sorting and tissue envelopment | 15 |
|    4.3. Origins of heterotypic adhesion energy | 16 |
|    4.4. Cadherin expression levels versus cytoskeleton remodeling | 18 |
|    4.5. Intercellular adhesion energy and polarity | 19 |
|    4.6. Cell interfacial tension refinements | 20 |
|    4.7. Loss of tissue liquid-like behavior | 21 |
|    4.8. Dynamics | 22 |
| 5. Conclusions | 24 |
| Acknowledgments | 25 |
| References | 26 |

* Department of Biology, Stanford University, Stanford, California, USA
† Department of Molecular and Cellular Physiology, Stanford University, Stanford, California, USA

**Abstract**

A major challenge in developmental biology is to understand how cellular processes that result from expression of the genetic program determine the material properties and shape transformations of tissues during morphogenesis. Cell/cell adhesion is critical in development, and it controls many aspects of tissue rearrangements that support morphogenesis. Intercellular adhesion not only allows cells to adhere together but also supports structure and function compartmentalization on the scale of cell assemblies, tissues, and organs. In metazoans, cadherins comprise a major class of cell/cell adhesion proteins. They form $Ca^{2+}$-dependent, homophilic adhesive contacts between neighboring cells that results in remodeling of the underlying cortical cytoskeleton with consequential changes in mechanical properties of cells. During development, programmed cues modulate cadherin levels and subtype expression, and downstream signaling to the cortical cytoskeleton resulting in a wide continuum of adhesive properties. A quantitative output from cell/cell adhesion is intercellular adhesion energy, which as a critical determinant of cell shape and position within the tissue, and tissue shape and position in the organism. We discuss molecular mechanisms underlying intercellular adhesion energy and its role in tissue morphogenesis.

## 1. INTRODUCTION

Cell/cell adhesion appeared concomitantly with the evolution of multicellular metazoans and is an essential, characteristic feature of organized multicellular life. In metazoans, cell/cell adhesion enabled the formation of epithelial tissues that separated different biological compartments—in its simplest form, the outside environment from the inside of the organism. This compartmentalization allowed organisms to regulate ionic homeostasis by vectorial transport of ions and solutes across a continuous epithelial cell barrier.

The life cycle of metazoans requires that an organism rebuilds itself from a single cell at each generation. During development, there are extensive cell and multicellular (tissue) reorganizations that require fine modulation of intercellular adhesion without complete dissociation of individual cells, including changes in overall tissue shape, tissues repositioning themselves relative to one another, complex cell rearrangements within groups of cells, and loosening of intercellular cohesion and cell delamination. A further complication is that these processes are not mutually exclusive. To maintain tissue cohesion, while allowing cell rearrangements, the strength of intercellular adhesion is controlled in a graded manner (i.e., it is not all or nothing), and this gradation is tightly regulated and can span a significant range that enables either subtle or more dramatic changes in cell or tissue shape. In extreme cases, intercellular adhesion is completely downregulated resulting in cohesive cells dissociating from each other and migrating as single cells, a process termed epithelial-to-mesenchyme transition (EMT) (Thiery and Sleeman, 2006).

The role of intercellular adhesion in multicellular organization has been intensively studied for many decades. Over 50 years ago, Townes and Holtfreter (1955) showed that cells from enzymatically dissociated embryos retained the ability to sort out *in vitro* according to their original organization in a fashion reminiscent of an immiscible fluid demixion. Steinberg (1963) later proposed that the gradation of intercellular adhesion energies between cells of the same and different types determined if, and how, cells sorted out from each other. Further *in vitro* experiments with embryo explants validated the idea that tissues have liquid-like properties and that their surface tension—a measure of intercellular adhesion energy between cells—was predicted by the sorting outcome when different tissues were mixed together (Foty *et al.*, 1996).

Molecular control of intercellular adhesion involves a large number of different adhesion proteins. However, the evolutionarily conserved cadherin superfamily of adhesion proteins is thought to be fundamental in morphogenesis. The complex developmental pattern of expression of different cadherin subtypes coupled with differences in their levels of expression account for differences in adhesion energies involved in sorting between different cell types (Takeichi, 1988). While contacts between the extracellular domain of cadherins on opposing cell surfaces control many aspects of intercellular adhesion, cadherin engagement is also accompanied by the reorganization of the cortical cytoskeleton (Mege *et al.*, 2006). This cytoskeletal response is essential for the rise of significant adhesion energy between cells (Chu *et al.*, 2004). Despite recent insight into how actin cytoskeleton architecture and dynamics are locally controlled by the cadherin–catenin adhesion complex (Drees *et al.*, 2005; Yamada *et al.*, 2005) and evidence for an important role of local control of actomyosin contractile activity in tissue remodeling during development (Lecuit and Lenne, 2007; Rauzi *et al.*, 2008), the contribution of cytoskeleton remodeling to intercellular adhesion is poorly understood.

The role of cadherins in intercellular adhesion energy and cell sorting has been confirmed in studies with reconstituted cell aggregates *in vitro* and to a lesser extent *in vivo* (Foty and Steinberg, 2005; Godt and Tepass, 1998; Hayashi and Carthew, 2004; Kafer *et al.*, 2007; Steinberg and Takeichi, 1994). However, *in vitro* binding assays between cadherin-coated surfaces and some *in vitro* cell sorting assays in which the level of cadherin expression was carefully controlled challenged the contribution of cadherin subtype-binding specificity in intercellular adhesion energy (Foty and Steinberg, 2005; Niessen and Gumbiner, 2002). More recently, the contribution of differences in cadherin expression levels in cell sorting in embryo explants has been questioned (Krieg *et al.*, 2008), which promoted the idea that adhesion energy was not the graded output that determined cell sorting in tissue morphogenesis (Ewald and Wallingford, 2008; Hammerschmidt and Wedlich, 2008; Krieg *et al.*, 2008; Montell, 2008; Schwartz and DeSimone, 2008).

Taken together, these observations prompt a clarification of intercellular adhesion energy definition to understand its relevance and its limitations in tissue morphogenesis. Therefore, this review will provide a brief summary of molecular mechanisms involved in intercellular adhesion, followed by an analysis of different mechanisms that modulate adhesion during development and how they may control-through spatial and temporal patterning of adhesion energy-tissue morphogenesis.

## 2. MOLECULAR BASIS OF INTERCELLULAR ADHESION

Though intercellular adhesion does not rely exclusively on cadherins, this protein family is the best understood of intercellular adhesion proteins, and plays a major role in adhesion-dependent morphogenetic processes. Cadherins were first identified as proteins responsible for intercellular adhesion during vertebrate development (Gallin *et al.*, 1983; Peyrieras *et al.*, 1983; Vestweber and Kemler, 1984; Yoshida and Takeichi, 1982), and were shown subsequently to be critical for development of invertebrates, and homologues have been found in simple organisms such as sponges and premetazoan single-celled choanoflagellates (Abedin and King, 2008; King *et al.*, 2003). Cadherins represent a large family of proteins comprising classical cadherins, protocadherins, and atypical cadherins involved in planar cell polarity (Halbleib and Nelson, 2006); here, we focus on classical cadherins that are the best described of the cadherin subfamilies. Classical cadherins are widely expressed in most cohesive tissues and comprise different subtypes (Halbleib and Nelson, 2006). They are transmembrane proteins—the extracellular domain forms homophilic interactions between neighboring cells (*trans*-binding), and the cytoplasmic domain signals cytoskeleton remodeling through a variety of cytoplasmic binding partners (Fig. 1.1).

### 2.1. Formation of *trans*-binding interactions by classical cadherins

The extracellular domain of classical cadherins is composed of five repeat domains. *Trans*-cadherin binding is mediated by the N-terminal repeat EC1 (Chen *et al.*, 2005; Tamura *et al.*, 1998). Force measurements using atomic force microscopy (AFM) between tethered cadherin extracellular domains also suggested a more extensive *trans*-binding between additional EC repeat domains, including an overlap between all five EC domains

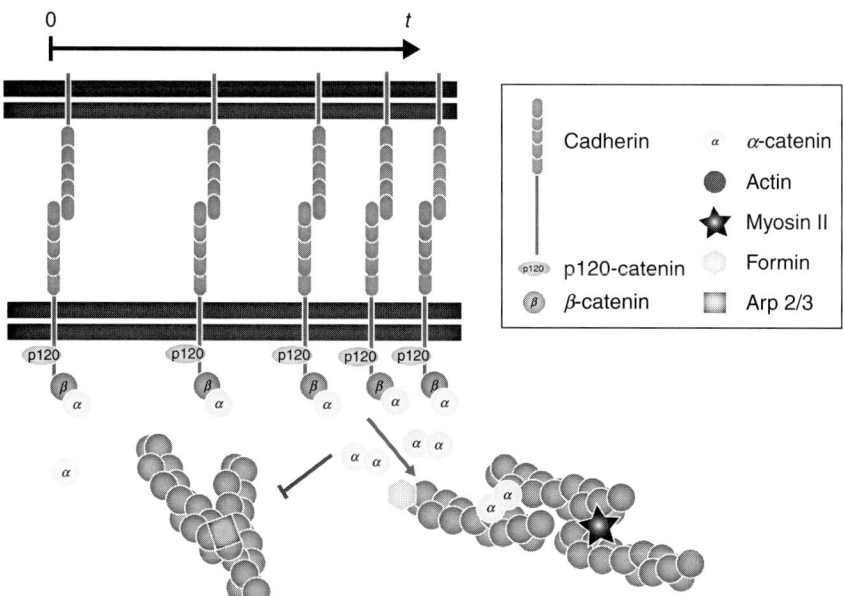

**Figure 1.1** Cadherin-mediated adhesion. Cadherin-mediated adhesion is a two-sided coin. On the extracellular side, cadherins engage homophilic bonds between adjacent cells. On the cytoplasmic side, they bind catenins that trigger cytoskeleton remodeling. As the intercellular contact matures (*t*), the concentration of cadherin/catenin complexes may allow α-E-catenin homodimerization. In turn, α-E-catenin dimer may bundle actomyosin filaments and recruit the actin nucleator formin, and prevent actin branching by competing with Arp2/3 (Drees *et al.*, 2005; Yamada *et al.*, 2005). (See Color Insert.)

(Chappuis-Flament *et al.*, 2001; Sivasankar *et al.*, 2001; Zhu *et al.*, 2003). X-ray crystallography and electron tomography of cadherins also support the existence of *cis*-dimers (Al-Amoudi *et al.*, 2007; Boggon *et al.*, 2002; He *et al.*, 2003) and a widely accepted model is that *cis*-dimerization is a prerequisite for *trans*-binding (Alberts, 2002). It was thought that *cis*-dimerization could induce lateral clustering (Chen *et al.*, 2005) which together with more extensive overlap in *trans*-dimerization along the length of the extracellular domain could strengthen the adhesive bond, which is weak between individual cadherins (Haussinger *et al.*, 2004). However, a recent study using single molecule FRET and AFM force measurements revealed that *cis*-dimerization is extremely slow and is dispensable for EC1 *trans*-dimerization, although increased density (*cis*-proximity) of cadherin extracellular domains enhanced *trans*-binding affinity threefold (Zhang *et al.*, 2009).

The extracellular domain of cadherins may also regulate cadherin subtype *trans*-binding specificity. There is strong evidence that the EC1 repeat domain regulates intercellular adhesion specificity *in vitro* (cell aggregation with exogenous cadherin expression; Nose et al., 1990) and *in vivo* (sorting of motor neurons in chick embryos; Patel et al., 2006). That the EC1 repeat domain is critical for *trans*-dimerization suggests that specificity of cadherin-subtype adhesion could be based on differences in binding affinity between homotypic and heterotypic EC1 interactions. However, a lack of subtype specificity has been observed in some *in vitro* binding assays involving cadherin-coated beads or surfaces (Niessen and Gumbiner, 2002). It is possible that EC1-mediated homotypic specificity arise only from an increase of *trans*-binding affinity upon lateral clustering at intercellular contacts (Chen et al., 2005).

## 2.2. Intracellular signaling by cadherins

The cytoplasmic domain of classical cadherins is highly conserved among different subtypes and directly binds to the cytoplasmic proteins p120 (Thoreson et al., 2000) and $\beta$-catenin. P120 regulates cytoskeleton dynamics through direct modulation of the Rho GTPase family proteins activity (Anastasiadis, 2007) by inhibiting RhoA and activating Rac1 and Cdc42 (Anastasiadis et al., 2000; Grosheva et al., 2001; Noren et al., 2000). Disruption of cadherin–p120 binding impairs Rac1 recruitment and activation at nascent intercellular contacts (Gavard et al., 2004). In addition, p120 binding to cadherins controls the stability and retention of cadherins at the plasma membrane by inhibiting cadherin endocytosis (see below) (Davis et al., 2003; Ireton et al., 2002; Thoreson et al., 2000).

$\beta$-Catenin binds directly to another cytoplasmic protein $\alpha$-catenin (Aberle et al., 1994), which, in turn, binds and bundles actin filaments and interacts with other actin partners (see below). Although indirect evidence supported a link, through the $\alpha/\beta$-catenin complex, between cadherins and the actin cytoskeleton, $\alpha$-E-catenin itself does not directly link cadherins and actin filaments (Drees et al., 2005; Yamada et al., 2005). *In vitro* binding studies with purified proteins showed that $\alpha$-E-catenin does not bind simultaneously to the cadherin/$\alpha$-catenin complex and F-actin. This mutually exclusive binding is explained by allosteric regulation of $\alpha$-E-catenin binding affinities by $\alpha$-E-catenin monomers and homodimers; $\alpha$-E-catenin monomers preferentially bind the cadherin/$\alpha$-catenin complex, whereas homodimers preferentially bind actin filaments and inhibit Arp2/3-mediated actin polymerization (Drees et al., 2005). It was proposed that $\alpha$-E-catenin homodimers formed locally from $\alpha$-E-catenin monomers at sites of cell–cell adhesion at which the concentration of cadherin/$\beta$-catenin/$\alpha$-E-catenin monomers is high (Fig. 1.1). However, the existence of high concentrations of $\alpha$-E-catenin homodimers at intercellular contact remains to be demonstrated. Indeed a theoretical analysis posited that, at steady state, the equilibrium between bound and unbound monomeric $\alpha$-E-catenin at the membrane may not suffice to locally increase monomer concentration and induce homodimerization (Dawes, 2009). A local increase in α-E-catenin dimer concentration would be possible if dimerization occurs from a specific monomer pool found at the membrane only, or dimerization kinetics from unbound monomer are increased in the vicinity of the membrane due to topological constraints. Further studies in cells are required to resolve these questions.

It remains possible that α-E-catenin interacts with other actin regulators to physically bind cadherins to the actin cytoskeleton. α-E-Catenin recruits another actin regulator, Formin-1, which favors unbranched actin nucleation (Kobielak et al., 2004) and Eplin, an inhibitor of actin depolymerization (Abe and Takeichi, 2008). Eplin binds simultaneously to the cadherin–catenin complex, through α-E-catenin, and F-actin *in vitro*, and localizes to intercellular contacts. However, Eplin can regulate actin organization at intercellular junctions in the absence of α-E-catenin (Abe and Takeichi, 2008) indicating that it may have a function that is independent of the cadherin–catenin complex. Even if α-E-catenin is not involved in a molecular link between cadherins and the actin cytoskeleton, its actin-binding ability may provide the actin cytoskeleton cortex an architecture that promote restricted lateral mobility of cadherin *trans*-dimer complexes at intercellular contacts (Cavey et al., 2008).

Although the mechanisms that link the cadherin–catenin complex to the underlying actin cytoskeleton remain poorly understood, tethering of the cytoskeleton to intercellular contacts is thought to be required for the control of intercellular adhesion by the cytoskeleton during tissue morphogenesis. Note that many developmental studies generalize cell–cell adhesion as a function of the "adherens junction," which contains not only cadherins, but also other intercellular adhesion proteins, such as nectins that bind the cytoplasmic protein afadin, a binding partner of actin and α-E-catenin (Pokutta et al., 2002; Sato et al., 2006; Takai and Nakanishi, 2003), or vezatin that links Myosin VII to the cadherins (Kussel-Andermann et al., 2000). Even though detailed molecular and physical description of mechanotransduction at intercellular contacts requires further investigation, cadherin-mediated adhesion is able to trigger a significant increase of membrane–cytoskeleton mechanical coupling, as evidenced by membrane tube extrusion (Tabdanov et al., 2009). Furthermore, cells are able to exert forces at cadherin-mediated adhesion sites that have a magnitude comparable to that at sites of integrin-mediated adhesion to the extracellular matrix (Dzamba et al., 2009; Ganz et al., 2006).

## 2.3. Cytoskeletal responses

The establishment of intercellular adhesion, from the early steps of cell–cell recognition to the formation of a mature intercellular junction, involves dramatic changes in the cortical cytoskeleton. Cadherin-dependent initial intercellular contacts between migrating simple epithelial cells in tissue culture result in a local increase in Rac1-mediated lamellipodial activity and a concomitant disruption of the cortical belt of bundled actin (Adams *et al.*, 1998; Ehrlich *et al.*, 2002). Rac1 and lamellipodial activities are transient, however, and decrease at established intercellular contacts (Ehrlich *et al.*, 2002; Perez *et al.*, 2008), but are reinitiated as new cell/cell contacts are formed at the periphery of the spreading intercellular adhesion zone (Yamada and Nelson, 2007). Concomitantly, RhoA is activated at the edges of the adhesion zone, where it promotes actomyosin contraction and contact expansion (Yamada and Nelson, 2007). Local regulation of Rac1 and RhoA activities may be through mechanisms involving p120, α-E-catenin or other proteins associated with the cadherin complex (see above).

Further maturation of intercellular contacts depends on specific tissue differentiation. In simple epithelia, E-cadherin-mediated adhesion sites become organized into a structure termed the adherens junction, which is localized in a ring around the cell at the apex of the lateral membrane. The actin cytoskeleton forms a ring of bundled filaments located close to the adherens junction. Different isoforms of myosins play roles in the proper organization of the adherens junctions, both at the level of recruitment of the cadherin/catenin complex and in the organization of the cortical actin cytoskeleton (Conti *et al.*, 2004; Ivanov *et al.*, 2005; Shewan *et al.*, 2005; Sousa *et al.*, 2005). Hence, in addition to their potential role in mechanotransduction (Kussel-Andermann *et al.*, 2000), myosins play multiple synergistic roles in organizing intercellular adhesions. The microtubule (MT) cytoskeleton also interacts with the adherens junction through MT minus and plus ends (Chausovsky *et al.*, 2000; Meng *et al.*, 2008; Stehbens *et al.*, 2006; Waterman-Storer *et al.*, 2000). MT anchoring to the adherens junction may depend on β-catenin binding to dynein (Ligon *et al.*, 2001) or p120 binding to two novel proteins Plekha7 and Nezha (Meng *et al.*, 2008).

## 2.4. Cell interfacial tension response and the rise of adhesion energy

Changes in the structure of the cell surface and underlying cortical cytoskeleton upon intercellular adhesion can be mechanically described as differences in cell interfacial tensions (the energy required to extend the cell surface of a unit area) between contacting and noncontacting cell surfaces. Two adhering cells exhibit lower cell interfacial tension at the intercellular interface ($\sigma_{cc}$) than at the free surface contacting the outer medium ($\sigma_{cm}$). Intercellular adhesion energy $W_{cc}$ is defined as the work per unit area to separate the two cells, or the energy release associated with the loss of a unit area of free cell surface to the benefit of half a unit of cell/cell interface: $W_{cc} = \sigma_{cm} - \sigma_{cc}/2$ (Fig. 1.2A); the greater the difference between cell/cell and cell/non-cell interfacial tensions, the higher the

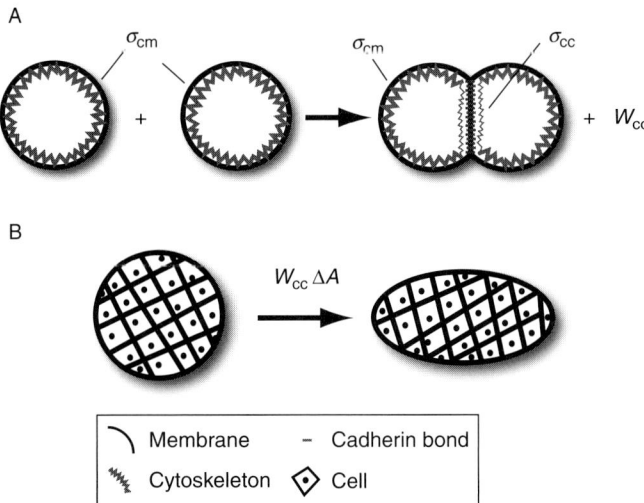

**Figure 1.2** Adhesion energy. The intercellular adhesion energy $W_{cc}$ is the work required to separate two cells (A). It results from differences in cell/cell and cell/non-cell interfacial tensions $\sigma_{cc}$ and $\sigma_{cm}$ that arise from cadherin recruitment (green) and subsequent cytoskeleton remodeling (red) (Chu et al., 2004) by unit area: $W_{cc} = \sigma_{cm} - \sigma_{cc}/2$. The intercellular adhesion energy can also be measured by tissue compression provided the tissue behaves as a liquid (B). As cells are free to translocate from the center of the aggregate to the surface, the work needed to expand the tissue surface of a unit area $\Delta A$ is the intercellular adhesion energy (Foty et al., 1994, 1996; Phillips and Davis, 1978). (See Color Insert.)

adhesion energy. This definition holds for living as well as nonliving matter and relates unambiguously adhesion energy and interfacial tensions. It does not discriminate between mechanisms that account for the difference between $\sigma_{cm}$ and $\sigma_{cc}$. In particular, it reflects the contributions of specific adhesion through interfacial bonds, and nonspecific adhesion (depletion forces, electrostatic, Van der Walls), repulsive forces (steric, membrane fluctuations) (Bell et al., 1984), and eventually any subsequent change in the tension of the supporting membrane and cortical cytoskeleton. From here, we use this definition which applies whenever adhesion *energy* (in the context of intercellular adhesion or not) is invoked.

Assuming that cells behave as a soft, slightly deformable material, measurement of the separation force between cell doublets provides a measure of intercellular adhesion energy (Brochard-Wyart and de Gennes, 2003; Chu et al., 2005). Intercellular adhesion energy increases for approximately an hour as the cytoskeleton underlying the forming intercellular contact reorganizes from the initial adhesion event (Chu et al., 2004). Note, however, that the initial minute of adhesion is weak and does not depend on cytoskeleton remodeling, whereas it takes at least half an hour for the adhesion energy to strengthen and plateau through Rac1 and Cdc42-dependent cytoskeleton remodeling. Using controlled exogenous cadherin expression in cadherin-deficient cells, Chu *et al.* showed that adhesion energy increased with higher cadherin expression. In addition, homotypic adhesion energy was different with different cadherin subtypes expressed at similar levels, whereas heterotypic adhesion energy was always low. These results quantitatively support a determinant role of cadherin subtype and subtype specificity in the extent of cytoskeleton remodeling and subsequent magnitude of adhesion energy (Chu *et al.*, 2006).

## 3. Modulation of Intercellular Adhesion in Development

From extracellular contacts between cadherins to remodeling the cytoskeleton, intercellular adhesion relies on a wide range of cellular events involving many molecular partners. Hence, there are likely multiple levels of regulation of cell/cell adhesion during embryonic development. Schematically, intercellular adhesion modulation can occur at the levels of proper localization and activation of the cadherin/catenin complex at the plasma membrane, as well as the organization and response of the underlying cortical cytoskeleton. Although the impact of these processes on adhesion energy has not always been assessed directly, they have been shown to regulate a number of morphogenetic events believed to involve intercellular adhesion.

### 3.1. Modulation of cadherin–catenin membrane levels

Cadherin expression is regulated at many levels, from gene transcription to protein exocytosis, endocytosis, and degradation. At the transcriptional level, cadherin expression is controlled by regulatory elements contained in the second intron present in several classical cadherin genes (Stemmler *et al.*, 2005), and by repression of transcription promoter activity by methylation or binding Zinc-finger proteins including Snail and Zeb family members (Conacci-Sorrell *et al.*, 2003; Strathdee, 2002).

Posttranslational regulation of cadherin expression occurs at different sites. Exit of newly synthesized E-cadherin from the endoplasmic reticulum requires $\beta$-catenin binding (Chen *et al.*, 1999). At the plasma membrane, the integrity of the cadherin–catenin complex is regulated by phosphorylation and endocytosis. Three serine residues in the cadherin cytoplasmic domain (S684, S686, and S692) are phosphorylated by CKII and GSK3$\beta$ kinases, which creates additional interactions with $\beta$-catenin, resulting in a large increase in the affinity of the interaction (Huber *et al.*, 2001).

In contrast, tyrosine phosphorylation of $\beta$-catenin at Y489 or Y654 disrupts binding to cadherin, and phosphorylation of $\beta$ catenin at Y142 disrupts binding to $\alpha$-catenin (Lilien and Balsamo, 2005).

Stability of the cadherin/catenin complex is regulated by endocytosis. Overall stability of the complex appears to depend on binding of p120 (Davis *et al.*, 2003). E-cadherin is also a target for ubiquitinylation by the E3 ligase Hakai, which like p120 binds to the juxtamembrane domain of cadherins (Fujita *et al.*, 2002). Cadherin endocytosis occurs via clathrin-coated vesicles (Bryant and Stow, 2004), but it remains poorly understood if this requires loss of p120 binding and concurrent ubiquitinylation. Finally, both the cadherin extracellular and intracellular domains are subject to proteolytic cleavage which may also lead to endocytosis, or the release or retention of fragments of cadherin that can subsequently compete with full length cadherin *trans*-binding or signal back to cadherin transcription, respectively (Marambaud *et al.*, 2002, 2003).

Constitutive turnover of plasma membrane cadherin is essential for cell rearrangements in cohesive tissues, as shown during hexagonal cell repacking in *Drosophila* wing epithelium (Classen *et al.*, 2005), indicating that cadherin unbinding can be finely tuned to allow cells to slip past each other and yet maintain overall tissue cohesion. During development, changes in cadherin membrane levels often trigger morphogenetic processes involved in loss of tissue cohesion such as EMT (Batlle *et al.*, 2000; Cano *et al.*, 2000). However, cadherin membrane levels may also vary between cells while tissue cohesion is maintained, which is thought to enable the oocyte, follicle cells and germline cells to sort in the *Drosophila* ovary (Godt and Tepass, 1998). Cadherin levels also vary over time and even between different interfaces of the same cell in a cohesive epithelium, which may drive cell rearrangements during *Drosophila* germband convergence–extension (Blankenship *et al.*, 2006; Zallen and Wieschaus, 2004). These differences may also be accompanied by local changes in cortical cytoskeleton response that could also modulate intercellular adhesion (see below).

## 3.2. Modulation of cortical cytoskeleton organization and response

There is growing evidence that modulation of cortical actomyosin activity at cell/cell interfaces is an important determinant of various aspects of morphogenesis. Actin dynamics and architecture are directly regulated by cadherin-mediated adhesion through $\alpha$-E-catenin signaling that may inhibit Arp2/3 activity (Drees *et al.*, 2005) or recruit formin (Kobielak *et al.*, 2004), or p120-mediated regulation of Rho family GTPases (Anastasiadis, 2007) Rho family GTPases are the nexus for control of many effectors of cytoskeleton activity, dynamics and organization. In general, Rac1 and Cdc42 target regulators of actin polymerization (e.g., Arp2/3, WAVE) (Jaffe and Hall, 2005), whereas Rho targets regulators of myosin activity.

Various isoforms of myosin are involved in several morphogenetic processes that depend on cadherin-mediated adhesion. For example, Myosin VI activity is required in *Drosophila* dorsal closure and ovary border cell migration (Geisbrecht and Montell, 2002; Millo *et al.*, 2004). Myosin II activity is important during convergence–extension and apical constriction, two morphogenetic processes where constriction of the apical actomyosin ring drives shortening of cell interfaces in epithelia (Lecuit and Lenne, 2007 and below). During *Drosophila* ventral furrow formation, Myosin II is activated upon phosphorylation by Rok, which is in turn activated by Rho and its regulator RhoGEF2; RhoGEF2 is localized to the apical membrane, thereby ensuring local activation of Myosin II at the correct site for apical constriction (Dawes-Hoang *et al.*, 2005; Kolsch *et al.*, 2007; Nikolaidou and Barrett, 2004). To achieve coordinated contraction, actin and myosin may also rely on common regulators, such as Abl or Dia (Fox and Peifer, 2007; Homem and Peifer, 2008; Stevens *et al.*, 2008). In vertebrate neural tube closure, localized Myosin II activation occurs similar to that in *Drosophila*, though the regulators may be different (Hildebrand, 2005; Nishimura and Takeichi, 2008).

## 3.3. Differential expression of cadherin subtypes

Distinct tissue types tend to express predominantly one cadherin subtype (Takeichi, 1988). Epithelia generally express E-cadherin, which is required for the maintenance of the typical apicobasal polarity of these tissues. Neural tissues and muscles express N-cadherins, forebrain and bone express R-cadherin, the epidermal basal layer expresses P-cadherin, endothelia express VE-cadherins, mesoderm expresses Cadherin-11 and kidney expresses Cadherin-6. Although there is generally preferential expression of one cadherin subtype, other cadherins may be expressed simultaneously in the same tissue: for example, E-cadherin and N-cadherin are coexpressed in the nervous system (Takeichi, 1988). In addition to tissue-specific expression patterns, different cadherins may exhibit patterns of chronological expression during development. For example, during *Drosophila* mesoderm invagination, cadherin-subtype expression switches from E- to N-cadherin (Oda *et al.*, 1998). Similarly, there is a switch from E- to N-cadherin during neurulation in chick embryos (Hatta *et al.*, 1987).

In cells expressing multiple types of cadherins, such as neural tissues, cadherins of a given subtype are localized at intercellular contacts provided that both neighbor cells express that subtype, regardless of the tissue origin of the cells. In contrast, when cells do not share expression of the same subtype, cadherin localization at cell/cell interfaces is compromised. This subcellular localization pattern was first shown *in vitro* (Hirano *et al.*, 1987), but was recently observed *in vivo* in *Drosophila* ommatidia (Hayashi and Carthew, 2004).

*In vivo* misexpression of cadherin subtypes causes cells to redistribute to regions that endogenously express the same cadherin subtype (Cortes *et al.*, 2003; Inoue *et al.*, 2001; Luo *et al.*, 2004; Patel *et al.*, 2006; Price *et al.*, 2002). Interestingly, even replacement of the EC1 domain of a cadherin subtype for that of a different subtype can induce cell missorting (Patel *et al.*, 2006; Price *et al.*, 2002); this supports the idea that the EC1 domain specifies subtype homotypic adhesion.

In addition to subtype-binding specificity provided by the cadherin extracellular (EC1) domain, differences in membrane levels and interface localization of different cadherin subtypes might be determined by subtype-dependent targeting or cytoskeleton response. Thus, combinations of different subtypes and membrane levels of cadherins in time and place, and different cytoskeleton rearrangements could multiply the vocabulary of the adhesion code to generate different adhesion energies on each cell interface to allow cell rearrangements while maintaining a degree of tissue integrity (Fig. 1.3).

## 4. Intercellular Adhesion Energy and Morphogenesis

The first attempts to understand the impact of intercellular adhesion on cell sorting were carried out on simplified systems comprising reconstituted cell aggregates from dissociated embryos or tissue culture cell lines. These aggregates had an interesting property that they behaved like liquids governed by their surface tension, which made the assessment of intercellular adhesion energy easier. It is believed, and in some cases verified, that this liquid behavior occurs to a large extent in living organisms during different stages of the development. This makes these developmental steps good candidates to understand how modulation of intercellular adhesion regulates morphogenesis.

### 4.1. Intercellular adhesion energy and tissue surface tension

A liquid governed by its surface tension tends to adopt a shape that minimizes its surface energy—a sphere. Similar behavior is observed in reconstituted embryonic cell aggregates as they eventually round up into a ball. Compression experiments on reconstituted aggregates showed that their surface tension relaxed after deformation to the same constant value regardless of the deformation amplitude (Foty *et al.*, 1994, 1996; Phillips and Davis, 1978). Although aggregates may exhibit some elasticity on a minute time

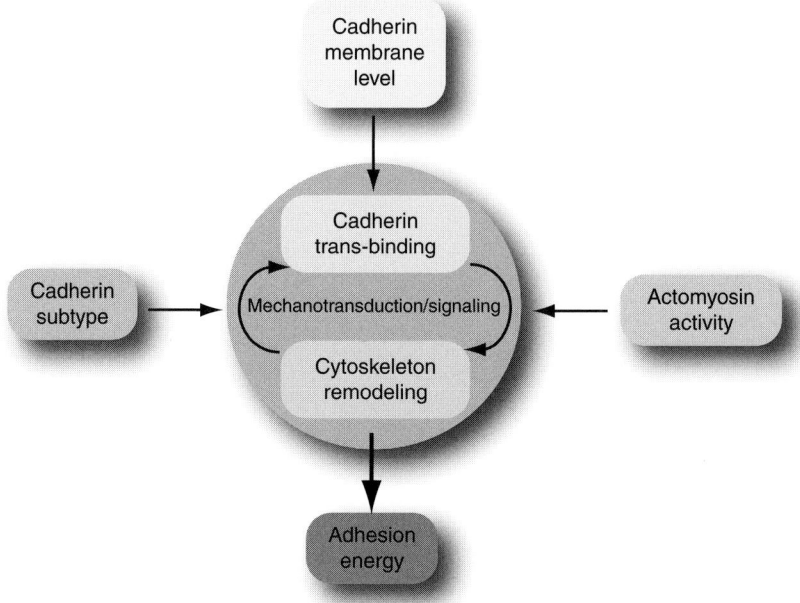

**Figure 1.3** Adhesion modulation. During development, cadherin-mediated adhesion is modulated by cadherin membrane levels, cadherin-subtype expression, and actomyosin activity. Cadherin membrane levels are regulated by cadherin expression, targeting and recycling. Differences in cadherin levels in cells through the tissue at a given time or throughout development support morphogenetic processes such as oocyte positioning in *Drosophila* oocyte (Godt and Tepass, 1998). Cadherin-subtype expression switching is observed at the onset of *Drosophila* gastrulation and during vertebrate neurulation (Hatta *et al.*, 1987; Oda *et al.*, 1998). Actomyosin activity is required in many processes including ventral furrow formation, germband convergence–extension, dorsal closure and ovary border cell migration in *Drosophila*, and neural tube closure in vertebrates.

scale, tension relaxation guarantees a liquid-like behavior for longer times (Forgacs *et al.*, 1998).

If, like molecules in a liquid, cells can be assumed to have no positional constraints in an aggregate, any deviation of shape from a sphere results in cell translocation from the interior of the aggregate to the surface. Subsequently, any increase in aggregate surface area is compensated by a loss of cell/cell interface area. Then, the *tissue* surface tension (the energy change for an increase of a unit of *tissue* surface area) can be written as $\gamma_{cm} = \sigma_{cm} - \sigma_{cc}/2$. Thus, *tissue* surface tension is a direct readout of intercellular adhesion energy (Fig. 1.2B).

Recently, the dependence of tissue surface tension on cadherin cell surface density was confirmed by compressing aggregates of cells expressing controlled levels of exogenous cadherins (Foty and Steinberg, 2005).

Contrary to Chu *et al.* (2004), tissue surface tension appeared to be independent of cadherin subtype, provided that the surface densities of the different cadherins were similar. Thus, the ability of a cadherin subtype to strengthen adhesion to a specific extent for a given surface density is likely to rely on other cellular factors. Nevertheless, typical tissue surface tension and separation force measurements give similar adhesion energy values of 1 to several 10 mN/m, depending on the cadherin expression level. Overall, these results provide further evidence for the relevance of liquid behavior to cell aggregates, and the use of tissue surface tension as a direct readout of intercellular adhesion energy.

## 4.2. Cell sorting and tissue envelopment

More than 50 years ago, Townes and Holtfreter (1955) observed that cell aggregates reconstituted *in vitro* from cells originating from two distinct embryonic tissues sorted out into subaggregates of the same cell type, one of which partially or totally enveloping the other. This phenomenon is reminiscent of an immiscible liquid demixion as well as morphogenetic processes observed at early stages of embryogenesis.

*Tissue* interfacial tension between subaggregates composed of two cell types, termed "a" and "b," involves three cell–cell interfacial tensions $\sigma_{aa}$, $\sigma_{bb}$, and $\sigma_{ab}$. A gain of a unit area of heterotypic contact is compensated by a loss of 2 half-units of homotypic contact. Therefore, the *tissue* interfacial tension is $\gamma_{ab} = \sigma_{ab} - (\sigma_{aa} + \sigma_{bb})/2$. If $\gamma_{ab} > 0$, it is energetically favorable to minimize heterotypic interfacial area and, therefore, cells will tend to sort and form large subaggregates of the same kind. If $\gamma_{ab} < 0$, heterotypic contacts are more stable that homotypic contacts and cells will not sort. Similar to homotypic intercellular contacts, heterotypic adhesion energy is defined as the energy released with the loss of one free cell surface of each kind to generate a heterotypic surface of the same area: $W_{ab} = (\sigma_{am} + \sigma_{bm} - \sigma_{ab})/2 = (W_{aa} + W_{bb})/2 - \gamma_{ab}/2$. Then, the sorting condition rewrites to $W_{ab} < (W_{aa} + W_{bb})/2$. If heterotypic adhesion energy is weaker than the average of homotypic adhesion energies, then cell sorting is energetically favorable. The envelopment order is predicted by the balance between tissue interfacial tensions. If $\gamma_{bm} > \gamma_{am} + \gamma_{ab}$, then subaggregates of cell-type "a" will envelop subaggregates of cell-type "b". This condition also rewrites to $W_{aa} < W_{ab}$, and since sorting occurs ($W_{ab} < (W_{aa} + W_{bb})/2$), it implies $W_{aa} < W_{bb}$. If $W_{aa} > W_{ab}$, partial envelopment occurs, and for the limit $W_{ab} = 0$ aggregates of cell-type "a" and "b" will separate (Steinberg, 1963) (Fig. 1.4).

These assumptions were first tested experimentally with chick embryo explants and revealed for a number of tissue combinations that if $W_{aa} < W_{bb}$ then aggregates of cell-type "a" enveloped aggregates of cell-type "b" (Foty *et al.*, 1996), thereby confirming the relevance of the differential adhesion

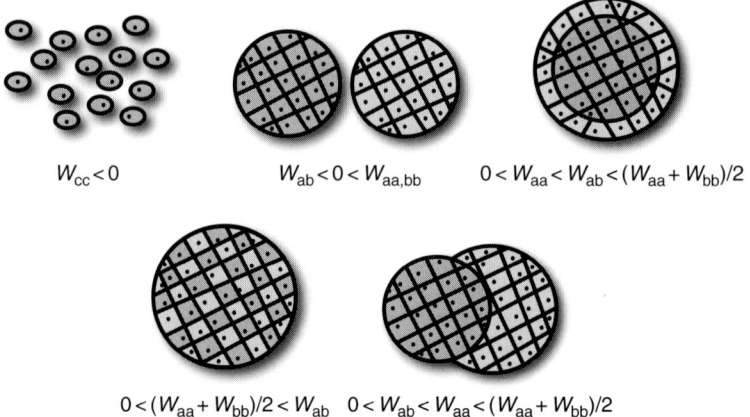

**Figure 1.4** Cell sorting and tissue envelopment. Depending on the values of homotypic and heterotypic adhesion energies, cells of different types, "a" (red) and "b" (blue) (according to their cadherin expression level, subtype, or cytoskeleton remodeling extent upon adhesion), can aggregate, sort, or unsort, and tissues can envelop each other partially or completely, as verified with tissue explants and culture cells (Foty and Steinberg, 2005; Foty et al., 1996; Friedlander et al., 1989; Krieg et al., 2008). (See Color Insert.)

hypothesis for cell sorting and tissue morphogenesis. The complete array of mixing/sorting, and partial and total envelopment behaviors has also been simulated *in silico* using the adhesion energy conditions formulated by Steinberg (Glazier and Graner, 1993; Graner and Glazier, 1992).

### 4.3. Origins of heterotypic adhesion energy

Although there is agreement between homotypic adhesion energy ranking and envelopment order, measurement of heterotypic adhesion energy $W_{ab}$ was generally missed in these aggregate envelopment studies. Indeed, whereas $\gamma_{aa}$ and $\gamma_{bb}$ (and hence $W_{aa}$ and $W_{bb}$) can be measured at the tissue scale by homotypic aggregate compression or centrifugation, $\gamma_{ab}$ can only be inferred from contact angle measurement between subaggregates in a partial envelopment configuration, which is not always experimentally feasible and even then has no predictive value. Hence, whether $W_{ab}$ falls in the appropriate range for cell sorting and total envelopment is usually not verified. On the other hand, some tissues comprised of two cell types, such as the avian oviduct epithelium, adopt a checkerboard pattern of cell organization (Honda et al., 1986), which can be explained for the unsorted condition $\gamma_{ab} < 0$ (or $W_{ab} > (W_{aa} + W_{bb})/2$).

How $W_{ab}$ is related to $W_{aa}$ and $W_{bb}$ remains an open question. For cell types differing only by the surface density of the same adhesion protein, and assuming that heterotypic adhesion $W_{ab}$ is limited by the lowest homotypic adhesion (e.g., $W_{ab}=W_{aa}$) expected for the lowest expressing population (Bell et al., 1984), $W_{ab}$ is strictly smaller than $(W_{aa} + W_{bb})/2$. This predicts that if homotypic adhesion energies are different, cell sorting is favored and corresponds to the limit between partial and total envelopment (Steinberg, 1963).

It has been experimentally verified that two cell populations sort and envelop one another when the only difference is the cell surface level of cadherins (Duguay et al., 2003; Foty and Steinberg, 2005; Friedlander et al., 1989; Steinberg and Takeichi, 1994). As expected, the highest expressing cells were the most adhesive and were enveloped (Foty and Steinberg, 2005); however, $W_{ab}$ was again not measured and hence $W_{ab} < (W_{aa}+W_{bb})/2$ has not been experimentally verified. Partial envelopment was observed occasionally, though total envelopment was the predominant outcome. This result suggests that heterotypic adhesion energy is often higher than the smallest homotypic adhesion energy. How cells bearing the lowest cadherin surface density manage to build up higher adhesion energy toward cells expressing higher amounts than themselves is unknown; uneven cadherin distribution between homotypic and heterotypic contacts is a possibility, but the underlying mechanism in reconstituted aggregates is unknown.

Cell sorting is also observed in aggregates of cell populations expressing different cadherin subtypes (Duguay et al., 2003; Friedlander et al., 1989; Jaffe et al., 1990; Niessen and Gumbiner, 2002; Nose et al., 1988). The fact that populations segregate but that subaggregates remain in contact, which corresponds to $W_{ab} > 0$, supports the hypothesis that non-null adhesion energy arises from binding between cells expressing different cadherin subtypes, and thus indicates some binding affinity between cadherin subtypes. In addition, the complete spectrum of cell sorting outcomes has been observed, in particular the random distribution in aggregates of cells expressed similar amounts of different cadherins subtypes (Duguay et al., 2003). Since there was no homotypic adhesion energy difference among cadherin subtypes expressed in these cells (i.e., $W_{aa} = W_{bb}$) (Foty and Steinberg, 2005), the random distribution of cells in the aggregate can be explained by $W_{ab} = W_{aa} = W_{bb}$. In other words, when the same amounts of different cadherin subtypes elicit similar homotypic adhesion energies, there is no cadherin-subtype *trans*-binding specificity. Thus, it was proposed that cell sorting and aggregate mutual envelopment would result from differences in cadherin surface density, regardless of cadherin subtype. However, this contradicts other sorting experiments that showed segregation between cells expressing equivalent amounts of different cadherin subtypes and even segregation of cells when the only difference was the cytoplasmic domain of cadherin (Friedlander et al., 1989; Jaffe et al., 1990). Cadherin-subtype *trans*-binding specificity is further supported by the observation of cadherin-subtype-specific localization at cell interfaces in mixed cell populations *in vitro* and *in vivo* (Hayashi and Carthew, 2004; Hirano *et al*., 1987) as well as heterotypic adhesion energy measurement by cell doublet separation (Chu *et al*., 2004). This suggests that other factors dependent on the cellular background are required to potentiate differences between homotypic and heterotypic cadherin binding and adhesion energies.

## 4.4. Cadherin expression levels versus cytoskeleton remodeling

Whereas cadherin expression levels have been shown to correlate with adhesion energy when expressed exogenously in the same cellular background in cultured cells (Chu *et al*., 2004; Foty and Steinberg, 2005), this has not been verified *in vivo*. As adhesion depends on the extent of cortical cytoskeleton response, which varies with cellular background, the adhesion energy outcome may not be simply correlated with the level of cadherin expression but depend on other factors affecting the actin cytoskeleton.

The respective contributions of cadherin expression level and cytoskeleton response to the intercellular adhesion energy have been revealed in recent studies with zebrafish embryos. At the onset of gastrulation, the three primary germ layers (ectoderm, mesoderm, and endoderm) have different levels of cadherins. Surprisingly, these levels ranked in the opposite order to homotypic adhesion energies, as inferred from the envelopment order (Krieg *et al*., 2008) or measured by tissue compression of reconstituted cell aggregates (Schötz *et al*., 2008). Analysis of cell mechanical indentation further revealed that cell/noncell interfacial tensions ($\sigma_{cm}$) for the three germ layers ranked in agreement with envelopment order. Direct imaging showed that the cortical cytoskeleton became thinner at intercellular contacts to a similar extent in all cell aggregates (Krieg *et al*., 2008). This suggests that even though cell/non-cell interfacial tensions are different, the largest cytoskeletal rearrangements upon adhesion occur in cells with the highest tension, and cell/cell interfacial tensions ($\sigma_{cc}$) fall in the same range for all cell/cell contact combinations. Thus, cells that expressed the lowest level of surface cadherins unexpectedly exhibited stronger cytoskeleton remodeling, which becomes the major contribution to the adhesion energy $W_{cc} = \sigma_{cm} - \sigma_{cc}/2$.

Interestingly, adhesion energies measured by doublet separation after short contact times (maximum 1 min) ranked in the same order as cadherin expression levels (Krieg *et al*., 2008). These adhesion energies are likely to account for the cadherin homophilic bonding contribution prior to cytoskeleton remodeling. Indeed, they are much smaller than mature adhesion energies. This is further supported by the independence of short contact adhesion energies on myosin inhibition or upregulation, whereas cell sorting was altered when myosin was inhibited or upregulated. This provides evidence for the requirement of actomyosin contraction for maintaining strong cell interfacial tension contrast between contacting and non-contacting cell surfaces.

Taken together, these studies reveal that differential cadherin expression among different tissues is not always a relevant indicator of their relative positioning within a cell aggregate since actin compliance to remodeling is tissue-type dependent, and therefore not related to the absolute cadherin expression level. However, differential adhesion is still an indicator of tissue positioning, but does not correlate with cadherin expression level.

## 4.5. Intercellular adhesion energy and polarity

The contribution of differential intercellular adhesion as a driving mechanism for overall spatial polarization of organisms during development was initially proposed to explain anteroposterior (AP) elongation during convergence–extension (C–E) of *Drosophila* germband (Irvine and Wieschaus, 1994). In this system, as well as in vertebrate models, segmented or graded gene expression in the tissue along the AP axis regulates elongation. It was proposed that gene expression patterns define dorsoventral (DV)-oriented segments of cell subpopulations with distinct adhesive properties (Irvine and Wieschaus, 1994). The tendency of cell subpopulations to minimize their DV interfaces would lead to cell intercalation and AP elongation. It was later shown that explants taken from different segments along the *Xenopus* chordamesoderm sort *in vitro*, and therefore have distinct adhesive properties, and undergo C–E at their interface (Ninomiya *et al.*, 2004).

Fine examination of cell intercalation during C–E in *Drosophila* germband revealed that cell rearrangements occurred by T1 transitions and rosette formations by which cell interfaces perpendicular to the AP axis disappeared and interfaces along the AP axis appeared (Bertet *et al.*, 2004; Blankenship *et al.*, 2006; Zallen and Wieschaus, 2004) However, these observations did not clearly identify distinct subpopulations of cells during these changes in tissue polarization. Conversely, correct cell rearrangements during of C–E could be simulated using differences in cell interfacial tensions according to their orientation relative to the AP axis (Zajac *et al.*, 2000, 2003).

It remains unclear how genetic or environmental cues elicit differences in adhesion energies according to an orientational pattern rather than a positional pattern. However, experimental observation revealed anisotropic recruitment of intercellular adhesion (E-cadherin) and cytoskeletal (actin, myosin) proteins according to interface orientation (Bertet *et al.*, 2004; Blankenship *et al.*, 2006; Zallen and Wieschaus, 2004). Recently, laser nanoablation experiments of intercellular cortex in the *Drosophila* germband during C–E confirmed that the surface tension of AP-oriented cell interfaces were on average twice as small as that at DV-oriented interfaces (Rauzi et al., 2008). Nanoablation provoked local disruption of cortical actin along a specific cell interface and was accompanied by redistribution of cadherin without membrane rupture. The magnitude of intercellular tension was found to correlate with myosin recruitment, again emphasizing the role of contractile actomyosin ring in adhesion energy (see above). In addition, a twofold anisotropy between DV and AP cell interfacial tensions quantitatively described cell geometry evolution along intercalation process and global tissue elongation during C–E completion (Fig. 1.5A).

## 4.6. Cell interfacial tension refinements

As the independence of *tissue* surface tensions $\gamma_{cc}$ and $\gamma_{cm}$ on tissue surface area is a good approximation for the description of aggregate sorting and envelopment, it is tempting to attribute the same property for *cell* interfacial tensions $\sigma_{cc}$ and $\sigma_{cm}$.

Indeed, the geometry and topology of cells in a tissue are often reminiscent of liquid foams (Hayashi and Carthew, 2004) in which bubbles are well described by a constant interfacial tension. In the case of foams, this is justified by the fact that interfaces are a two-dimensional liquid throughout the entire foam (i.e., the surface material is not bound to a specific bubble). In contrast, within a tissue, cells do not exchange their surface material with each other. Instead, each cell has a given amount of surface material

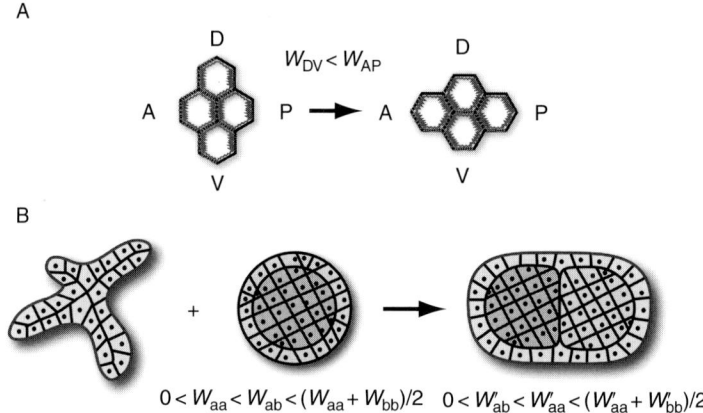

**Figure 1.5** Cell and tissue patterns and polarity. Adhesion energy differences between the dorsoventral and anteroposterior interfaces of *Drosophila* germband cells drive convergence–extension (A) (Rauzi et al., 2008). The combination of liquid-like induced chordamesoderm and mesendoderm and non-liquid-like coated epithelium (yellow) can promote linear ordering, instead of concentric organization, and mimic body elongation in *Xenopus* (B) (Ninomiya and Winklbauer, 2008). (See Color Insert.)

available (both plasma membrane and cortical cytoskeleton) that behaves as an elastic sheet. To reflect this aspect, the simplest *cell* interfacial tension expression consists of the sum of a constant term and terms dependent on cell surface areas, $\sigma_{ab} = \sigma_{ab}^0 + K_a \Delta S_a + K_b \Delta S_b$, where $\sigma_{ab}^0$ is the cell interfacial tension at rest (no stretch of cell surfaces), $K_{a/b}$ is a surface elastic modulus of the cell "a/b" and $\Delta S_{a/b}$ the surface area increase of the cell "a/b" (Farhadifar et al., 2007; Kafer et al., 2007; Rauzi et al., 2008). $\sigma^0$ and $K$ each reflect mechanical properties of the plasma membrane and the cytoskeleton combined. If there are differences in $\sigma^0$ and/or $K$ among interfaces of the same cell, this is the result of differential adhesion. There is no reason to attribute differences in the first term $\sigma^0$ to cadherin binding only and in the second $K$ to cytoskeleton response only.

Such an cell surface area-dependent tension model describes the geometry and topology (interfaces lengths and contact angles at cell vertices) of *Drosophila* ommatidia better than a constant tension model, in the context of the wild-type phenotype and a variety of mutants with over-numerous cells or when cells misexpressed a specific cadherin subtype (Hilgenfeldt et al., 2008; Kafer et al., 2007). It also better describes vertex relaxation upon nanoablation of cell cortex in the *Drosophila* germband (Rauzi et al., 2008). Eventually, an area-dependent tension model also allows for the description of cell geometry and topology (number of neighbors, cross-section area as a function of neighbor number) in the growing *Drosophila* wing epithelium (Farhadifar et al., 2007).

## 4.7. Loss of tissue liquid-like behavior

While predictions from differences in intercellular adhesion energies have always been found compatible with *in vitro* aggregate sorting and envelopment order, envelopment patterns from tissues of embryo explants *in vitro* do not always reproduce the order observed *in vivo* (Krieg et al., 2008; Phillips and Davis, 1978). Despite this apparent discrepancy, differences in intercellular adhesion energies could still account for the observed phenomena, for example by the presence of an additional tissue that does not display a liquid-like behavior but with which adhesion of liquid-like tissues would perturb envelopment order.

In *Rana pipiens* gastrulation, germ layer explants sort *in* vitro such that the endoderm envelops mesoderm, which in turn envelops ectoderm, as predicted by adhesion energy differences (Phillips and Davis, 1978). However, this outcome is observed only if the external single-cell "coated" layer of ectoderm is removed from the late blastula embryo during dissection of the explants. This external layer has a "shiny" nonadherent apical side, and a basal side that directly adheres to underlying ectodermal cells. Being non adhesive, the external "coated" apical side preferentially faces the external medium, whereas the basal side seeks intercellular contact. Hence, it was proposed that by its constrained localization at the embryo surface, the "coated" ectoderm drives the ectoderm outward and inverts the envelopment order (Phillips and Davis, 1978). Similarly, in zebrafish embryos, the extraembryonic tissues such as the enveloping cell layer (EVL) and the yolk syncytial layer (YSL) are assumed to adhere strongly with the ectoderm and the endoderm, respectively, forcing *in vivo* envelopment in an order opposite to that of germ layer explants *in vitro* (Krieg et al., 2008).

Although an epithelium often exhibits extensive neighbor rearrangements that suggest it has retained most of the mechanical features of a liquid in the plane of the tissue, it clearly does not behave as a liquid in the transverse direction. Indeed, "coated" epithelial layers isolated from *Xenopus* gastrula do not round up but fold irregularly. Nevertheless, impairment of apical–basal polarity allows epithelium rounding (Ninomiya and Winklbauer, 2008). When added to underlying tissues *in vitro*, the epithelium alters envelopment by lowering the embryo/non-contact interfacial tension. However, instead of inverting the envelopment order it forces partial envelopment such that a series of *in vitro*-induced anteroposterior tissues, which should sort concentrically in the absence of "coated" epithelium, sorts along an axis in its presence, thereby mimicking their *in vivo* counterparts (Fig. 1.5B). Hence, the association of solid-like to liquid-like tissues combined with a fine tuning of adhesion energy differences allows for the elaboration from the simple sphere to complex tissue shapes and organizations.

The solid-like behavior of epithelia is probably responsible for some the most spectacular morphogenetic processes that occur in embryonic development, particularly because it has different mechanical properties on each side. For example, inward bending of an epithelium is observed in vertebrate neurulation, mesoderm invagination in *Drosophila* or blastopore lip initiation in *Xenopus*. During these processes, the apical side of the epithelium constricts whereas the basal side extends. Apical constriction requires the contractile actomyosin ring that underlies the adherens junction (see above). This suggests that a decrease of adhesion energy upon actomyosin contraction, potentially accompanied by perturbation of adherens junction components, forces cells to reduce their apical perimeter to maintain intercellular cohesion. Here, the localization of the intercellular adhesion protein complex and cytoskeletal proteins to the apex of the lateral membrane is critical for proper morphogenesis (Dawes-Hoang *et al.*, 2005; Hildebrand, 2005; Homem and Peifer, 2008; Kolsch *et al.*, 2007).

## 4.8. Dynamics

Contrary to the stable configurations that result from sorting and envelopment of cell aggregates *in vitro*, the dynamics of cell sorting and envelopment of reconstituted aggregates cannot be explained solely from equilibrium properties, such as differences in intercellular adhesion energies. Conversely, they should reveal their dynamic mechanical properties, which may be useful to describe the dynamics of morphogenetic processes observed *in vivo*.

One may consider first that time scales of adhesion energy changes are larger than those involved in cell rearrangements. For *in vitro* experiments on engineered cells lines or tissue explants, it is likely that cell adhesive properties are stable and do not evolve with time. Hence, during the rearrangement process, the aggregate travels through a predetermined energy landscape. Provided that cells are motile enough to escape from local surface energy minima, the system evolves toward its most stable configuration, without the need for any directed migration.

These predictions concern situations such as the rounding-up of a nonspherical aggregate, the fusion of two miscible aggregates or the sorting and envelopment immiscible aggregates. By analogy with liquids, the dynamics of rounding-up, fusion, demixion, and envelopment in cell aggregates are governed by the balance of a driving force, the surface tension $\gamma$, resisted by an effective viscosity, $\eta$. The ratio $\gamma/\eta = V^\star$ defines the characteristic velocity of the tissue flow that arises during the process, which, together with the specific geometry involved in the rearrangement, determines its time scale.

In the simple case of fusion of two cell aggregates of the same kind, the contact area $x^2$ grows as $aV^\star t$, where $a$ is the initial size of the aggregate and $t$ is the time (Gordon *et al.*, 1972). Knowing $\gamma$ by independent measurement, such as aggregate compression, allows the determination of tissue viscosity. Typically, tissue viscosity ranges between $10^4$ and $10^7$ Pa s. In the case of aggregates of mixed cell populations, the aggregate goes through a first phase during which all cells of the most cohesive population internalize, so that the aggregate is surrounded by a layer of less cohesive cells. Then, the most cohesive cells form subaggregates that coalesce and grow, driven by tissue interfacial tension $\gamma_{ab}$. During this second coalescence phase, the typical size of a subaggregate $L$ grows as $V^\star t$. This linear relationship has been experimentally verified (Beysens *et al.*, 2000) and determination of $V^\star$ is compatible with typical values of tissue surface tensions and tissue viscosity measured earlier.

The tissue viscosities determined by such methods are much higher than that of either water or cytoplasm. They reflect the dissipative processes that take place during cell rearrangements, which could occur through intercellular bond rupture (Brochard-Wyart and de Gennes, 2003), cortical cytoskeleton relaxation or remodeling (Cuvelier *et al.*, 2007), or membrane recycling. Hence, a quantitative investigation of reconstituted aggregates dynamics combined with visualization and manipulation of molecular actors of intercellular adhesion promises valuable information on the molecular mechanisms of intercellular adhesion dynamics in morphogenesis.

In vivo, cell interfacial tensions are likely to evolve concomitantly with cell rearrangements. For instance, cell intercalation in the *Drosophila* germband exhibits temporal changes in relevant protein recruitment at the membrane throughout the process (Bertet *et al.*, 2004; Blankenship *et al.*, 2006; Zallen and Wieschaus, 2004). This suggests temporal changes of cell–cell interfacial tension, as revealed by nanoablation of cell cortex (Rauzi *et al.*, 2008). Therefore, the dynamics are governed by the timing of the cellular cues that determine tension evolution, which might complicate the explanation.

However, there may be some instances where, according to the schedule of the development plan, rapid changes in adhesion energies occur that far exceed in amplitude their further evolution during the longer morphological process that they trigger. In this case, it may be possible to consider that adhesion energies measured at the onset of the event, by tissue explants compression for instance, are those that actually drive the process to completion. This is the assumption adopted by Schötz *et al.* (2008) to describe the dynamics of mesoderm ingression between the yolk and ectoderm in the zebrafish shield. Using tissue viscosity from aggregate fusion dynamics measurements, the estimated ingression velocity was in the same range albeit slightly smaller than the observed velocity ($\sim 1$ $\mu$m/min). Hence, constant adhesion energies and tissue viscosity might be a reasonable approximation in this case.

## 5. Conclusions

Intercellular adhesion is a complex phenomenon that relies on a multitude of molecular mechanisms. A major, but not the sole component that provides support for a structural link between membranes of neighboring cells in metazoans is the transmembrane protein cadherin. However, molecular *trans*-binding between cells is only one aspect of intercellular adhesion. Cadherins, in cooperation with a variety of molecular partners, elicit dramatic changes in the architecture and dynamics of the cortical cytoskeleton that underlies the intercellular contact. That adhesion energy strongly depends on cytoskeleton remodeling implies that any cue that specifically affects cytoskeleton organization and activity at a given cell/cell interface will impact adhesion energy. Hence, mechanisms involved in intercellular adhesion cannot be simply reduced to the parameters of cadherin expression levels and *trans*-binding affinities. This perspective reconciles apparent controversies concerning the contribution of adhesion in the description of cell sorting equilibrium states and morphogenesis. It also provides new directions to explain discrepancies between molecular- and cellular-scale adhesion data.

Nevertheless, many questions remain unanswered. First, the molecular mechanisms that underlie cortical cytoskeleton organization at intercellular contacts remain poorly understood. Despite recent progress in understanding mechanisms that drive adhesion maturation and the role of actomyosin activity in mature intercellular contacts during specific morphogenetic processes, how mechanical forces are transmitted through the cadherin/catenin complex to the cortical cytoskeleton is poorly understood. A number of candidates may serve as a mechanical link but further investigation is needed to provide evidence that the putative simultaneous interactions occur *in vivo*. In addition, it is not known how mechanotransduction occurs in the presence of constitutive turnover of molecular components, though this turnover is important to allow adhesion plasticity during morphogenesis. Combinations of cell-scale biophysical adhesion assays with visualization of protein turnover and interactions might provide valuable insight.

Second, adhesion mechanisms specific to cadherin subtype require further investigation. This is of particular relevance for nervous tissues in which E- and N-cadherins, which have very similar cytoplasmic domains, are coexpressed in the same cells yet do not colocalized in the same subcellular regions. In addition, whereas adhesion energy differences provide a powerful predictive framework for cell sorting *in vitro* and *in vivo*, heterotypic adhesion is always assumed to fulfill the appropriate condition. However, nothing is known about the molecular mechanisms of heterotypic adhesion, especially when it involves different cadherin subtypes. Systematic cell-scale adhesion assays combined with pharmacological and genetic manipulation may provide new insight.

Finally, it remains unclear to what extent intercellular adhesion suffices to explain morphogenetic processes observed *in vivo*. Adhesion energy is only the steady state mechanical manifestation of intercellular adhesion and nothing is known how the same underlying processes regulate tissue viscosities and the dynamics of cell rearrangements and tissue morphogenesis. Moreover, if solid-like tissues are involved, their overall mechanical properties will not simply reflect intercellular adhesion energies as tissue surface tension does for a liquid-like tissue. Cells will have positional constraints and their elastic properties will impact tissue behavior. Conversely, if cells exhibit directed migration, simple liquid behavior will be lost. Here, new tools are needed to assess the mechanical properties of tissues and control cell environments to understand which cues determine transitions between noncohesiveness and the different forms of tissue cohesiveness.

## *ACKNOWLEDGMENTS*

This work was supported by grants from the NIH (GM35527 and 78270) to WJN, a postdoctoral fellowship to NB from the FRM (Fondation Recherche Médicale, SPE20060407147) and a grant from the Bio-X Program at Stanford University.

# REFERENCES

Abe, K., and Takeichi, M. (2008). Inaugural article: EPLIN mediates linkage of the cadherin catenin complex to F-actin and stabilizes the circumferential actin belt. *Proc. Natl. Acad. Sci. USA* **105,** 13–19.

Abedin, M., and King, N. (2008). The premetazoan ancestry of cadherins. *Science* **319,** 946–948.

Aberle, H., Butz, S., Stappert, J., Weissig, H., Kemler, R., and Hoschuetzky, H. (1994). Assembly of the cadherin–catenin complex *in vitro* with recombinant proteins. *J. Cell. Sci.* **107**(Pt 12), 3655–3663.

Adams, C. L., Chen, Y.-T., Smith, S. J., and James Nelson, W. (1998). Mechanisms of epithelial cell–cell adhesion and cell compaction revealed by high-resolution tracking of e-cadherin—Green fluorescent protein. *J. Cell Biol.* **142,** 1105–1119.

Al-Amoudi, A., Diez, D. C., Betts, M. J., and Frangakis, A. S. (2007). The molecular architecture of cadherins in native epidermal desmosomes. *Nature* **450,** 832–837.

Alberts, B. (2002). "Molecular Biology of the Cell." Garland Science, New York.

Anastasiadis, P. Z. (2007). p120-ctn: A nexus for contextual signaling via Rho GTPases. *Bioch. Biophys. Acta (BBA)—Mol. Cell Res.* **1773,** 34–46.

Anastasiadis, P. Z., Moon, S. Y., Thoreson, M. A., Mariner, D. J., Crawford, H. C., Zheng, Y., and Reynolds, A. B. (2000). Inhibition of RhoA by p120 catenin. *Nat. Cell Biol.* **2,** 637–644.

Batlle, E., Sancho, E., Franci, C., Dominguez, D., Monfar, M., Baulida, J., and Garcia De Herreros, A. (2000). The transcription factor snail is a repressor of E-cadherin gene expression in epithelial tumour cells. *Nat. Cell Biol.* **2,** 84–89.

Bell, G. I., Dembo, M., and Bongrand, P. (1984). Cell adhesion. Competition between nonspecific repulsion and specific bonding. *Biophys. J.* **45,** 1051–1064.

Bertet, C., Sulak, L., and Lecuit, T. (2004). Myosin-dependent junction remodelling controls planar cell intercalation and axis elongation. *Nature* **429,** 667–671.

Beysens, D. A., Forgacs, G., and Glazier, J. A. (2000). Cell sorting is analogous to phase ordering in fluids. *Proc. Natl. Acad. Sci. USA* **97,** 9467–9471.

Blankenship, J. T., Backovic, S. T., Sanny, Justina, S. P., Weitz, O., and Zallen, J. A. (2006). Multicellular rosette formation links planar cell polarity to tissue morphogenesis. *Dev. Cell* **11,** 459–470.

Boggon, T. J., Murray, J., Chappuis-Flament, S., Wong, E., Gumbiner, B. M., and Shapiro, L. (2002). C-cadherin ectodomain structure and implications for cell adhesion mechanisms. *Science* **296,** 1308–1313.

Brochard-Wyart, F., and de Gennes, P.-G. (2003). Unbinding of adhesive vesicles. *Comptes Rendus Phys.* **4,** 281–287.

Bryant, D. M., and Stow, J. L. (2004). The ins and outs of E-cadherin trafficking. *Trends Cell Biol.* **14,** 427–434.

Cano, A., Perez-Moreno, M. A., Rodrigo, I., Locascio, A., Blanco, M. J., del Barrio, M. G., Portillo, F., and Nieto, M. A. (2000). The transcription factor snail controls epithelial–mesenchymal transitions by repressing E-cadherin expression. *Nat. Cell Biol.* **2,** 76–83.

Cavey, M., Rauzi, M., Lenne, P. F., and Lecuit, T. (2008). A two-tiered mechanism for stabilization and immobilization of E-cadherin. *Nature* **453,** 751–756.

Chappuis-Flament, S., Wong, E., Hicks, L. D., Kay, C. M., and Gumbiner, B. M. (2001). Multiple cadherin extracellular repeats mediate homophilic binding and adhesion. *J. Cell Biol.* **154,** 231–243.

Chausovsky, A., Bershadsky, A. D., and Borisy, G. G. (2000). Cadherin-mediated regulation of microtubule dynamics. *Nat. Cell Biol.* **2,** 797–804.

Chen, Y. T., Stewart, D. B., and Nelson, W. J. (1999). Coupling assembly of the E-cadherin/beta-catenin complex to efficient endoplasmic reticulum exit and basal–lateral membrane targeting of E-cadherin in polarized MDCK cells. *J. Cell Biol.* **144,** 687–699.

Chen, C. P., Posy, S., Ben-Shaul, A., Shapiro, L., and Honig, B. H. (2005). Specificity of cell–cell adhesion by classical cadherins: Critical role for low-affinity dimerization through beta-strand swapping. *Proc. Natl. Acad. Sci. USA* **102,** 8531–8536.

Chu, Y.-S., Thomas, W. A., Eder, O., Pincet, F., Perez, E., Thiery, J. P., and Dufour, S. (2004). Force measurements in E-cadherin-mediated cell doublets reveal rapid adhesion strengthened by actin cytoskeleton remodeling through Rac and Cdc42. *J. Cell Biol.* **167,** 1183–1194.

Chu, Y.-S., Dufour, S., Thiery, J. P., Perez, E., and Pincet, F. (2005). Johnson–Kendall–Roberts theory applied to living cells. *Phys. Rev. Lett.* **94,** 028102–028104.

Chu, Y.-S., Eder, O., Thomas, W. A., Simcha, I., Pincet, F., Ben-Ze'ev, A., Perez, E., Thiery, J. P., and Dufour, S. (2006). Prototypical type I E-cadherin and Type II cadherin-7 mediate very distinct adhesiveness through their extracellular domains. *J. Biol. Chem.* **281,** 2901–2910.

Classen, A.-K., Anderson, K. I., Marois, E., and Eaton, S. (2005). Hexagonal packing of *Drosophila* wing epithelial cells by the planar cell polarity pathway. *Dev. Cell* **9,** 805–817.

Conacci-Sorrell, M., Simcha, I., Ben-Yedidia, T., Blechman, J., Savagner, P., and Ben-Ze'ev, A. (2003). Autoregulation of E-cadherin expression by cadherin–cadherin interactions: The roles of beta-catenin signaling, slug, and MAPK. *J. Cell Biol.* **163,** 847–857.

Conti, M. A., Even-Ram, S., Liu, C., Yamada, K. M., and Adelstein, R. S. (2004). Defects in cell adhesion and the visceral endoderm following ablation of nonmuscle myosin heavy chain II-A in mice. *J. Biol. Chem.* **279,** 41263–41266.

Cortes, F., Daggett, D., Bryson-Richardson, R. J., Neyt, C., Maule, J., Gautier, P., Hollway, G. E., Keenan, D., and Currie, P. D. (2003). Cadherin-mediated differential cell adhesion controls slow muscle cell migration in the developing zebrafish myotome. *Dev. Cell* **5,** 865–876.

Cuvelier, D., Thery, M., Chu, Y.-S., Dufour, S., Thiery, J.-P., Bornens, M., Nassoy, P., and Mahadevan, L. (2007). The universal dynamics of cell spreading. *Curr. Biol.* **17,** 694–699.

Davis, M. A., Ireton, R. C., and Reynolds, A. B. (2003). A core function for p120-catenin in cadherin turnover. *J. Cell Biol.* **163,** 525–534.

Dawes, A. T. (2009). A mathematical model of alpha-catenin dimerization at adherens junctions in polarized epithelial cells. *J. Theor. Biol.* **257,** 480–488.

Dawes-Hoang, R. E., Parmar, K. M., Christiansen, A. E., Phelps, C. B., Brand, A. H., and Wieschaus, E. F. (2005). Folded gastrulation, cell shape change and the control of myosin localization. *Development* **132,** 4165–4178.

Drees, F., Pokutta, S., Yamada, S., Nelson, W. J., and Weis, W. I. (2005). Alpha-catenin is a molecular switch that binds E-cadherin–beta-catenin and regulates actin-filament assembly. *Cell* **123,** 903–915.

Duguay, D., Foty, R. A., and Steinberg, M. S. (2003). Cadherin-mediated cell adhesion and tissue segregation: Qualitative and quantitative determinants. *Dev. Biol.* **253,** 309–323.

Dzamba, B. J., Jakab, K. R., Marsden, M., Schwartz, M. A., and DeSimone, D. W. (2009). Cadherin adhesion, tissue tension, and noncanonical Wnt signaling regulate fibronectin matrix organization. *Dev. Cell* **16,** 421–432.

Ehrlich, J. S., Hansen, M. D. H., and Nelson, W. J. (2002). Spatio-temporal regulation of rac1 localization and lamellipodia dynamics during epithelial cell–cell adhesion. *Dev. Cell* **3,** 259–270.

Ewald, A. J., and Wallingford, J. B. (2008). Vertebrate gastrulation: Separation is sticky and tense. *Curr. Biol.* **18,** R615–R617.

Farhadifar, R., Röper, J.-C., Aigouy, B., Eaton, S., and Jülicher, F. (2007). The influence of cell mechanics, cell–cell interactions, and proliferation on epithelial packing. *Curr. Biol.* **17,** 2095–2104.

Forgacs, G., Foty, R. A., Shafrir, Y., and Steinberg, M. S. (1998). Viscoelastic properties of living embryonic tissues: A quantitative study. *Biophys. J.* **74,** 2227–2234.

Foty, R. A., and Steinberg, M. S. (2005). The differential adhesion hypothesis: A direct evaluation. *Dev. Biol.* **278,** 255–263.

Foty, R. A., Forgacs, G., Pfleger, C. M., and Steinberg, M. S. (1994). Liquid properties of embryonic tissues: Measurement of interfacial tensions. *Phys. Rev. Lett.* **72,** 2298.

Foty, R. A., Pfleger, C. M., Forgacs, G., and Steinberg, M. S. (1996). Surface tensions of embryonic tissues predict their mutual envelopment behavior. *Development* **122,** 1611–1620.

Fox, D. T., and Peifer, M. (2007). Abelson kinase (Abl) and RhoGEF2 regulate actin organization during cell constriction in *Drosophila. Development* **134,** 567–578.

Friedlander, D. R., Mège, R. M., Cunningham, B. A., and Edelman, G. M. (1989). Cell sorting-out is modulated by both the specificity and amount of different cell adhesion molecules (CAMs) expressed on cell surfaces. *Proc. Natl. Acad. Sci. USA* **86,** 7043–7047.

Fujita, Y., Krause, G., Scheffner, M., Zechner, D., Leddy, H. E., Behrens, J., Sommer, T., and Birchmeier, W. (2002). Hakai, a c-Cbl-like protein, ubiquitinates and induces endocytosis of the E-cadherin complex. *Nat. Cell Biol.* **4,** 222–231.

Gallin, W. J., Edelman, G. M., and Cunningham, B. A. (1983). Characterization of L-CAM, a major cell adhesion molecule from embryonic liver cells. *Proc. Natl. Acad. Sci. USA* **80,** 1038–1042.

Ganz, A., Mireille, L., Saez, A., Silberzan, P., Buguin, A., Mège, R.-M., and Ladoux, B. (2006). Traction forces exerted through N-cadherin contacts. *Biol. Cell* **98,** 721–730.

Gavard, J., Lambert, M., Grosheva, I., Marthiens, V., Irinopoulou, T., Riou, J.-F., Bershadsky, A., and Mege, R.-M. (2004). Lamellipodium extension and cadherin adhesion: Two cell responses to cadherin activation relying on distinct signalling pathways. *J. Cell Sci.* **117,** 257–270.

Geisbrecht, E. R., and Montell, D. J. (2002). Myosin VI is required for E-cadherin-mediated border cell migration. *Nat. Cell Biol.* **4,** 616–620.

Glazier, J. A., and Graner, F. (1993). Simulation of the differential adhesion driven rearrangement of biological cells. *Phys. Rev. E Stat. Phys. Plasmas Fluids Relat. Interdiscipl. Top.* **47,** 2128–2154.

Godt, D., and Tepass, U. (1998). *Drosophila* oocyte localization is mediated by differential cadherin-based adhesion. *Nature* **395,** 387–391.

Gordon, R., Goel, N. S., Steinberg, M. S., and Wiseman, L. L. (1972). A rheological mechanism sufficient to explain the kinetics of cell sorting. *J. Theor. Biol.* **37,** 43–73.

Graner, F., and Glazier, J. A. (1992). Simulation of biological cell sorting using a two-dimensional extended Potts model. *Phys. Rev. Lett.* **69,** 2013–2016.

Grosheva, I., Shtutman, M., Elbaum, M., and Bershadsky, A. (2001). p120 catenin affects cell motility via modulation of activity of Rho-family GTPases: A link between cell–cell contact formation and regulation of cell locomotion. *J. Cell Sci.* **114,** 695–707.

Halbleib, J. M., and Nelson, W. J. (2006). Cadherins in development: Cell adhesion, sorting, and tissue morphogenesis. *Genes Dev.* **20,** 3199–3214.

Hammerschmidt, M., and Wedlich, D. (2008). Regulated adhesion as a driving force of gastrulation movements. *Development* **135,** 3625–3641.

Hatta, K., Takagi, S., Fujisawa, H., and Takeichi, M. (1987). Spatial and temporal expression pattern of N-cadherin cell adhesion molecules correlated with morphogenetic processes of chicken embryos. *Dev. Biol.* **120,** 215–227.

Haussinger, D., Ahrens, T., Aberle, T., Engel, J., Stetefeld, J., and Grzesiek, S. (2004). Proteolytic E-cadherin activation followed by solution NMR and X-ray crystallography. *EMBO J.* **23,** 1699–1708.

Hayashi, T., and Carthew, R. W. (2004). Surface mechanics mediate pattern formation in the developing retina. *Nature* **431,** 647–652.

He, W., Cowin, P., and Stokes, D. L. (2003). Untangling desmosomal knots with electron tomography. *Science* **302,** 109–113.

Hildebrand, J. D. (2005). Shroom regulates epithelial cell shape via the apical positioning of an actomyosin network. *J. Cell Sci.* **118,** 5191–5203.

Hilgenfeldt, S., Erisken, S., and Carthew, R. W. (2008). Physical modeling of cell geometric order in an epithelial tissue. *Proc. Natl. Acad. Sci. USA* **105,** 907–911.

Hirano, S., Nose, A., Hatta, K., Kawakami, A., and Takeichi, M. (1987). Calcium-dependent cell–cell adhesion molecules (cadherins): Subclass specificities and possible involvement of actin bundles. *J. Cell Biol.* **105,** 2501–2510.

Homem, C. C., and Peifer, M. (2008). Diaphanous regulates myosin and adherens junctions to control cell contractility and protrusive behavior during morphogenesis. *Development* **135,** 1005–1018.

Honda, H., Yamanaka, H., and Eguchi, G. (1986). Transformation of a polygonal cellular pattern during sexual maturation of the avian oviduct epithelium: Computer simulation. *J. Embryol. Exp. Morphol.* **98,** 1–19.

Huber, A. H., Stewart, D. B., Laurents, D. V., Nelson, W. J., and Weis, W. I. (2001). The cadherin cytoplasmic domain is unstructured in the absence of beta-catenin. A possible mechanism for regulating cadherin turnover. *J. Biol. Chem.* **276,** 12301–12309.

Inoue, T., Tanaka, T., Takeichi, M., Chisaka, O., Nakamura, S., and Osumi, N. (2001). Role of cadherins in maintaining the compartment boundary between the cortex and striatum during development. *Development* **128,** 561–569.

Ireton, R. C., Davis, M. A., van Hengel, J., Mariner, D. J., Barnes, K., Thoreson, M. A., Anastasiadis, P. Z., Matrisian, L., Bundy, L. M., Sealy, L., Gilbert, B., van Roy, F., *et al.* (2002). A novel role for p120 catenin in E-cadherin function. *J. Cell Biol.* **159,** 465–476.

Irvine, K., and Wieschaus, E. (1994). Cell intercalation during *Drosophila* germband extension and its regulation by pair-rule segmentation genes. *Development* **120,** 827–841.

Ivanov, A. I., Hunt, D., Utech, M., Nusrat, A., and Parkos, C. A. (2005). Differential roles for actin polymerization and a myosin II motor in assembly of the epithelial apical junctional complex. *Mol. Biol. Cell.* **16,** 2636–2650.

Jaffe, A. B., and Hall, A. (2005). Rho GTPases: Biochemistry and biology. *Annu. Rev. Cell Dev. Biol.* **21,** 247–269.

Jaffe, S. H., Friedlander, D. R., Matsuzaki, F., Crossin, K. L., Cunningham, B. A., and Edelman, G. M. (1990). Differential effects of the cytoplasmic domains of cell adhesion molecules on cell aggregation and sorting-out. *Proc. Natl. Acad. Sci. USA* **87,** 3589–3593.

Kafer, J., Hayashi, T., Maree, A. F. M., Carthew, R. W., and Graner, F. (2007). Cell adhesion and cortex contractility determine cell patterning in the *Drosophila* retina. *Proc. Natl. Acad. Sci. USA* **104,** 18549–18554.

King, N., Hittinger, C. T., and Carroll, S. B. (2003). Evolution of key cell signaling and adhesion protein families predates animal origins. *Science* **301,** 361–363.

Kobielak, A., Pasolli, H. A., and Fuchs, E. (2004). Mammalian formin-1 participates in adherens junctions and polymerization of linear actin cables. *Nat. Cell Biol.* **6,** 21–30.

Kolsch, V., Seher, T., Fernandez-Ballester, G. J., Serrano, L., and Leptin, M. (2007). Control of *Drosophila* gastrulation by apical localization of adherens junctions and Rho-GEF2. *Science* **315,** 384–386.

Krieg, M., Arboleda-Estudillo, Y., Puech, P.-H., Kafer, J., Graner, F., Muller, D. J., and Heisenberg, C.-P. (2008). Tensile forces govern germ-layer organization in zebrafish. *Nat. Cell Biol.* **10,** 429–436.

Kussel-Andermann, P., El-Amraoui, A., Safieddine, S., Nouaille, S., Perfettini, I., Lecuit, M., Cossart, P., Wolfrum, U., and Petit, C. (2000). Vezatin, a novel transmembrane protein, bridges myosin VIIA to the cadherin-catenins complex. *EMBO J.* **19,** 6020–6029.

Lecuit, T., and Lenne, P.-F. (2007). Cell surface mechanics and the control of cell shape, tissue patterns and morphogenesis. *Nat. Rev. Mol. Cell Biol.* **8,** 633–644.

Ligon, L. A., Karki, S., Tokito, M., and Holzbaur, E. L. (2001). Dynein binds to beta-catenin and may tether microtubules at adherens junctions. *Nat. Cell Biol.* **3,** 913–917.

Lilien, J., and Balsamo, J. (2005). The regulation of cadherin-mediated adhesion by tyrosine phosphorylation/dephosphorylation of beta-catenin. *Curr. Opin. Cell Biol.* **17,** 459–465.

Luo, J., Treubert-Zimmermann, U., and Redies, C. (2004). Cadherins guide migrating Purkinje cells to specific parasagittal domains during cerebellar development. *Mol. Cell Neurosci.* **25,** 138–152.

Marambaud, P., Shioi, J., Serban, G., Georgakopoulos, A., Sarner, S., Nagy, V., Baki, L., Wen, P., Efthimiopoulos, S., Shao, Z., Wisniewski, T., and Robakis, N. K. (2002). A presenilin-1/gamma-secretase cleavage releases the E-cadherin intracellular domain and regulates disassembly of adherens junctions. *EMBO J.* **21,** 1948–1956.

Marambaud, P., Wen, P. H., Dutt, A., Shioi, J., Takashima, A., Siman, R., and Robakis, N. K. (2003). A CBP binding transcriptional repressor produced by the PS1/epsilon-cleavage of N-cadherin is inhibited by PS1 FAD mutations. *Cell* **114,** 635–645.

Mège, R.-M., Gavard, J., and Lambert, M. (2006). Regulation of cell–cell junctions by the cytoskeleton. *Curr. Opin. Cell Biol.* **18,** 541–548.

Meng, W., Mushika, Y., Ichii, T., and Takeichi, M. (2008). Anchorage of microtubule minus ends to adherens junctions regulates epithelial cell–cell contacts. *Cell* **135,** 948–959.

Millo, H., Leaper, K., Lazou, V., and Bownes, M. (2004). Myosin VI plays a role in cell–cell adhesion during epithelial morphogenesis. *Mech. Dev.* **121,** 1335–1351.

Montell, D. J. (2008). Morphogenetic cell movements: Diversity from modular mechanical properties. *Science* **322,** 1502–1505.

Niessen, C. M., and Gumbiner, B. M. (2002). Cadherin-mediated cell sorting not determined by binding or adhesion specificity. *J. Cell Biol.* **156,** 389–400.

Nikolaidou, K. K., and Barrett, K. (2004). A Rho GTPase signaling pathway is used reiteratively in epithelial folding and potentially selects the outcome of Rho activation. *Curr. Biol.* **14,** 1822–1826.

Ninomiya, H., and Winklbauer, R. (2008). Epithelial coating controls mesenchymal shape change through tissue-positioning effects and reduction of surface-minimizing tension. *Nat. Cell Biol.* **10,** 61–69.

Ninomiya, H., Elinson, R. P., and Winklbauer, R. (2004). Antero-posterior tissue polarity links mesoderm convergent extension to axial patterning. *Nature* **430,** 364–367.

Nishimura, T., and Takeichi, M. (2008). Shroom3-mediated recruitment of Rho kinases to the apical cell junctions regulates epithelial and neuroepithelial planar remodeling. *Development* **135,** 1493–1502.

Noren, N. K., Liu, B. P., Burridge, K., and Kreft, B. (2000). p120 catenin regulates the actin cytoskeleton via Rho family GTPases. *J. Cell Biol.* **150,** 567–580.

Nose, A., Nagafuchi, A., and Takeichi, M. (1988). Expressed recombinant cadherins mediate cell sorting in model systems. *Cell* **54,** 993–1001.

Nose, A., Tsuji, K., and Takeichi, M. (1990). Localization of specificity determining sites in cadherin cell adhesion molecules. *Cell* **61,** 147–155.

Oda, H., Tsukita, S., and Takeichi, M. (1998). Dynamic behavior of the cadherin-based cell–cell adhesion system during *Drosophila* gastrulation. *Dev. Biol.* **203,** 435–450.

Patel, S. D., Ciatto, C., Chen, C. P., Bahna, F., Rajebhosale, M., Arkus, N., Schieren, I., Jessell, T. M., Honig, B., Price, S. R., and Shapiro, L. (2006). Type II cadherin ectodomain structures: Implications for classical cadherin specificity. *Cell* **124,** 1255–1268.

Perez, T. D., Tamada, M., Sheetz, M. P., and Nelson, W. J. (2008). Immediate-early signaling induced by E-cadherin engagement and adhesion. *J. Biol. Chem.* **283,** 5014–5022.

Peyrieras, N., Hyafil, F., Louvard, D., Ploegh, H. L., and Jacob, F. (1983). Uvomorulin: A nonintegral membrane protein of early mouse embryo. *Proc. Natl. Acad. Sci. USA* **80,** 6274–6277.

Phillips, H. M., and Davis, G. S. (1978). Liquid-tissue mechanics in amphibian gastrulation: Germ-layer assembly in *Rana pipiens. Am. Zool.* **18,** 81–93.

Pokutta, S., Drees, F., Takai, Y., Nelson, W. J., and Weis, W. I. (2002). Biochemical and structural definition of the L-afadin- and actin-binding sites of alpha-catenin. *J. Biol. Chem.* **277,** 18868–18874.

Price, S. R., De Marco Garcia, N. V., Ranscht, B., and Jessell, T. M. (2002). Regulation of motor neuron pool sorting by differential expression of type II cadherins. *Cell* **109,** 205–216.

Rauzi, M., Verant, P., Lecuit, T., and Lenne, P. F. (2008). Nature and anisotropy of cortical forces orienting *Drosophila* tissue morphogenesis. *Nat. Cell Biol.* **10,** 1401–1410.

Sato, T., Fujita, N., Yamada, A., Ooshio, T., Okamoto, R., Irie, K., and Takai, Y. (2006). Regulation of the assembly and adhesion activity of E-cadherin by nectin and afadin for the formation of adherens junctions in Madin-Darby canine kidney cells. *J. Biol. Chem.* **281,** 5288–5299.

Schötz, E.-M., Burdine, R. D., Jülicher, F., Steinberg, M. S., Heisenberg, C.-P., and Foty, R. A. (2008). Quantitative differences in tissue surface tension influence zebrafish germ layer positioning. *HFSP J.* **2,** 42–56.

Schwartz, M. A., and DeSimone, D. W. (2008). Cell adhesion receptors in mechanotransduction. *Curr. Opin. Cell Biol.* **20,** 551–556.

Shewan, A. M., Maddugoda, M., Kraemer, A., Stehbens, S. J., Verma, S., Kovacs, E. M., and Yap, A. S. (2005). Myosin 2 is a key Rho kinase target necessary for the local concentration of E-cadherin at cell–cell contacts. *Mol. Biol. Cell.* **16,** 4531–4542.

Sivasankar, S., Gumbiner, B., and Leckband, D. (2001). Direct measurements of multiple adhesive alignments and unbinding trajectories between cadherin extracellular domains. *Biophys. J.* **80,** 1758–1768.

Sousa, S., Cabanes, D., Archambaud, C., Colland, F., Lemichez, E., Popoff, M., Boisson-Dupuis, S., Gouin, E., Lecuit, M., Legrain, P., and Cossart, P. (2005). ARHGAP10 is necessary for [alpha]-catenin recruitment at adherens junctions and for *Listeria* invasion. *Nat. Cell Biol.* **7,** 954–960.

Stehbens, S. J., Paterson, A. D., Crampton, M. S., Shewan, A. M., Ferguson, C., Akhmanova, A., Parton, R. G., and Yap, A. S. (2006). Dynamic microtubules regulate the local concentration of E-cadherin at cell–cell contacts. *J. Cell Sci.* **119,** 1801–1811.

Steinberg, M. S. (1963). Reconstruction of tissues by dissociated cells. *Science* **141,** 401–408.

Steinberg, M. S., and Takeichi, M. (1994). Experimental specification of cell sorting, tissue spreading, and specific spatial patterning by quantitative differences in cadherin expression. *Proc. Natl. Acad. Sci. USA* **91,** 206–209.

Stemmler, M. P., Hecht, A., and Kemler, R. (2005). E-cadherin intron 2 contains cis-regulatory elements essential for gene expression. *Development* **132,** 965–976.

Stevens, T. L., Rogers, E. M., Koontz, L. M., Fox, D. T., Homem, C. C., Nowotarski, S. H., Artabazon, N. B., and Peifer, M. (2008). Using Bcr-Abl to examine mechanisms by which abl kinase regulates morphogenesis in *Drosophila. Mol. Biol Cell.* **19,** 378–393.

Strathdee, G. (2002). Epigenetic versus genetic alterations in the inactivation of E-cadherin. *Semin. Cancer Biol.* **12,** 373–379.

Tabdanov, E., Borghi, N., Brochard-Wyart, F., Dufour, S., and Thiery, J. P. (2009). Role of E-cadherin in membrane–cortex interaction probed by nanotube extrusion. *Biophys. J.* **96,** 2457–2465.

Takai, Y., and Nakanishi, H. (2003). Nectin and afadin: Novel organizers of intercellular junctions. *J. Cell Sci.* **116,** 17–27.

Takeichi, M. (1988). The cadherins: Cell–cell adhesion molecules controlling animal morphogenesis. *Development* **102,** 639–655.

Tamura, K., Shan, W. S., Hendrickson, W. A., Colman, D. R., and Shapiro, L. (1998). Structure function analysis of cell adhesion by neural (N-) cadherin. *Neuron* **20,** 1153–1163.

Thiery, J. P., and Sleeman, J. P. (2006). Complex networks orchestrate epithelial–mesenchymal transitions. *Nat. Rev. Mol. Cell Biol.* **7,** 131–142.

Thoreson, M. A., Anastasiadis, P. Z., Daniel, J. M., Ireton, R. C., Wheelock, M. J., Johnson, K. R., Hummingbird, D. K., and Reynolds, A. B. (2000). Selective uncoupling of p120(ctn) from E-cadherin disrupts strong adhesion. *J. Cell Biol.* **148,** 189–202.

Townes, P. L., and Holtfreter, J. (1955). Directed movements and selective adhesion of embryonic amphibian cells. *J. Exp. Zool.* **128,** 53–120.

Vestweber, D., and Kemler, R. (1984). Rabbit antiserum against a purified surface glycoprotein decompacts mouse preimplantation embryos and reacts with specific adult tissues. *Exp. Cell Res.* **152,** 169–178.

Waterman-Storer, C. M., Salmon, W. C., and Salmon, E. D. (2000). Feedback interactions between cell–cell adherens junctions and cytoskeletal dynamics in newt lung epithelial cells. *Mol. Biol Cell.* **11,** 2471–2483.

Yamada, S., and Nelson, W. J. (2007). Localized zones of Rho and Rac activities drive initiation and expansion of epithelial cell cell adhesion. *J. Cell Biol.* doi:10.1083/jcb.200701058.

Yamada, S., Pokutta, S., Drees, F., Weis, W. I., and Nelson, W. J. (2005). Deconstructing the cadherin–catenin–actin complex. *Cell* **123,** 889–901.

Yoshida, C., and Takeichi, M. (1982). Teratocarcinoma cell adhesion: Identification of a cell-surface protein involved in calcium-dependent cell aggregation. *Cell* **28,** 217–224.

Zajac, M., Jones, G. L., and Glazier, J. A. (2000). Model of Convergent Extension in Animal Morphogenesis. *Phys. Rev. Lett.* **85,** 2022–2025.

Zajac, M., Jones, G. L., and Glazier, J. A. (2003). Simulating convergent extension by way of anisotropic differential adhesion. *J. Theor. Biol.* **222,** 247–259.

Zallen, J. A., and Wieschaus, E. (2004). Patterned gene expression directs bipolar planar polarity in *Drosophila*. *Dev. Cell* **6,** 343–355.

Zhang, Y., Sivasankar, S., Nelson, W. J., and Chu, S. (2009). Resolving cadherin interactions and binding cooperativity at the single-molecule level. *Proc. Natl. Acad. Sci. USA* **106,** 109–114.

Zhu, B., Chappuis-Flament, S., Wong, E., Jensen, I. E., Gumbiner, B. M., and Leckband, D. (2003). Functional analysis of the structural basis of homophilic cadherin adhesion. *Biophys. J.* **84,** 4033–4042.

# CHAPTER TWO

# Remodeling of the Adherens Junctions During Morphogenesis

Tamako Nishimura *and* Masatoshi Takeichi

## Contents

| | |
|---|---|
| 1. Introduction | 34 |
| 2. Basic Machinery of the Adherens Junction | 36 |
| 3. Remodeling by Small GTPase | 37 |
| 4. Remodeling by Cadherin Turnover and Endocytosis | 39 |
|     4.1. Basic processes of cadherin recycling | 39 |
|     4.2. p120-catenin-dependent cadherin stabilization | 39 |
|     4.3. MicroRNA-dependent turnover | 40 |
|     4.4. Other mechanisms | 40 |
| 5. Remodeling by Nonclassic Cadherins and Nectins | 41 |
|     5.1. Protocadherins | 41 |
|     5.2. Fat cadherins | 43 |
|     5.3. Nectins | 44 |
| 6. Junctional Remodeling During Morphogenesis | 44 |
|     6.1. Remodeling through actin modulation | 44 |
|     6.2. Cadherin endocytosis during epithelial cell packing | 46 |
|     6.3. Classic cadherin regulation during gastrulation and neurulation | 47 |
| 7. Perspectives | 48 |
| Acknowledgments | 49 |
| References | 49 |

## Abstract

Morphogenesis of epithelial tissues involves various forms of reshaping of cell layers, such as invagination or bending, convergent extension, and epithelial–mesenchymal transition. At the cellular level, these processes include changes in the shape, position, and assembly pattern of cells. During such morphogenetic processes, epithelial sheets in general maintain their multicellular architecture, implying that they must engage the mechanisms to change the spatial relationship with their neighbors without disrupting the junctions. A major junctional structure in epithelial tissues is the "adherens junction," which is composed of cadherin adhesion receptors and associated proteins including F-actin. The adherens junctions are required for the firm associations between cells, as disruption of them causes disorganization of the epithelial architecture. The adherens junctions, however, appear to be a dynamic entity, allowing the rearrangement of cells within cell sheets. This dynamic nature of the adherens junctions seems to be supported by various mechanisms, such as the interactions of cadherins with actin cytoskeleton, endocytosis and recycling of cadherins, and the cooperation of cadherins with other adhesion receptors. In this chapter, we provide an overview of these mechanisms analyzed *in vitro* and *in vivo*.

## 1. INTRODUCTION

Epithelial cell layers dynamically change their morphology via a number of processes, such as folding, invagination, and convergent extension, during development. Epithelial remodeling continues into adulthood, occurring on various occasions such as during cell renewal, tissue regeneration, and wound healing. Cancer growth and invasion also involve epithelial remodeling. Since an important structural function of the epithelia is to cover and seal the tissues or organs, they need to maintain persistently their cell–cell junctions during the remodeling processes. Accordingly, the epithelia must have elaborate mechanisms to coordinate the stability and yet flexibility of their junctions.

The cell–cell junctions in the simple epithelia typically comprise the tight and adherens junctions and the desmosome. These three junctional components are clustered together at the apical-most portion of the lateral cell–cell contacts, forming the apical "junctional complex" (Farquhar and Palade, 1963), although the adherens junctions (AJs) and desmosomes are also distributed throughout the cell–cell contacts. The AJ located at the junctional complex is specifically termed the "zonula adherens" or "adhesion belt," which encloses the cells at a site near their apical surface, along with the circumferential actin belt.

The major adhesion receptors constituting the AJ are the classic cadherins, which are $Ca^{2+}$-dependent, homophilic cell–cell adhesion molecules (Fig. 2.1). They are a group of single-pass transmembrane proteins, consisting of about 20 subtypes. The classic cadherins are conserved among vertebrates and invertebrates, although the size of their extracellular domain varies between the species (Takeichi, 2007). Their cytoplasmic domain binds p120-catenin and $\beta$-catenin (or plakoglobin) at its N-terminal and C-terminal side, respectively; and $\beta$-catenin/plakoglobin, in turn, binds $\alpha$-catenin (Gumbiner, 2005). Many other proteins are also associated with the cadherin–catenin complex. In addition to the classic cadherins, there are a

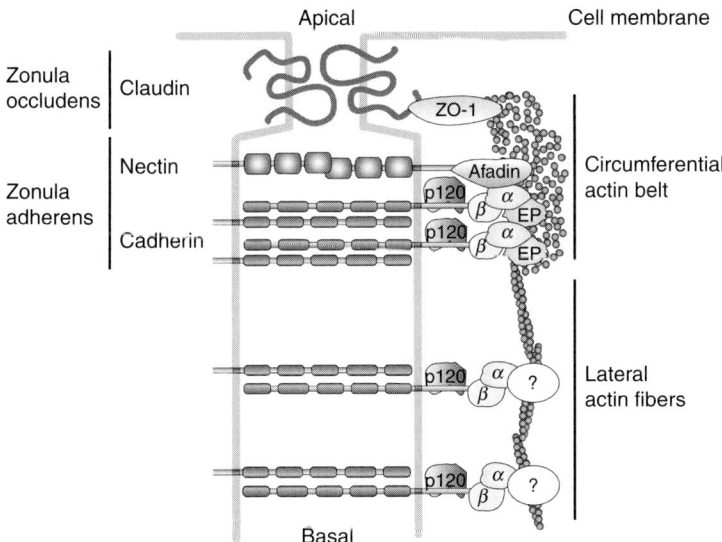

**Figure 2.1** Basic molecular elements organizing the adherens junction in simple epithelial cells of the vertebrates. Tight junction (zonula occludens) is also depicted. The adherens junction adjacent to the tight junction is called zonula adherens, which is specialized by the association with the circumferential actin belt. The cadherin–catenin complexes themselves are in general distributed throughout the cell–cell contacts in vertebrate cells. Vertical section of the junctions is shown. EP, EPLIN; p120, p120-catenin; $\alpha$, $\alpha$-catenin; $\beta$, $\beta$-catenin; ?, unknown molecules.

large number of nonclassic cadherins, constituting the cadherin superfamily (Takeichi, 2007). As far as studied, the nonclassic cadherins seem not to be components of the adherens junction. Many of them, however, interact homophilically as the classic cadherins do, thereby being concentrated at cell–cell contacts; and they often affect the classic cadherin-based AJs. Besides the cadherins, nectins, a family of immunoglobulin-domain membrane proteins, are also localized in the AJs (Takai and Nakanishi, 2003).

The classic cadherin-based AJs play a major role in the physical association of epithelial cells in both vertebrates and invertebrates, and therefore the regulation of them is assumed to be critical for epithelial sheet remodeling. There appear to be multiple ways to reshape the AJs. For example, in many morphogenetic processes, epithelial cells undergo shape changes or movement, not disrupting the zonula adherens (ZA), the most organized part of the AJ. In this form of remodeling, the regulation of the contractility of the ZA seems to be important. The other ways of remodeling include the downregulation of cadherins within a range that can keep the cohesion of cells, but yet allow their rearrangement within the mass. (Cells forming solid tissues always express some types of classic cadherins, even in loose tissues such as mesenchymal ones.) These mechanisms enable an epithelial layer to rearrange its cellular elements for assuming new architecture. In this chapter, we discuss how these forms of junctional remodeling are molecularly controlled.

##  2. BASIC MACHINERY OF THE ADHERENS JUNCTION

The adhesive function of cadherins requires cytoplasmic partners, that is, the catenins. Among the catenins, α-catenin is essential for the firm associations of epithelial cells, as it is well known that α-catenin-deficient cells cannot organize the typical epithelial junctions (Hirano et al., 1992; Vasioukhin et al., 2001; Watabe-Uchida et al., 1998). This is also the case even for neuronal contacts; the synaptic contacts in α-catenin-deficient hippocampal neurons are unstable and turn over more rapidly than those in the wild-type ones (Abe et al., 2004). Thus, given that the actions of α-catenin may be regulated physiologically, this catenin might join the processes for junctional remodeling (Takeichi and Abe, 2005).

Early studies showed that α-catenin could directly bind F-actin (Rimm et al., 1995). It has, thus, long been thought that a role of α-catenin is to tether the cadherin–β-catenin complex to F-actin. However, recent studies have challenged this idea, by demonstrating that the cadherin–β-catenin–α-catenin complex cannot associate with F-actin in vitro but that only free α-catenins can do so (Drees et al., 2005; Yamada et al., 2005). Despite these findings, it is widely accepted that normally functioning cadherins always colocalize with actin fibers, bringing up the question of how these in vitro and in vivo observations can be reconciled. One answer has been given by the finding of a molecule that mediates the linkage between the cadherin–β-catenin–α-catenin complex and F-actin (Abe and Takeichi, 2008). This is EPLIN, which is known to be an actin-binding and -stabilizing protein (Maul et al., 2003). EPLIN can also bind α-catenin associated with the cadherin–β-catenin complex; and through this interaction with α-catenin, EPLIN serves to link this complex to the actin fibers. Depletion experiments have indicated that EPLIN is required not only for the linkage between cadherin and F-actin but also for maintaining the circumferential actin belt. These studies also suggested that there would be another mechanism(s) to link cadherin and F-actin, because these two molecules still colocalized to each other in the absence of EPLIN, although their association patterns were dramatically altered (Fig. 2.2).

A recent work on Drosophila E-cadherin has strengthened the concept that α-catenin functions to tether the cadherin to actin fibers (Cavey et al., 2008). Only in the presence of α-catenin are homotypic cadherin clusters mobile along the cell junctions. Another work demonstrated that α-catenin is required for cadherin molecules to move together with the actin filaments

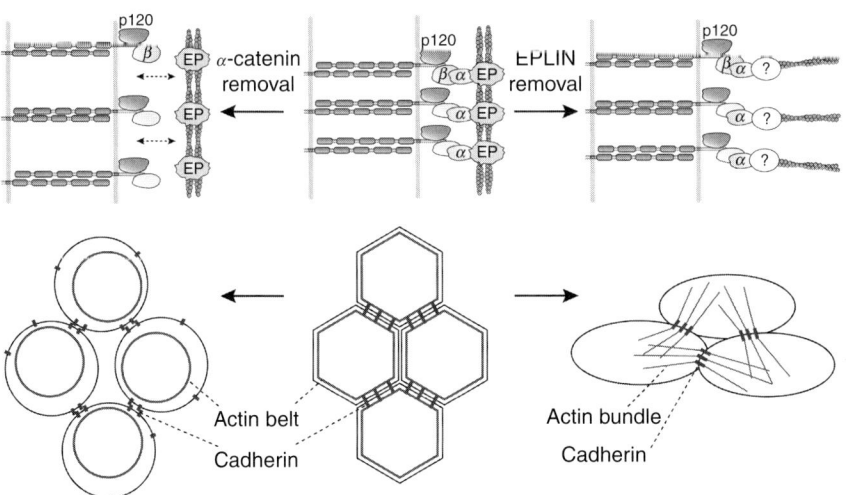

**Figure 2.2** Models to explain the roles of α-catenin (α) and EPLIN (EP) in cadherin–actin interactions. Horizontal sections of the junction at the zonula adherens level are shown. See Abe and Takeichi (2008) and Watabe-Uchida *et al.* (1998) for details of EPLIN and α-catenin functions, respectively.

undergoing a retrograde flow, again supporting the idea of its tethering role (Kametani and Takeichi, 2007). The detailed mechanisms of how the α-catenin tethered to actin fibers functions to maintain the AJs still remain unclear. The actin cytoskeleton is important for the shaping, contraction, and movement of cells, while the AJs are essential for the zippering of them. The structural and functional association between these two molecular systems must be a critical way to control the complex behavior of cells forming a sheet.

It should additionally be noted that, in cells treated with actin-depolymerizing reagents, the junctional cadherins still associate with residual actin clusters (Cavey *et al.*, 2008), and this association of cadherin and actin seems to be α-catenin independent. Actually, the AJs also have other actin-binding proteins such as afadin (see below). The entire story on the interaction between the AJ and actin systems is thus still incomplete.

## 3. Remodeling by Small GTPase

Rho-family small GTPases, as well as their regulators such as guanine nucleotide exchange factors (GEFs) and GTPase-activating proteins (GAPs), are known to be crucial in establishing and maintaining the AJs (Braga and Yap, 2005; Fukata and Kaibuchi, 2001), suggesting that they may also participate in junctional remodeling. A recent addition to the list of this class of AJ regulators is a Cdc42-specific GEF known as Tuba (Otani et al., 2006). Tuba is localized along the apical-most region of the cell junctions interacting with ZO-1; and at this site, Tuba activates Cdc42 and in turn regulates N-WASP, a Cdc42 effector. Depletion of Tuba and N-WASP equally disrupts the network-like organization of E-cadherin and F-actin distributed at the lateral cell–cell contacts, but not at the ZA. Concomitantly, the loss of Tuba causes the junctions to have less tension. Thus, the Tuba–Cdc42–N-WASP pathway seems to regulate the junctional tension via the regulation of actin polymerization at the lateral cell–cell contacts (Fig. 2.3). An interesting point in these observations is that Tuba is required for the lateral portions of the junction, but not for the apical-most ZA, despite the localization of Tuba in the close vicinity of the ZA. The ZA may function to regulate the other portions of the junction.

Another small GTPase, Rap1, first identified as a repressor for cell transformation by Ras, also has attracted attention as a regulator for cell junction formation (Bos, 2005; Kooistra et al., 2007). In *Drosophila*, Rap1 is enriched at the AJs. In Rap1-depleted *Drosophila* cells, the AJs become condensed to one side of the cells; and the cohesion between cells is lost (Knox and Brown, 2002). C3G, a Rap1 GEF, interacts with the cytoplasmic region of E-cadherin, and then activates Rap1 (Hogan et al., 2004). E-cadherin-mediated adhesion is required for the Rap1 activation; and, conversely, Rap1 activity is necessary for the localization of E-cadherin at cell–cell contacts. Rap1 rescues the Ras-transformed or HGF-induced

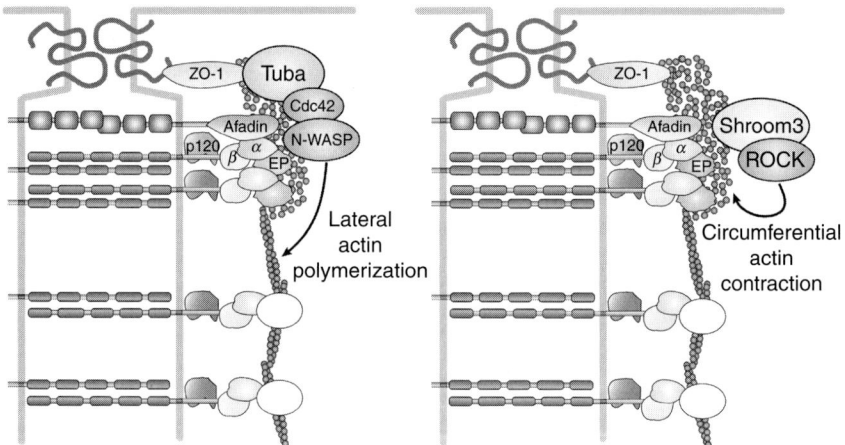

**Figure 2.3** Models to explain the roles of Tuba and Shroom3 in junctional remodeling. Tuba regulates the polymerization of lateral actin fibers (Otani et al., 2006), whereas Shroom3 regulates the contraction of circumferential actin belts (Nishimura and Takeichi, 2008). Vertical sections are shown.

downregulation of E-cadherin, and the effect is enhanced by the activation of another Rap GEF, Epac1 (Price et al., 2004). The potential role of Rap1 in more dynamic morphogenetic systems, however, remains to be investigated.

## 4. Remodeling by Cadherin Turnover and Endocytosis

One of the mechanisms to modulate the AJs would be to remove/add cadherin molecules from/to the junctional sites. Several mechanisms for cadherin trafficking have been investigated and elucidated, as outlined below.

### 4.1. Basic processes of cadherin recycling

Cell-surface cadherins seem to be actively turned over, which would affect the adhesion activity between cells and facilitate junctional remodeling (Bryant and Stow, 2004; Yap et al., 2007). Newly synthesized E-cadherin is transferred from the Golgi to specific sites on the plasma membrane with the Sec6/8 exocyst complex (Yeaman et al., 2004). Once having attained the cell surface, on the other hand, E-cadherin is internalized by clathrin-mediated endocytosis, especially when cells are not in stable contacts (Le et al., 1999). Internalized E-cadherin is transported to recycling endosomes colocalizing with Rab11, and then is recycled back to the cell surface with the exocyst (Langevin et al., 2005) or trafficked to late endosomes and lysosomes for degradation (Palacios et al., 2005). This E-cadherin trafficking from recycling endosomes to the cell adhesion sites appears to depend on the interaction of Rab11 and $\beta$-catenin with the exocyst components Sec15 and Sec10, respectively (Langevin et al., 2005). The homophilic interaction of E-cadherin molecules makes them resistant to endocytosis, via the activation of Cdc42 and Rac, and then IQGAP, which enhances F-actin crosslinking (Izumi et al., 2004).

### 4.2. p120-catenin-dependent cadherin stabilization

p120-catenin (p120), which binds the juxtamembrane region of the E-cadherin tail, plays a critical role in cadherin turnover (Xiao et al., 2007). A pioneering study (Ireton et al., 2002) demonstrated that exogenous expression of p120 in SW40 carcinoma cells upregulated the surface level of E-cadherin, leading them to acquire a typical epithelial configuration, which cells were otherwise loosely associated. This finding suggested that p120 is required for the stabilizing of cadherins on the cell surface, and this idea was confirmed by the experiments to deplete p120 by siRNA, which caused elimination of cadherins from the junctions (Davis et al., 2003; Xiao et al., 2003). In vivo depletion of p120 in the salivary gland also causes E-cadherin deficiency as well as severe defects in epithelial cell adhesion and polarity (Davis and Reynolds, 2006). These effects of p120 expression on cadherin turnover are dependent on its binding to cadherins via the armadillo domain (Liu et al., 2007; Xiao et al., 2005). A dileucine motif on the juxtamembrane domain of E-cadherin (Miranda et al., 2001) is required for its clathrin-mediated endocytosis (Miyashita and Ozawa, 2007a,b), and this motif is masked by p120 binding, explaining the mechanism of p120-dependent stabilization of E-cadherin.

## 4.3. MicroRNA-dependent turnover

MicroRNAs (miRNAs) are small, noncoding RNAs that modulate gene expression of target molecules. Recent screening of miRNAs identified miR-200, a family of microRNAs, whose expression was inversely correlated with E-cadherin downregulation or the epithelial-to-mesenchymal transition (EMT). Ectopic expression of miR-200 in cancer cell lines causes upregulation of E-cadherin; and conversely, inhibition of miR-200 reduces E-cadherin expression in epithelial cells, enhancing their migration (Gregory et al., 2008; Park et al., 2008). This action of miR-200 is elicited via suppression of the expression of E-cadherin repressors, ZEB1 ($\delta$EF1) and ZEB2 (SIP2), which are thought to be involved in EMT. In addition, a significant correlation between E-cadherin and miR-200 expression was found in primary human cancer specimens, confirming the association of miR-200 with the E-cadherin status in in vivo situations.

## 4.4. Other mechanisms

E-cadherin internalization is enhanced by tyrosine kinase activation via various pathways, such as those mediated by the HGF receptor c-Met, EGF receptor, and Src, causing the EMT featured by cell scattering and fibroblast-like morphology. Upon HGF treatment or v-Src activation, two tyrosine residues within the juxtamembrane domain of E-cadherin are phosphorylated, and then Hakai, an E3 ubiquitin ligase, ubiquitinates E-cadherin (Fujita et al., 2002). Ubiquitinated E-cadherin binds to the HGF-regulated tyrosine kinase substrate and is shuttled to lysosomes in a Src-activated Rab5- and Rab7-dependent manner (Palacios et al., 2005). In addition, the signaling mechanisms of HGF-induced Rab5 activation have been revealed (Kimura et al., 2006): HGF treatment activates Ras, which in turn activates RIN2, a Rab5-GEF localizing at cell–cell junctions. This process leads to Rab5 activation and E-cadherin endocytosis.

## 5. Remodeling by Nonclassic Cadherins and Nectins

Many of nonclassic cadherins also show the activity of homophilic binding, and thereby are concentrated at cell–cell contacts. Although some of these molecules function to sustain cell–cell adhesion, others seem to serve as a modulator of the classic cadherins, or even as an inhibitor of cell adhesion, rather than as adhesion receptors. Nectins, a subfamily of the immunoglobulin superfamily, is localized in the AJs, and cooperate with the classic cadherin in junctional remodeling. Examples of the action of these molecules are outlined below.

### 5.1. Protocadherins

Protocadherin represents a subfamily of the cadherin superfamily, whose cytoplasmic domain is not identical to that of the classic cadherin (Redies *et al.*, 2005; Vanhalst *et al.*, 2005), thus suggesting their distinct functions. One of them, paraxial protocadherin (PAPC, protocadherin-8), has been shown to be important for gastrulation. In the developing *Xenopus* embryo, this protocadherin is expressed first in Spemann's organizer and then in the paraxial mesoderm. The expression of a dominant-negative form of PAPC or its depletion by morpholino oligos inhibits the convergent extension movement of the mesoderm (Kim *et al.*, 1998; Medina *et al.*, 2004; Unterseher *et al.*, 2004). The phenotypes observed here are similar to those for cells defective in planar cell polarity (PCP), which is regulated by the Wnt/Frizzled 7 signaling pathway. Indeed, PAPC was shown to activate RhoA and c-Jun N-terminal kinase (JNK) (Medina *et al.*, 2004; Unterseher *et al.*, 2004), which are the effectors of the Wnt/Frizzled 7 signals. Interaction between PAPC and Frizzled 7 at their extracellular domain was also observed, although these two molecules appeared to regulate the PCP signals separately. A recent report shows that ankyrin repeats domain protein 5 (ANR5) interacts with PAPC (Chung *et al.*, 2007). Depletion of ANR5 causes defects in the elongation of activin-treated animal caps and tissue separation, critical for gastrulation movement, and also inhibits PAPC-induced activation of RhoA and JNK, suggesting that ANR5 is a functional partner for PAPC. PAPC was also shown to bind Sprouty (Wang *et al.*, 2008), a receptor tyrosine kinase inhibitor protein, which has the ability to inhibit the convergent extension movement (Sivak *et al.*, 2005). It seems that PAPC promotes gastrulation by sequestering and inactivating Sprouty.

One of the biological functions of PAPC seems to be downregulation of C-cadherin activity. Overexpression of PAPC decreases the adhesive activity of C-cadherin without changing its expression level (Chen and Gumbiner, 2006). Activin treatments induced PAPC expression, and simultaneously decreased C-cadherin activity during elongation of the animal cap. Depletion of PAPC interferes with this activin-induced downregulation of C-cadherin activity, inhibiting animal cap elongation. To elucidate how this activity of PAPC is linked with its role in PCP signaling is an intriguing future subject. It is noteworthy that arcadlin, the mammalian homolog of PAPC, also downregulates another classic cadherin, N-cadherin (Yasuda et al., 2007). N-cadherin, which is localized at synaptic junctions in neurons, is endocytosed in an activity-dependent manner. The expression of arcadlin is upregulated by excitatory stimuli of hippocampal neurons, and this promotes N-cadherin internalization. During this process, arcadlin interacts with the cytoplasmic region of N-cadherin, and also interacts with TAO2$\beta$, a MAPKKK. Homophilic interaction of arcadlin molecules activates the TAO2$\beta$–MEK3 MAPKK–p38 MAPK pathway, which in turn phosphorylates the arcadlin and then accelerates the internalization of N-cadherin. These observations suggest that downregulation of classic cadherins might be a conserved function of PAPC/arcadlin.

OL-protocadherin (OL-pc, protocadherin-10) has also been shown to interfere with the classic cadherin function. In the knockout mice for OL-pc, the growth cones of striatal neurons do not normally migrate, as they lump together (Uemura et al., 2007). OL-pc interacts with the Nap1–WAVE complex (Nakao et al., 2008), a regulator of actin assembly, which functions downstream of Rac signaling. Although the Nap–WAVE complex is generally localized at the lamellipodia to sustain cell migration, this complex becomes redistributed to cell–cell contacts when OL-pc is expressed, because the OL-pc is concentrated there due to its homophilic interactions. As a consequence, the cell–cell contacts become unstable, and the action of classic cadherins to hold the apposing cell membranes is abrogated. Based on these observations, it has been proposed that a function of OL-pc is to upregulate the cell motile machinery at cell–cell contact sites, interfering with the classic cadherin-dependent contact inhibition of cell movement (Fig. 2.4).

Another protocadherin, NF-protocadherin (BH-protocadherin, protocadherin-7), is involved in the integrity of the deep ectoderm layer in *Xenopus* embryos (Bradley et al., 1998). NF-protocadherin interacts with TAF1 via the cytoplasmic domain of the former (Heggem and Bradley, 2003). Depletion of either this protocadherin or TAF1 in embryos results in neural tube closure defects, influencing the columnar epithelial morphology as well as convergent extension movement (Rashid et al., 2006). Thus, NF-protocadherin appears to contribute to maintaining the epithelial architecture and remodeling. Detailed molecular mechanisms of its action on cell–cell adhesion are not known.

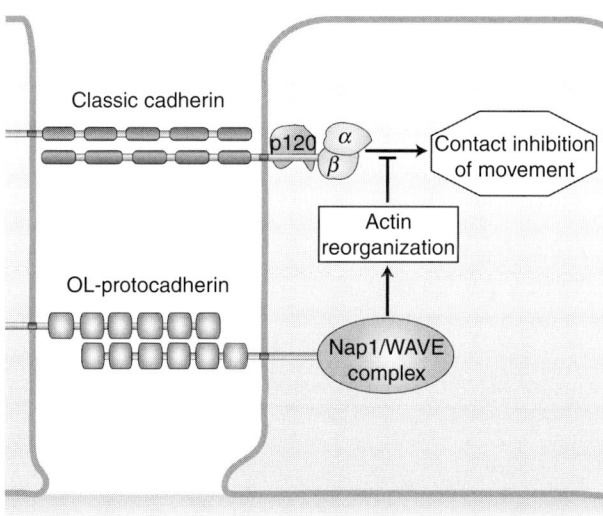

**Figure 2.4** Hypothetical role of OL-protocadherin in regulation of cell motility. Classic cadherins are known to induce contact inhibition of cell movement (Bracke et al., 1997; Chen and Obrink, 1991; Huttenlocher et al., 1998), and this activity of cadherins might be blocked by the OL-protocadherin–associated Nap1/WAVE complex, resulting in an increase in the motility of cells within their sheets (Nakao et al., 2008).

## 5.2. Fat cadherins

Fat cadherins, constituting another subfamily of the cadherin superfamily, are characterized by their unusually large extracellular domain (Tanoue and Takeichi, 2005). In *Drosophila*, although Fat is known to regulate the proliferation of cells, as well as planar cell polarity, the other subfamily member "Fat-like" is involved in tubular morphogenesis (Castillejo-Lopez et al., 2004), suggesting that the latter might play a role in epithelial remodeling. Cytological studies of a vertebrate homologue of Fat-like, termed Fat1, showed that and this molecule was in abundance at the basal regions of cell–cell contacts, segregated from the apically concentrated classic cadherins. Nevertheless, depletion of Fat1 loosens the AJ-based junctions (Tanoue and Takeichi, 2004). Biochemical studies showed that Fat1 interacts with Ena/VASP proteins at its cytoplasmic tail, promoting the polymerization of actin fibers such as stress fibers (Moeller et al., 2004; Tanoue and Takeichi, 2004). Interestingly, Fat1 loss in epithelial cells causes a reduction in not only the number of basally located stress fibers but also the amount of AJ associated F-actin. This finding explains why the cell junctions are widely disrupted by Fat depletion. These observations suggest that Fat1 indirectly control the AJ integrity via promotion of cytoplasmic actin polymerization. How Fat1-dependent actin polymerization affects the AJ-associated actin networks remains unknown.

## 5.3. Nectins

Nectins are immunoglobulin (Ig)-like cell adhesion molecules (Takai and Nakanishi, 2003). Nectin–nectin trans-interactions cause the activation of Cdc42 and Rac (Kawakatsu *et al.*, 2002), facilitating cadherin-mediated AJ formation (Fukuhara *et al.*, 2003, 2004). Nectin binds afadin at the cytoplasmic region of the former. Afadin interacts with Rap1 and further with p120-catenin, resulting in the strengthening of the binding of p120 to E-cadherin (Hoshino *et al.*, 2005), a process critical for E-cadherin stabilization. This pathway is considered to be one of the mechanisms of Rap1-dependent AJ remodeling.

An important biological function of the nectins is to recruit classic cadherins to specific junctional sites through their cross-interactions via afadin. Members of the nectin subfamily interact with each other in a heterophilic fashion more strongly than in their homophilic interactions; for example, the binding between nectin-1 and nectin-3 is stronger than that between 1 and 1, or 3 and 3 (Takai and Nakanishi, 2003). As a result, when cells expressing nectin-1 and nectin-3 are mixed, these nectins are selectively concentrated at the heterotypic cell-contact sites. This leads classic cadherins to accumulate preferentially to the heterotypic cell boundaries where the binding of nectin-1 to nectin-3 is taking place (Togashi *et al.*, 2006). This way of cooperation between classic cadherins and nectins (or other Ig-superfamily members) might play an important role in polarizing the AJ distribution, as well as serve to selectively link a specific pair of cells, in which the nectins and cadherins would function as a recognition receptor and adhesion stabilizer, respectively.

## 6. Junctional Remodeling During Morphogenesis

For the actual morphogenetic processes, various mechanisms of AJ remodeling, some of which have been discussed above, are assumed to operate singly or coordinately. Some examples are described below.

## 6.1. Remodeling through actin modulation

Given the critical role of actin filaments at the AJs, the regulation of F-actin polymerization or contractility is expected to have significant impacts on junctional remodeling. A typical example can be seen in the action of

Shroom3 (Hildebrand and Soriano, 1999). This actin-binding protein is localized along the circumferential actin belts (Fig. 2.3), which are also associated with myosin II (Hildebrand, 2005). Depending on the activity of Rho kinase (ROCK), the Shroom3-associated actin filaments contract, resulting in the apical constriction of the epithelial layers. The role of Shroom3 in this system is to recruit ROCKs to the junctions via its direct binding to them (Nishimura and Takeichi, 2008).

A more dynamic feature of junctional actin remodeling was observed in cells undergoing intercalation during the germ-band extension (GBE) in *Drosophila* embryos (Bertet *et al.*, 2004; Zallen and Wieschaus, 2004). In the intercalating cells, their junctions shrink selectively at the sides oriented toward the dorsoventral axis, resulting in the formation of vertices composed of four cells. Subsequently, the vertices elongate toward the anteroposterior direction to create new horizontal junctions. Myosin specifically localizes at the junctions to be shrunken, mediating the contraction of the actomyosin-associated junctions. Additional observations revealed that many of these intercalating cells are transiently organized into a rosette configuration (Blankenship *et al.*, 2006). These rosettes are then resolved so as to organize the anteroposteriorly oriented junctions. These ways of cellular rearrangement have been proposed to generate a driving force for the intercalation of germ-band cells (Lecuit and Lenne, 2007).

Rosette formation was also found in vertebrate ectodermal cells undergoing neural tube (Nishimura and Takeichi, 2008) or primitive streak (Wagstaff *et al.*, 2008) formation, implying that this would be a general mechanism for the epithelial cell rearrangement involving intercalation. Detailed observations of the apical surface of the closing neural tubes have clarified additional molecular events. The neural tube closure is a Shroom/ROCK dependent event (Haigo *et al.*, 2003; Hildebrand, 2005; Hildebrand and Soriano, 1999; Wei *et al.*, 2001), and these molecules are concentrated along the AJs located close to the apical surface of the columnar neuroepithelial cells (Nishimura and Takeichi, 2008), suggesting that the ROCK-dependent contraction of the AJ-associated actomyosin is responsible for the epithelial bending. However, the neuroepithelial bending is a polarized phenomenon; that is, the invaginating neuroepithelial cells bend only along the dorsoventral axis to form a tube. Correlating with this polarity, the phosphorylation of myosin light chain (MLC), which represents the active form of myosin II, is detectable preferentially along the AJs distributed toward the dorsoventral direction, which corresponds to the direction of bending (Fig. 2.5). ROCK inhibitors, or ROCK fragments which can interfere with the interaction of ROCK and Shroom3, abolish the rosette formation as well as the dorsoventrally polarized distribution of phosphorylated MLC. Concomitantly, neural tube closure fails. These findings suggest that the junctional actomyosin is locally activated in a

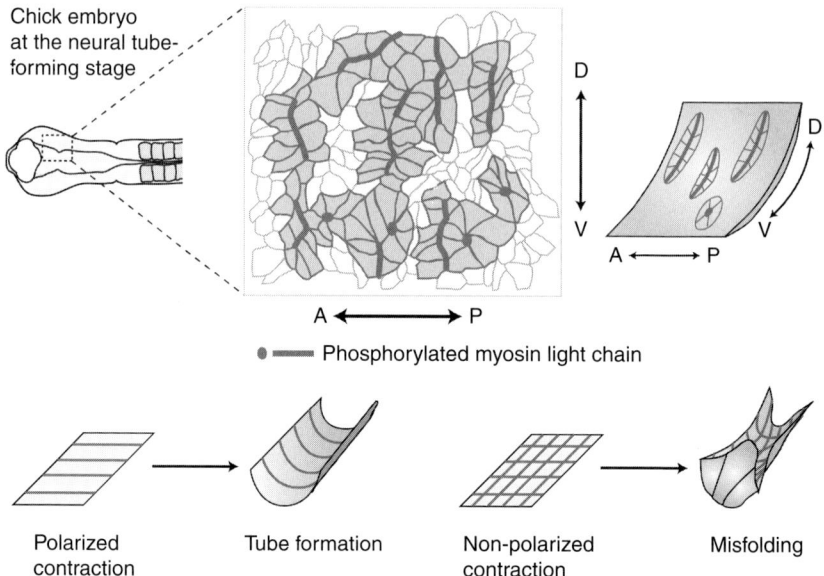

**Figure 2.5** Distribution of phosphorylated myosin light chain at the apical (ventricular) surface of the closing neural tube, with reference to the polarity of neuroepithelial bending. See Nishimura and Takeichi (2008) for the original data. In the lower panels, two types of the apical junctional contraction in an epithelial sheet are compared. The contraction toward the dorsoventral axis alone would be necessary for proper tube formation; and that along random directions may result in misfolding of the neuroepithelial sheet. D, dorsal; V, ventral; A, anterior; P, posterior.

planar-polarity fashion, so as to bend the neuroepithelial layer in a restricted direction, which leads to correct neural tube formation. How the actomyosin is locally activated remains to be elucidated.

## 6.2. Cadherin endocytosis during epithelial cell packing

Another type of epithelial remodeling has been discovered from experiments using the developing *Drosophila* wings (Classen et al., 2005). Wing epithelial cells are irregularly shaped in larvae, but most of them are reshaped to become hexagonal during the pupal stage before hair formation begins. At this time, E-cadherin is vigorously endocytosed with Rab11-positive endosomes and recycled, which may contribute to the active junctional remodeling and hexagonal repacking. E-cadherin recycling is affected by the mutation of PCP signaling molecules such as Flamingo cadherin, a cadherin superfamily member, and Frizzled, suggesting the involvement of PCP signaling in this type of junctional remodeling.

## 6.3. Classic cadherin regulation during gastrulation and neurulation

In *Xenopus* embryos, cadherins are involved in gastrulation movements. C-cadherin is a classical cadherin expressed in *Xenopus* embryos, and acts as a major mediator of intercellular adhesion in the blastula (Heasman *et al.*, 1994). Injection of mRNA encoding a dominant-negative form of C-cadherin into the prospective dorsal marginal zone causes gastrulation defects, such as incomplete involution and an open blastopore (Lee and Gumbiner, 1995). In the zebrafish, E-cadherin depletion impairs gastrulation movement (Babb and Marrs, 2004); and the E-cadherin-mediated cell adhesion between the deep cells and enveloping layer seems to be important for the epiboly movement (Shimizu *et al.*, 2005). On the other hand, downregulation of C-cadherin activity in *Xenopus* embryos is required for the convergent extension movement of gastrulation, as animal cap extension by activin is inhibited by a C-cadherin-activating antibody (Zhong *et al.*, 1999). Thus classic cadherins are likely essential for the collective cell migration (Friedl and Wolf, 2003) necessary for gastrulation, but their adhesive activity has to be downregulated to allow the convergent extension of the cell mass.

A number of mechanisms to downregulate the classic cadherins during gastrulation have been suggested. C-cadherin is downregulated by a TGF$\beta$ signaling pathway (Ogata *et al.*, 2007): Activin/nodal members of the TGF$\beta$ superfamily induce the expression of two genes, fibronectin leucine-rich repeat transmembrane 3 (FLRT3), a type I transmembrane protein containing extracellular leucine-rich repeats, and the small GTPase Rnd1. FLRT-3 physically interacts with Rnd1. Depletion of FLRT-3 or Rnd1 blocks the activin-induced animal cap elongation, upregulating C-cadherin-mediated cell adhesion. FLRT3 mediates C-cadherin internalization via Rab5- and dynamin-dependent endocytosis.

Wnt pathways are also involved in E-cadherin regulation during zebrafish gastrulation (Ulrich *et al.*, 2005). In Wnt11 mutants, the coordinated movement of the prechordal plate is disturbed. The mutant cells show cell cohesion defects, and lack Rab5-mediated endocytosis of E-cadherin. Enhancing Rab5c activity rescues the zebrafish from the mutant phenotypes. Thus, Wnt11 signaling plays a role in gastrulation movement through Rab5c-mediated E-cadherin endocytosis.

In the mouse, E-cadherin is transcriptionally downregulated by Snail (Cano *et al.*, 2000). This E-cadherin downregulation is accompanied by N-cadherin upregulation during gastrulation and neurulation (Takeichi, 1988); and E-cadherin proteins sharply disappear during this process. The p38 MAP kinase and a p38-interacting protein (p38IP) have been reported to downregulate the level of E-cadherin protein during gastrulation (Zohn *et al.*, 2006). In mice having a mutation in their p38IP, which interacts with and activates p38, both the downregulation of E-cadherin protein and cell migration during gastrulation are inhibited at the posterior primitive streak; although E-cadherin transcription is unaffected. Thus, the p38 pathway regulates gastrulation by acting on E-cadherin protein levels and, in turn, cell movement. As mentioned above, since microRNAs also regulate E-cadherin expression at EMT (Gregory et al., 2008; Park et al., 2008), they might be involved in the regulation of E-cadherin in epithelial remodeling during gastrulation.

## 7. Perspectives

We have outlined various mechanisms of AJ remodeling, as summarized in Fig. 2.6. Many of these mechanisms have been revealed by cell biological studies. One of the next important challenges is to test how they are utilized *in vivo*. We need to understand how the individual events affecting the AJs are networked together in developing embryos, and how these are linked to other signal pathways controlling morphogenesis. For example, MLC in the AJs of the closing neural tube is phosphorylated at

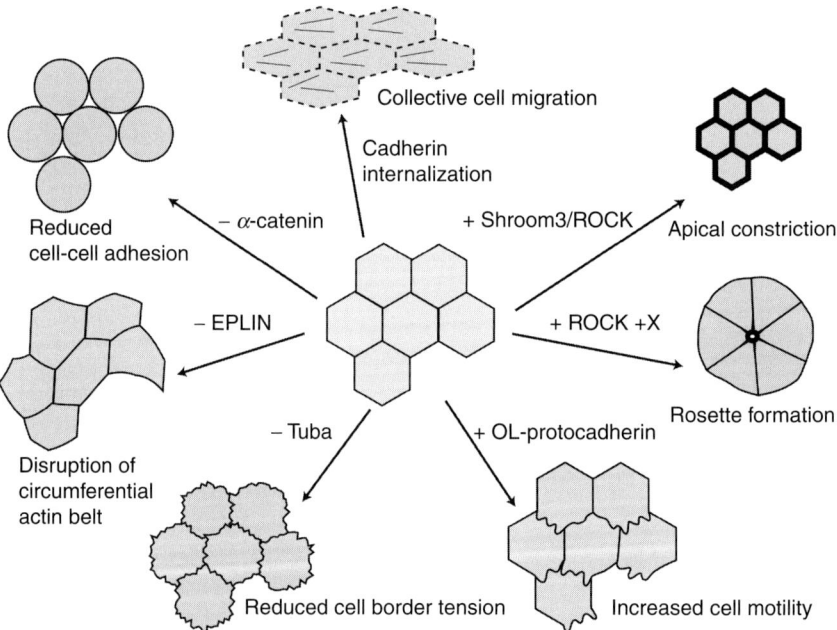

**Figure 2.6** Reshaping of epithelial sheets, which can be regulated by various forms of adherens junction remodeling.

restricted parts of the junctions. This restriction strongly suggests the involvement of PCP signaling, which widely controls tissue patterns, in this regulation of phosphorylation. In this case, our goal should be to clarify the linkage between AJ modulation and PCP signals.

On the other hand, the regulatory mechanisms of cell–cell adhesion are still unclear even at the cell biological level. New components of AJ are still being identified, and the entire molecular complex organizing the AJ appears to be more complex than ever thought. The detailed functions of each component remain unresolved for many of them. Even concerning the most classic component of AJ, α-catenin, its role is not perfectly understood yet. New technologies such as proteomics analysis of the AJ components and live-cell imaging of individual components should facilitate our eventual understanding of the AJ structure and remodeling mechanisms.

One of the final goals of developmental biology is to understand the mechanisms of how the complex body structures are organized by the cells. Remodeling of the epithelial cell junctions likely plays a key role in this aspect of development. In addition, misregulation of the AJ remodeling is very likely involved in pathogenetic behavior of cells, such as cancer metastasis. Understanding the regulation of junctional remodeling is thus critical for understanding both the basic morphogenetic processes and malignant cell behavior.

## ACKNOWLEDGMENTS

Work in our laboratory was supported by the program Grants-in-Aid for Specially Promoted Research of the Ministry of Education, Science, Sports, and Culture of Japan.

## REFERENCES

Abe, K., and Takeichi, M. (2008). EPLIN mediates linkage of the cadherin catenin complex to F-actin and stabilizes the circumferential actin belt. *Proc. Natl. Acad. Sci. USA* **105,** 13–19.

Abe, K., Chisaka, O., Van Roy, F., and Takeichi, M. (2004). Stability of dendritic spines and synaptic contacts is controlled by alpha N-catenin. *Nat. Neurosci.* **7,** 357–363.

Babb, S. G., and Marrs, J. A. (2004). E-cadherin regulates cell movements and tissue formation in early zebrafish embryos. *Dev. Dyn.* **230,** 263–277.

Bertet, C., Sulak, L., and Lecuit, T. (2004). Myosin-dependent junction remodelling controls planar cell intercalation and axis elongation. *Nature* **429,** 667–671.

Blankenship, J. T., Backovic, S. T., Sanny, J. S., Weitz, O., and Zallen, J. A. (2006). Multicellular rosette formation links planar cell polarity to tissue morphogenesis. *Dev. Cell* **11,** 459–470.

Bos, J. L. (2005). Linking Rap to cell adhesion. *Curr. Opin. Cell Biol.* **17,** 123–128.

Bracke, M. E., Depypere, H., Labit, C., Van Marck, V., Vennekens, K., Vermeulen, S. J., Maelfait, I., Philippe, J., Serreyn, R., and Mareel, M. M. (1997). Functional down-regulation of the E-cadherin/catenin complex leads to loss of contact inhibition of

motility and of mitochondrial activity, but not of growth in confluent epithelial cell cultures. *Eur. J. Cell Biol.* **74,** 342–349.
Bradley, R. S., Espeseth, A., and Kintner, C. (1998). NF-protocadherin, a novel member of the cadherin superfamily, is required for *Xenopus* ectodermal differentiation. *Curr. Biol.* **8,** 325–334.
Braga, V. M., and Yap, A. S. (2005). The challenges of abundance: Epithelial junctions and small GTPase signalling. *Curr. Opin. Cell Biol.* **17,** 466–474.
Bryant, D. M., and Stow, J. L. (2004). The ins and outs of E-cadherin trafficking. *Trends Cell Biol.* **14,** 427–434.
Cano, A., Perez-Moreno, M. A., Rodrigo, I., Locascio, A., Blanco, M. J., del Barrio, M. G., Portillo, F., and Nieto, M. A. (2000). The transcription factor snail controls epithelial–mesenchymal transitions by repressing E-cadherin expression. *Nat. Cell Biol.* **2,** 76–83.
Castillejo-Lopez, C., Arias, W. M., and Baumgartner, S. (2004). The fat-like gene of *Drosophila* is the true orthologue of vertebrate fat cadherins and is involved in the formation of tubular organs. *J. Biol. Chem.* **279,** 24034–24043.
Cavey, M., Rauzi, M., Lenne, P. F., and Lecuit, T. (2008). A two-tiered mechanism for stabilization and immobilization of E-cadherin. *Nature* **453,** 751–756.
Chen, X., and Gumbiner, B. M. (2006). Paraxial protocadherin mediates cell sorting and tissue morphogenesis by regulating C-cadherin adhesion activity. *J. Cell Biol.* **174,** 301–313.
Chen, W. C., and Obrink, B. (1991). Cell–cell contacts mediated by E-cadherin (uvomorulin) restrict invasive behavior of L-cells. *J. Cell Biol.* **114,** 319–327.
Chung, H. A., Yamamoto, T. S., and Ueno, N. (2007). ANR5, an FGF target gene product, regulates gastrulation in *Xenopus*. *Curr. Biol.* **17,** 932–939.
Classen, A. K., Anderson, K. I., Marois, E., and Eaton, S. (2005). Hexagonal packing of *Drosophila* wing epithelial cells by the planar cell polarity pathway. *Dev. Cell* **9,** 805–817.
Davis, M. A., and Reynolds, A. B. (2006). Blocked acinar development, E-cadherin reduction, and intraepithelial neoplasia upon ablation of p120-catenin in the mouse salivary gland. *Dev. Cell* **10,** 21–31.
Davis, M. A., Ireton, R. C., and Reynolds, A. B. (2003). A core function for p120-catenin in cadherin turnover. *J. Cell Biol.* **163,** 525–534.
Drees, F., Pokutta, S., Yamada, S., Nelson, W. J., and Weis, W. I. (2005). Alpha-catenin is a molecular switch that binds E-cadherin–beta-catenin and regulates actin-filament assembly. *Cell* **123,** 903–915.
Farquhar, M. G., and Palade, G. E. (1963). Junctional complexes in various epithelia. *J. Cell Biol.* **17,** 375–412.
Friedl, P., and Wolf, K. (2003). Tumour-cell invasion and migration: Diversity and escape mechanisms. *Nat. Rev. Cancer* **3,** 362–374.
Fujita, Y., Krause, G., Scheffner, M., Zechner, D., Leddy, H. E., Behrens, J., Sommer, T., and Birchmeier, W. (2002). Hakai, a c-Cbl-like protein, ubiquitinates and induces endocytosis of the E-cadherin complex. *Nat. Cell Biol.* **4,** 222–231.
Fukata, M., and Kaibuchi, K. (2001). Rho-family GTPases in cadherin-mediated cell–cell adhesion. *Nat. Rev. Mol. Cell Biol.* **2,** 887–897.
Fukuhara, A., Shimizu, K., Kawakatsu, T., Fukuhara, T., and Takai, Y. (2003). Involvement of nectin-activated Cdc42 small G protein in organization of adherens and tight junctions in Madin–Darby canine kidney cells. *J. Biol. Chem.* **278,** 51885–51893.
Fukuhara, T., Shimizu, K., Kawakatsu, T., Fukuyama, T., Minami, Y., Honda, T., Hoshino, T., Yamada, T., Ogita, H., Okada, M., and Takai, Y. (2004). Activation of Cdc42 by trans interactions of the cell adhesion molecules nectins through c-Src and Cdc42-GEF FRG. *J. Cell Biol.* **166,** 393–405.
Gregory, P. A., Bert, A. G., Paterson, E. L., Barry, S. C., Tsykin, A., Farshid, G., Vadas, M. A., Khew-Goodall, Y., and Goodall, G. J. (2008). The miR-200 family and

miR-205 regulate epithelial to mesenchymal transition by targeting ZEB1 and SIP1. *Nat. Cell Biol.* **10,** 593–601.

Gumbiner, B. M. (2005). Regulation of cadherin-mediated adhesion in morphogenesis. *Nat. Rev. Mol. Cell Biol.* **6,** 622–634.

Haigo, S. L., Hildebrand, J. D., Harland, R. M., and Wallingford, J. B. (2003). Shroom induces apical constriction and is required for hingepoint formation during neural tube closure. *Curr. Biol.* **13,** 2125–2137.

Heasman, J., Ginsberg, D., Geiger, B., Goldstone, K., Pratt, T., Yoshida-Noro, C., and Wylie, C. (1994). A functional test for maternally inherited cadherin in *Xenopus* shows its importance in cell adhesion at the blastula stage. *Development* **120,** 49–57.

Heggem, M. A., and Bradley, R. S. (2003). The cytoplasmic domain of *Xenopus* NF-protocadherin interacts with TAF1/set. *Dev. Cell* **4,** 419–429.

Hildebrand, J. D. (2005). Shroom regulates epithelial cell shape via the apical positioning of an actomyosin network. *J. Cell Sci.* **118,** 5191–5203.

Hildebrand, J. D., and Soriano, P. (1999). Shroom, a PDZ domain-containing actin-binding protein, is required for neural tube morphogenesis in mice. *Cell* **99,** 485–497.

Hirano, S., Kimoto, N., Shimoyama, Y., Hirohashi, S., and Takeichi, M. (1992). Identification of a neural alpha-catenin as a key regulator of cadherin function and multicellular organization. *Cell* **70,** 293–301.

Hogan, C., Serpente, N., Cogram, P., Hosking, C. R., Bialucha, C. U., Feller, S. M., Braga, V. M., Birchmeier, W., and Fujita, Y. (2004). Rap1 regulates the formation of E-cadherin-based cell–cell contacts. *Mol. Cell Biol.* **24,** 6690–6700.

Hoshino, T., Sakisaka, T., Baba, T., Yamada, T., Kimura, T., and Takai, Y. (2005). Regulation of E-cadherin endocytosis by nectin through afadin, Rap1, and p120ctn. *J. Biol. Chem.* **280,** 24095–24103.

Huttenlocher, A., Lakonishok, M., Kinder, M., Wu, S., Truong, T., Knudsen, K. A., and Horwitz, A. F. (1998). Integrin and cadherin synergy regulates contact inhibition of migration and motile activity. *J. Cell Biol.* **141,** 515–526.

Ireton, R. C., Davis, M. A., van Hengel, J., Mariner, D. J., Barnes, K., Thoreson, M. A., Anastasiadis, P. Z., Matrisian, L., Bundy, L. M., Sealy, L., Gilbert, B., van Roy, F., *et al.* (2002). A novel role for p120 catenin in E-cadherin function. *J. Cell Biol.* **159,** 465–476.

Izumi, G., Sakisaka, T., Baba, T., Tanaka, S., Morimoto, K., and Takai, Y. (2004). Endocytosis of E-cadherin regulated by Rac and Cdc42 small G proteins through IQGAP1 and actin filaments. *J. Cell Biol.* **166,** 237–248.

Kametani, Y., and Takeichi, M. (2007). Basal-to-apical cadherin flow at cell junctions. *Nat. Cell Biol.* **9,** 92–98.

Kawakatsu, T., Shimizu, K., Honda, T., Fukuhara, T., Hoshino, T., and Takai, Y. (2002). Trans-interactions of nectins induce formation of filopodia and Lamellipodia through the respective activation of Cdc42 and Rac small G proteins. *J. Biol. Chem.* **277,** 50749–50755.

Kim, S. H., Yamamoto, A., Bouwmeester, T., Agius, E., and Robertis, E. M. (1998). The role of paraxial protocadherin in selective adhesion and cell movements of the mesoderm during *Xenopus* gastrulation. *Development* **125,** 4681–4690.

Kimura, T., Sakisaka, T., Baba, T., Yamada, T., and Takai, Y. (2006). Involvement of the Ras–Ras-activated Rab5 guanine nucleotide exchange factor RIN2-Rab5 pathway in the hepatocyte growth factor-induced endocytosis of E-cadherin. *J. Biol. Chem.* **281,** 10598–10609.

Knox, A. L., and Brown, N. H. (2002). Rap1 GTPase regulation of adherens junction positioning and cell adhesion. *Science* **295,** 1285–1288.

Kooistra, M. R., Dube, N., and Bos, J. L. (2007). Rap1: A key regulator in cell–cell junction formation. *J. Cell Sci.* **120,** 17–22.

Langevin, J., Morgan, M. J., Sibarita, J. B., Aresta, S., Murthy, M., Schwarz, T., Camonis, J., and Bellaiche, Y. (2005). *Drosophila* exocyst components Sec5, Sec6, and Sec15 regulate DE-Cadherin trafficking from recycling endosomes to the plasma membrane. *Dev. Cell* **9,** 355–376.

Le, T. L., Yap, A. S., and Stow, J. L. (1999). Recycling of E-cadherin: A potential mechanism for regulating cadherin dynamics. *J. Cell Biol.* **146,** 219–232.

Lecuit, T., and Lenne, P. F. (2007). Cell surface mechanics and the control of cell shape, tissue patterns and morphogenesis. *Nat. Rev. Mol. Cell Biol.* **8,** 633–644.

Lee, C. H., and Gumbiner, B. M. (1995). Disruption of gastrulation movements in *Xenopus* by a dominant-negative mutant for C-cadherin. *Dev. Biol.* **171,** 363–373.

Liu, H., Komiya, S., Shimizu, M., Fukunaga, Y., and Nagafuchi, A. (2007). Involvement of p120 carboxy-terminal domain in cadherin trafficking. *Cell Struct. Funct.* **32,** 127–137.

Maul, R. S., Song, Y., Amann, K. J., Gerbin, S. C., Pollard, T. D., and Chang, D. D. (2003). EPLIN regulates actin dynamics by cross-linking and stabilizing filaments. *J. Cell Biol.* **160,** 399–407.

Medina, A., Swain, R. K., Kuerner, K. M., and Steinbeisser, H. (2004). *Xenopus* paraxial protocadherin has signaling functions and is involved in tissue separation. *EMBO J.* **23,** 3249–3258.

Miranda, K. C., Khromykh, T., Christy, P., Le, T. L., Gottardi, C. J., Yap, A. S., Stow, J. L., and Teasdale, R. D. (2001). A dileucine motif targets E-cadherin to the basolateral cell surface in Madin–Darby canine kidney and LLC-PK1 epithelial cells. *J. Biol. Chem.* **276,** 22565–22572.

Miyashita, Y., and Ozawa, M. (2007a). A dileucine motif in its cytoplasmic domain directs beta-catenin-uncoupled E-cadherin to the lysosome. *J. Cell Sci.* **120,** 4395–4406.

Miyashita, Y., and Ozawa, M. (2007b). Increased internalization of p120-uncoupled E-cadherin and a requirement for a dileucine motif in the cytoplasmic domain for endocytosis of the protein. *J. Biol. Chem.* **282,** 11540–11548.

Moeller, M. J., Soofi, A., Braun, G. S., Li, X., Watzl, C., Kriz, W., and Holzman, L. B. (2004). Protocadherin FAT1 binds Ena/VASP proteins and is necessary for actin dynamics and cell polarization. *EMBO J.* **23,** 3769–3779.

Nakao, S., Platek, A., Hirano, S., and Takeichi, M. (2008). Contact-dependent promotion of cell migration by the OL-protocadherin–Nap1 interaction. *J. Cell Biol.* **182,** 395–410.

Nishimura, T., and Takeichi, M. (2008). Shroom3-mediated recruitment of Rho kinases to the apical cell junctions regulates epithelial and neuroepithelial planar remodeling. *Development* **135,** 1493–1502.

Ogata, S., Morokuma, J., Hayata, T., Kolle, G., Niehrs, C., Ueno, N., and Cho, K. W. (2007). TGF-beta signaling-mediated morphogenesis: Modulation of cell adhesion via cadherin endocytosis. *Genes Dev.* **21,** 1817–1831.

Otani, T., Ichii, T., Aono, S., and Takeichi, M. (2006). Cdc42 GEF Tuba regulates the junctional configuration of simple epithelial cells. *J. Cell Biol.* **175,** 135–146.

Palacios, F., Tushir, J. S., Fujita, Y., and D'souza-Schorey, C. (2005). Lysosomal targeting of E-cadherin: A unique mechanism for the down-regulation of cell–cell adhesion during epithelial to mesenchymal transitions. *Mol. Cell Biol.* **25,** 389–402.

Park, S. M., Gaur, A. B., Lengyel, E., and Peter, M. E. (2008). The miR-200 family determines the epithelial phenotype of cancer cells by targeting the E-cadherin repressors ZEB1 and ZEB2. *Genes Dev.* **22,** 894–907.

Price, L. S., Hajdo-Milasinovic, A., Zhao, J., Zwartkruis, F. J., Collard, J. G., and Bos, J. L. (2004). Rap1 regulates E-cadherin-mediated cell–cell adhesion. *J. Biol. Chem.* **279,** 35127–35132.

Rashid, D., Newell, K., Shama, L., and Bradley, R. (2006). A requirement for NF-protocadherin and TAF1/Set in cell adhesion and neural tube formation. *Dev. Biol.* **291,** 170–181.

Redies, C., Vanhalst, K., and Roy, F. (2005). delta-Protocadherins: Unique structures and functions. *Cell Mol. Life Sci.* **62**, 2840–2852.

Rimm, D. L., Koslov, E. R., Kebriaei, P., Cianci, C. D., and Morrow, J. S. (1995). Alpha 1 (E)-catenin is an actin-binding and -bundling protein mediating the attachment of F-actin to the membrane adhesion complex. *Proc. Natl. Acad. Sci. USA* **92**, 8813–8817.

Shimizu, T., Yabe, T., Muraoka, O., Yonemura, S., Aramaki, S., Hatta, K., Bae, Y. K., Nojima, H., and Hibi, M. (2005). E-cadherin is required for gastrulation cell movements in zebrafish. *Mech. Dev.* **122**, 747–763.

Sivak, J. M., Petersen, L. F., and Amaya, E. (2005). FGF signal interpretation is directed by Sprouty and Spred proteins during mesoderm formation. *Dev. Cell* **8**, 689–701.

Takai, Y., and Nakanishi, H. (2003). Nectin and afadin: Novel organizers of intercellular junctions. *J. Cell Sci.* **116**, 17–27.

Takeichi, M. (1988). The cadherins: Cell–cell adhesion molecules controlling animal morphogenesis. *Development* **102**, 639–655.

Takeichi, M. (2007). The cadherin superfamily in neuronal connections and interactions. *Nat. Rev. Neurosci.* **8**, 11–20.

Takeichi, M., and Abe, K. (2005). Synaptic contact dynamics controlled by cadherin and catenins. *Trends Cell Biol.* **15**, 216–221.

Tanoue, T., and Takeichi, M. (2004). Mammalian Fat1 cadherin regulates actin dynamics and cell–cell contact. *J. Cell Biol.* **165**, 517–528.

Tanoue, T., and Takeichi, M. (2005). New insights into Fat cadherins. *J. Cell Sci.* **118**, 2347–2353.

Togashi, H., Miyoshi, J., Honda, T., Sakisaka, T., Takai, Y., and Takeichi, M. (2006). Interneurite affinity is regulated by heterophilic nectin interactions in concert with the cadherin machinery. *J. Cell Biol.* **174**, 141–151.

Uemura, M., Nakao, S., Suzuki, S. T., Takeichi, M., and Hirano, S. (2007). OL-protocadherin is essential for growth of striatal axons and thalamocortical projections. *Nat. Neurosci.* **10**, 1151–1159.

Ulrich, F., Krieg, M., Schotz, E. M., Link, V., Castanon, I., Schnabel, V., Taubenberger, A., Mueller, D., Puech, P. H., and Heisenberg, C. P. (2005). Wnt11 functions in gastrulation by controlling cell cohesion through Rab5c and E-cadherin. *Dev. Cell* **9**, 555–564.

Unterseher, F., Hefele, J. A., Giehl, K., De Robertis, E. M., Wedlich, D., and Schambony, A. (2004). Paraxial protocadherin coordinates cell polarity during convergent extension via Rho A and JNK. *EMBO J.* **23**, 3259–3269.

Vanhalst, K., Kools, P., Stacs, K., van Roy, F., and Redies, C. (2005). delta-Protocadherins: A gene family expressed differentially in the mouse brain. *Cell Mol. Life Sci.* **62**, 1247–1259.

Vasioukhin, V., Bauer, C., Degenstein, L., Wise, B., and Fuchs, E. (2001). Hyperproliferation and defects in epithelial polarity upon conditional ablation of alpha-catenin in skin. *Cell* **104**, 605–617.

Wagstaff, L. J., Bellett, G., Mogensen, M. M., and Munsterberg, A. (2008). Multicellular rosette formation during cell ingression in the avian primitive streak. *Dev. Dyn.* **237**, 91–96.

Wang, Y., Janicki, P., Koster, I., Berger, C. D., Wenzl, C., Grosshans, J., and Steinbeisser, H. (2008). Xenopus paraxial protocadherin regulates morphogenesis by antagonizing Sprouty. *Genes Dev.* **22**, 878–883.

Watabe-Uchida, M., Uchida, N., Imamura, Y., Nagafuchi, A., Fujimoto, K., Uemura, T., Vermeulen, S., van Roy, F., Adamson, E. D., and Takeichi, M. (1998). alpha-Catenin–vinculin interaction functions to organize the apical junctional complex in epithelial cells. *J. Cell Biol.* **142**, 847–857.

Wei, L., Roberts, W., Wang, L., Yamada, M., Zhang, S., Zhao, Z., Rivkees, S. A., Schwartz, R. J., and Imanaka-Yoshida, K. (2001). Rho kinases play an obligatory role in vertebrate embryonic organogenesis. *Development* **128**, 2953–2962.

Xiao, K., Allison, D. F., Buckley, K. M., Kottke, M. D., Vincent, P. A., Faundez, V., and Kowalczyk, A. P. (2003). Cellular levels of p120 catenin function as a set point for cadherin expression levels in microvascular endothelial cells. *J. Cell Biol.* **163,** 535–545.

Xiao, K., Garner, J., Buckley, K. M., Vincent, P. A., Chiasson, C. M., Dejana, E., Faundez, V., and Kowalczyk, A. P. (2005). p120-Catenin regulates clathrin-dependent endocytosis of VE-cadherin. *Mol. Biol. Cell* **16,** 5141–5151.

Xiao, K., Oas, R. G., Chiasson, C. M., and Kowalczyk, A. P. (2007). Role of p120-catenin in cadherin trafficking. *Biochim. Biophys. Acta* **1773,** 8–16.

Yamada, S., Pokutta, S., Drees, F., Weis, W. I., and Nelson, W. J. (2005). Deconstructing the cadherin–catenin–actin complex. *Cell* **123,** 889–901.

Yap, A. S., Crampton, M. S., and Hardin, J. (2007). Making and breaking contacts: The cellular biology of cadherin regulation. *Curr. Opin. Cell Biol.* **19,** 508–514.

Yasuda, S., Tanaka, H., Sugiura, H., Okamura, K., Sakaguchi, T., Tran, U., Takemiya, T., Mizoguchi, A., Yagita, Y., Sakurai, T., De Robertis, E. M., and Yamagata, K. (2007). Activity-induced protocadherin arcadlin regulates dendritic spine number by triggering N-cadherin endocytosis via TAO2beta and p38 MAP kinases. *Neuron* **56,** 456–471.

Yeaman, C., Grindstaff, K. K., and Nelson, W. J. (2004). Mechanism of recruiting Sec6/8 (exocyst) complex to the apical junctional complex during polarization of epithelial cells. *J. Cell Sci.* **117,** 559–570.

Zallen, J. A., and Wieschaus, E. (2004). Patterned gene expression directs bipolar planar polarity in *Drosophila*. *Dev. Cell* **6,** 343–355.

Zhong, Y., Brieher, W. M., and Gumbiner, B. M. (1999). Analysis of C-cadherin regulation during tissue morphogenesis with an activating antibody. *J. Cell Biol.* **144,** 351–359.

Zohn, I. E., Li, Y., Skolnik, E. Y., Anderson, K. V., Han, J., and Niswander, L. (2006). p38 and a p38-interacting protein are critical for downregulation of E-cadherin during mouse gastrulation. *Cell* **125,** 957–969.

CHAPTER THREE

# HOW THE CYTOSKELETON HELPS BUILD THE EMBRYONIC BODY PLAN: MODELS OF MORPHOGENESIS FROM *DROSOPHILA*

Tony J. C. Harris,* Jessica K. Sawyer,[†] *and* Mark Peifer[†,‡]

## Contents

| | |
|---|---|
| 1. Introduction | 56 |
| 2. Establishing Epithelial Structure | 57 |
|    2.1. Forming the first epithelial cells of the *Drosophila* embryo | 57 |
|    2.2. Connecting cells by assembling adherens junctions | 60 |
| 3. Epithelial Morphogenesis | 63 |
|    3.1. Internalizing mesoderm: Actomyosin contractility drives apical constriction | 63 |
|    3.2. Germband extension: Actomyosin contractility driving cell intercalation | 67 |
|    3.3. Microtubule arrays contributing to gastrulation | 69 |
|    3.4. Dorsal closure: Integrating signaling, the cytoskeleton, and adhesion to drive epithelial sheet migration and sealing | 70 |
|    3.5. Adding complexity to the JNK and Dpp pathways | 71 |
|    3.6. An embarrassment of riches: Five nonreceptor tyrosine kinases regulate dorsal closure | 74 |
|    3.7. Roles for novel regulators of adhesion and the cytoskeleton | 75 |
|    3.8. Powering closure by old and new mechanisms | 76 |
|    3.9. Reaching out to close the gap—Key roles for cell protrusions in dorsal closure | 77 |
| 4. Coordinating Cytoskeletal Machinery in Epithelial Cells | 78 |
| 5. Concluding Statement | 79 |
| References | 80 |

---

\* Department of Cell and Systems Biology, University of Toronto, Toronto, Ontario, Canada
[†] Department of Biology, University of North Carolina at Chapel Hill, Chapel Hill, North Carolina, USA
[‡] Lineberger Comprehensive Cancer Center, University of North Carolina at Chapel Hill, Chapel Hill, North Carolina, USA

## Abstract

One key challenge for cell and developmental biologists is to determine how the cytoskeletal toolkit is used to build embryonic tissues and organs. Here, we review recent progress in meeting this challenge, focusing on epithelial morphogenesis in the *Drosophila* embryo as a model. We outline how actin and microtubule networks are regulated by embryonic patterning systems, and how they affect cell shape, cell behavior, and cell–cell interactions to shape epithelial structures. We focus on the formation of the first epithelium at cellularization, the assembly of junctions, apical constriction of cells in the ventral furrow, cell intercalation in the germband, and epithelial sheet migration during dorsal closure. These events provide models for uncovering the cell biological basis of morphogenesis.

## 1. INTRODUCTION

Morphogenesis is the remarkable process by which the body's cells self-assemble into complex tissues and organs. The most common form of tissue organization in the animal body plan is the epithelium. Epithelia are sheets of adherent cells that form boundaries between the body's compartments. Each epithelial cell has an apical domain facing the surface of the sheet and a basolateral domain facing underlying tissue. This polarity is tightly linked to epithelial structure. In mature epithelia, cell–cell junctions form around the circumference of the apical domain—these include the adherens junctions (AJs) that connect neighboring cells and the tight junctions or septate junctions that seal cell–cell contacts. Each of these junctions interact with apical actin, but actin also localizes laterally and basally where it binds integrins engaged with the substratum. Similarly, microtubules (MTs) are organized in specific apical, lateral and basal networks, and intracellular trafficking pathways direct specific cargo to the apical or basolateral domains. This polarized organization of epithelial cells controls transport between body compartments, and is critical for the development and maintenance of epithelial structure. Its loss is a hallmark of cancer (reviewed in Gumbiner, 2005; Nelson, 2003; Tepass *et al.*, 2001).

Textbook models of epithelial structure often appear static, but establishing, remodeling, and maintaining epithelial tissues requires continual, dynamic interactions between AJs, cytoskeletal proteins, polarity cues, and intracellular trafficking. The *Drosophila* embryo has become a major model for understanding these processes and how they are regulated by embryonic patterning mechanisms to generate the animal body plan. In this chapter, we focus on the control and effects of cytoskeletal networks during the initial establishment of epithelial structure in *Drosophila*, and during epithelial remodeling at gastrulation and dorsal closure.

## 2. Establishing Epithelial Structure

### 2.1. Forming the first epithelial cells of the *Drosophila* embryo

The first epithelium of the *Drosophila* embryo forms through a process called cellularization (reviewed in Lecuit, 2004; Mazumdar and Mazumdar, 2002). Unlike most animals, *Drosophila* development begins in a syncytium. The zygotic nucleus undergoes 13 rounds of nuclear divisions, leading to $\sim 6000$ nuclei lining the embryo periphery just below the plasma membrane. The plasma membrane then forms furrows that invaginate from the embryo surface and synchronously compartmentalize the nuclei into $\sim 6000$ individual columnar cells. This process forms a tightly packed array of columnar hexagonal cells. It is organized primarily by cytoskeletal networks and membrane trafficking, and occurs prior to the establishment of AJ-based adhesion (Fig. 3.1).

The MT network acts as an important framework for organizing the basic internal structure of the cells and their precursors, the initial nuclear compartments. Each cell's MT organization was described over 20 years ago (Warn and Warn, 1986). Two centrosomes localize above the nucleus. MT bundles emanate from these sites and run basally around each nucleus to form "inverted baskets" (Fig. 3.1A). These MT baskets form before the plasma membrane invaginates, but it is unclear how these MT networks are recruited to the cortex and assembled. They function in clearing yolk away from the embryo periphery where the cells will form, and in organizing each cell's endoplasmic reticulum and Golgi apparatus around the cell nucleus (Frescas *et al.*, 2006; Fig. 3.1B). In this way, the MT cytoskeleton sets up basic cell architecture before the plasma membrane has even formed.

Membrane trafficking generates lateral plasma membrane invaginations (Fig. 3.1C and D). Photobleaching studies revealed that each cell organizes its own secretory system to deliver components to the plasma membrane (Frescas *et al.*, 2006). Membrane material is first trafficked to the apical surface of the cells, and then continually displaced to form lateral cell membranes around the MT baskets and nuclei (Lecuit and Wieschaus, 2000). MTs and dynein are required to transport Golgi vesicles into the apical domain (Lecuit and Wieschaus, 2000; Papoulas *et al.*, 2005), and disrupting the link between these vesicles and MTs (by mutating the Golgi associated protein Lava lamp) disrupts apical Golgi localization and lateral membrane formation (Papoulas *et al.*, 2005; Sisson *et al.*, 2000). Thus, Golgi vesicles may supply materials to the apical surface, which then flow basally to form the lateral membranes. Blocking membrane scission events through disruption of dynamin also blocks furrow formation, and this coincides with a block in post-Golgi trafficking at apical recycling endosomes (Pelissier *et al.*, 2003). Blocking the recycling endosome regulator Rab11 function

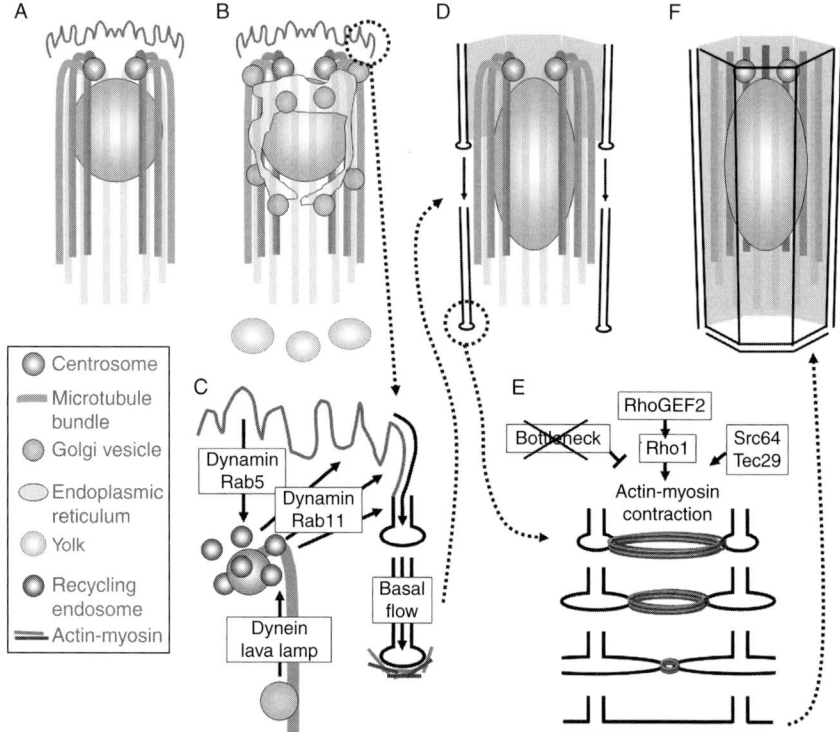

**Figure 3.1** Forming the first epithelial cells of the *Drosophila* embryo through cellularization. (A) Microtubule organization in a cellularization compartment before the formation of lateral and basal plasma membranes. The apical plasma is at the top and covers microvilli. The centrosomes are just above the nucleus. (B) The positioning of the endoplasmic reticulum, Golgi vesicles, and yolk in a cellularization compartment before the formation of lateral and basal plasma membranes. (C) Membrane trafficking pathways supplying the growth of lateral plasma membrane furrows (right). (D) Growth of lateral plasma membrane furrows to their full length around a cell compartment. (E) Signaling pathways regulating the contractile actomyosin ring that forms the basal membrane of the cell and the yolk membrane beneath. (F) A hexagonal cell at the end of cellularization. (See Color Insert.)

inhibits furrows as well (Pelissier *et al.*, 2003; Riggs *et al.*, 2003), further suggesting that material travels from the Golgi via the recycling endosomes to the plasma membrane. Interestingly, Discs Large and Strabismus, proteins normally associated with cortical cell polarity, have also been implicated in this post-Golgi trafficking (Lee *et al.*, 2003), although links to dynamin and Rab11 have not been made. The massive growth of plasma membranes during cellularization must require substantial trafficking from internal membranes, but additional mechanisms may also reorganize the plasma membranes as this occurs. For example, disruption of the early endosome regulator Rab5 also blocks plasma membrane invagination suggesting a role for plasma membrane endocytosis (Pelissier et al., 2003). Indeed, plasma membrane endocytosis was recently documented during the invagination process (Sokac and Wieschaus, 2008a). Moreover, remodeling of actin-based apical microvilli might also contribute to lateral membrane growth—as the density of the apical microvilli decreases, and thus the amount of apical membrane decreases, the lateral membrane grows (Grevengoed et al., 2003; Turner and Mahowald, 1976).

Contractile actin–myosin rings are critical for completing the formation of epithelial cells at the end of cellularization. These rings form at the base of the plasma membranes (the furrow canals) as they initiate invagination (Warn and Magrath, 1983; Warn et al., 1980; Young et al., 1991; Fig. 3.1C). The ring of one cell is connected to rings of each neighboring cell, creating a large network covering the base of all cells. Once the lateral membranes reach their full length, the actin–myosin rings detach from each other and constrict around the base of each cell (Royou et al., 2004). These contractile rings drive a specialized form of cell division that separates the base of the lateral membranes into the basal membrane of the cell and the yolk membrane beneath (Fig. 3.1E and F). Actomyosin contractility is often regulated the RhoGEF–Rho–Rho kinase–Myosin II regulatory light chain–Myosin II heavy chain regulatory cassette (Matsumura, 2005). Indeed, Rho is required for cellularization (Crawford et al., 1998). During cellularization, RhoGEF2 is required for recruiting Rho1 to the furrow canals (Padash Barmchi et al., 2005), and the final constriction of the rings is blocked by either full loss of RhoGEF2 function (Grosshans et al., 2005; Padash Barmchi et al., 2005) or disruptions of Myosin II (Royou et al., 2004). Other data suggest that RhoGEF2 may work with the formin Dia, which is also required for this process (Afshar et al., 2000; Grosshans et al., 2005). Additionally, the nonreceptor kinases Src64 and Tec29 are required for this final constriction (Thomas and Wieschaus, 2004). Significantly, actomyosin constriction does not fully separate the base of the cells from the yolk membrane in wild type, and actin-lined "yolk channels" are left connecting the yolk membrane and the cells during gastrulation (Rickoll and Counce, 1980); stability of these yolk channels during gastrulation requires the septin Peanut (Adam et al., 2000).

The timing of the basal constriction and the organization of basal actin is tightly regulated. For example, the protein Bottleneck prevents premature constriction of the rings during earlier plasma membrane invagination (Schejter and Wieschaus, 1993). Interestingly, it remains uncertain whether actomyosin contractility plays a major role in membrane invagination. Invagination is inhibited with disruption of Rho1, but invagination appears normal with full loss of RhoGEF2 (Padash Barmchi et al., 2005), disruption of Myosin II (Royou et al., 2004), or disruption of Src64 or Tec29 (Thomas and Wieschaus, 2004). However, RhoGEF2 is required to stabilize Bottleneck and *RhoGEF2, bottleneck* double mutants display major disorganization of the basal actin network during early invagination (Padash Barmchi *et al.*, 2005). Furthermore, an intact actin network appears to be important for invagination, as Nullo mutants have a diminished basal actin network and display sporadic loss of plasma membrane invaginations (Simpson and Wieschaus, 1990). Similarly, Abelson kinase mutants show reduced basal actin as a result of abnormally elevated apical actin, and also display plasma membrane furrow defects (Grevengoed *et al.*, 2003). Thus, the role for actin in plasma membrane invagination appears complex and may involve multiple, possibly redundant, functions. For example, actin also has a role in regulating endocytosis events at the furrow canals (Sokac and Wieschaus, 2008a). Although actomyosin contractility may not be essential for membrane invagination, it appears to be critical for constricting and pinching off the base of each cell to complete cellularization.

## 2.2. Connecting cells by assembling adherens junctions

Remarkably, cellularization does not require AJs. It occurs normally in strong mutants of AJ components in which AJ assembly is undetectable by either immunofluorescence or electron microscopy (Cox *et al.*, 1996; Harris and Peifer, 2004; Sokac and Wieschaus, 2008b). One explanation for this may be the tight packing of the cellular array produced by the cytoskeletal and trafficking activities described above. This tight packing may make cell–cell adhesion dispensable for holding the cells together. However, AJs are assembled during cellularization and become essential for epithelial structure as the tissue undergoes large-scale tissue remodeling at gastrulation (Cox *et al.*, 1996; Harris and Peifer, 2004). Intriguingly, MTs have a major influence on AJ assembly during cellularization (when AJs are dispensable) and then major links with actin arise during gastrulation (when AJs become essential; Fig. 3.2A–C).

The MT network plays a central role in directing AJ assembly during cellularization. Since AJs are dispensable at this stage, cellularization can be thought of as a staging ground for establishing AJ-based adhesion. Apical junctions initially form as discontinuous clusters called spot AJs (Tepass and Hartenstein, 1994). These spot AJs are positioned at the same apical–basal position as centrosomes, placing them in proximity to the MT minus ends. This positioning and the basal mislocalization of AJ components in mutants of the minus-end-directed motor Dynein suggest that apical MT minus ends act as landmarks for spot AJ assembly (Harris and Peifer, 2005). The polarity regulator Bazooka (Baz, the *Drosophila* homolog of PAR-3) appears to coordinate AJ–MT interactions during cellularization. Baz (PAR-3) localizes to spot AJ assembly sites next to centrosomes, and in *baz* mutants AJ components fail to accumulate at these assembly sites. Remarkably, in AJ

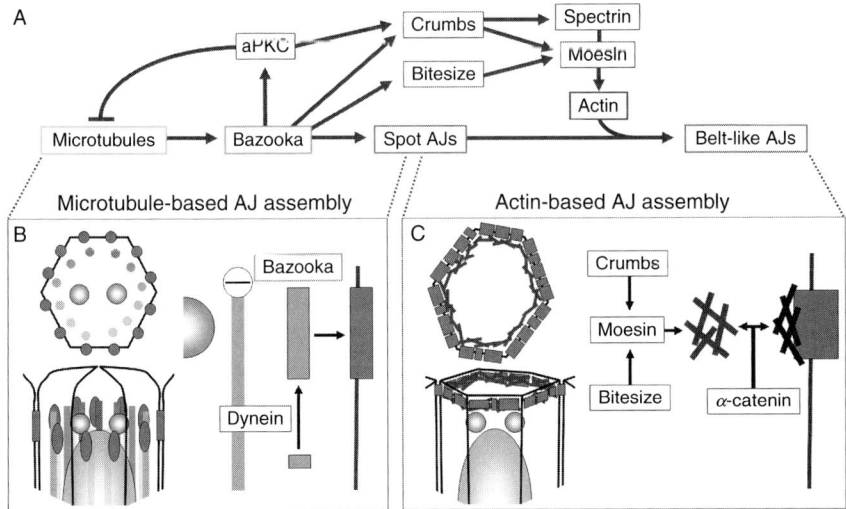

**Figure 3.2** The adherens junction assembly pathway during cellularization and gastrulation of the *Drosophila* embryo. (A) Pathways regulating the assembly of spot AJs and more mature belt-like AJs. (B) A MT-based assembly mechanism plays a major role in forming and positioning spot AJs during cellularization. Left: surface and side views of the top half of a cell show spot AJs (purple) in the subapical lateral region at the same apical–basal position as centrosomes (red) and in proximity to lateral MT bundles (green). Right: molecular interactions regulating this positioning. (C) An actin-based assembly mechanism plays a major role in forming and positioning belt-like AJs during gastrulation. Left: surface and side views of the top half of a cell show belt-like AJs (purple) at the apex of the lateral region well above the centrosomes but in close association with actin. Right: molecular interactions regulating this association with actin (light and dark blue). (See Color Insert.)

mutants, Baz (PAR-3) can accumulate at spot AJ assembly sites despite the absence of spot AJs (Harris and Peifer, 2004). This suggests a model in which MT polarity positions Baz (PAR-3) to organize spot AJ assembly sites (Fig. 3.2B), and more specifically that Baz (PAR-3) and AJ proteins may be transported and/or maintained apically by Dynein.

The PAR-6/aPKC complex is an important regulator of the switch from cellularization AJ–MT interactions to gastrulation AJ–actin interactions. The complex is recruited to the apical domain via undefined steps downstream of Baz (PAR-3) (Harris and Peifer, 2005). In many contexts, PAR-6/aPKC forms complexes with Baz (PAR-3) (reviewed in Macara, 2004; Nance, 2005; Ohno, 2001), but during *Drosophila* cellularization and gastrulation, PAR-6/aPKC complexes localize above Baz (PAR-3) in the apical domain (Harris and Peifer, 2005). In *par-6* and *apkc* mutants, AJs are normally positioned through the end of cellularization, and remain apically positioned during gastrulation. However, in the absence of Par-6 or aPKC, the normally symmetric spot AJs relocalize to two large puncta during early gastrulation, usually at the dorsal and ventral sides of the cells (Blankenship et al., 2006; Harris and Peifer, 2007). In WT embryos, AJs become organized into more continuous belt-like AJs by the end gastrulation (stage 7–8; Tepass and Hartenstein, 1994) and localize to a more extreme apical position above the position of the centrosomes (Harris and Peifer, 2005). Without PAR-6/aPKC, both of these changes fail. Closer examination of *apkc* mutants showed that the abnormal formation of the two large Baz (PAR-3)-associated AJ puncta is due to abnormally persistent associations with two overactive centrosomal MT asters (Harris and Peifer, 2007). Thus, PAR-6/aPKC activity appears to downregulate AJ–MT interactions to allow the next steps of AJ maturation. PAR-6/aPKC may also actively promote AJ maturation by phosphorylating and enhancing the activity of the apical determinant Crumbs (Sotillos et al., 2004).

AJ maturation requires complex links to the actin cytoskeleton. The larger cytoskeletal network of the apical domain is regulated by the Crumbs complex and Bitesize. Crumbs and Bitesize are recruited to AJs via poorly defined steps downstream of Baz (PAR-3) (Bilder et al., 2003; Pilot et al., 2006)—the Crumbs recruitment appears to be independent of AJs (Harris and Peifer, 2004), and the Bitesize recruitment involves localized PIP2 (Pilot et al., 2006). Crumbs is a transmembrane protein that localizes apical to AJs at the core of a conserved apical polarity complex (reviewed in Knust and Bossinger, 2002). Bitesize is a synaptotagmin-like protein that colocalizes with AJs (Pilot et al., 2006). Loss of either Crumbs or Bitesize function leads to discontinuous AJs by the end of gastrulation (Pilot et al., 2006; Tepass, 1996). For Crumbs, these effects are likely indirect, because Crumbs localizes above AJs. Crumbs may act via effects on the apical cytoskeleton, as Crumbs can form a complex with $\beta$-Heavy-spectrin and Moesin and is required for proper organization of the spectrin-based cytoskeleton in the apical domain (Medina et al., 2002). Interestingly, Bitesize also appears to affect AJ organization by regulating the apical cytoskeleton. Bitesize can also form a complex with Moesin, and with Bitesize disruption, AJ defects are accompanied by abnormal actin organization and defects in Moesin recruitment to junctions (Pilot et al., 2006). Thus, both Bitesize and Crumbs appear to recruit regulators of the actin cytoskeleton to promote belt-like AJs (Fig. 3.2A and C).

We are beginning to gain insights into the actin–AJs connections. Live imaging and FRAP analyses have revealed that even belt-like junctions are actually a mosaic of small foci (Cavey et al., 2008). These foci associate with stable actin patches, which appear to engage the larger actin network of the apical domain to stabilize the foci along the length of belt-like junctions. The actin-binding protein and AJ protein $\alpha$-catenin is apparently dispensable for assembling the local actin patches, but links the AJ foci to the larger actin network of the apical domain (Cavey et al., 2008). Thus, $\alpha$-catenin may connect the stable actin patches at AJ foci to the larger apical actin network (Fig. 3.2C), a mechanism that would be consistent with evidence that α-catenin cannot bind AJs and actin simultaneously (Yamada et al., 2005).

## 3. Epithelial Morphogenesis

At the end of cellularization, a simple epithelial monolayer of ~ 6000 cells covers the embryo surface. To generate the animal body plan, this epithelial sheet is dramatically remodeled by cytoskeletal activities directed by embryonic patterning systems. During gastrulation four major changes occur in the embryo. The ventral furrow and posterior midgut invaginate to form the mesoderm and endoderm, respectively. The lateral germband extends to elongate the body axis, and a patch of extraembryonic cells forms on the dorsal side. Another round of major epithelial rearrangements occurs later in development during dorsal closure, when the epidermis displaces the amnioserosa from the dorsal side to fully envelope the embryo so it can hatch as a larva. These events provide valuable models for a range of morphogenetic processes.

### 3.1. Internalizing mesoderm: Actomyosin contractility drives apical constriction

Internalization of the ventral furrow cells (mesoderm) and posterior midgut (endoderm) are the first steps in *Drosophila* gastrulation (reviewed in Lecuit and Lenne, 2007; Leptin, 1999). The ventral furrow involves a stripe of cells along the ventral midline that is 18 cells wide and 60 cells long. These cells will invaginate and form a tube in the interior of the embryo. Eventually, cells in the tube will disperse into a single layer of cells beneath the ectoderm and become mesoderm (Costa et al., 1993). This process of internalization is characterized by four distinct phases. First, cells apically flatten and display random cell constrictions. Second, cells spanning a 12-cell width begin apically constricting in a coordinated fashion, resulting in a bend in the epithelium. As cells constrict, small membrane protrusions/blebs form on the apical surface, which may be a response to, or possibly aid in, the reduction of the apical surface area. At the same time, cells elongate along the apical–basal axis to 1.7 times their original length. Additionally, their nuclei shift basally and their basal surfaces expand. Third, after the ventral furrow cells have reached their maximum length, they begin to shorten back to their original length, while remaining constricted apically. This shortening results in a wedge shape and may help to move the furrow beneath the epidermis. Finally, the lateral epidermis on either side of the furrow covers the tube of mesoderm, separating it from the overlaying ectoderm (Costa et al., 1993; Leptin and Grunewald, 1990; Sweeton et al., 1991).

Almost 20 years ago, apical constriction of ventral furrow cells was proposed to be a result of contraction of the actin cytoskeleton underlying the apices of the cells (Young et al., 1991). A pathway that instructs cells to constrict has begun to emerge, starting with specification and ending with cell shape change (Fig. 3.3A). First, specification and internalization of the mesoderm is controlled by the transcription factors Twist and Snail. Both genes are required for mesodermal fates. Loss of function of these genes results in elimination of ventral furrow formation and an expansion of lateral fates into the ventral domain (Costa et al., 1993).

Twist and Snail have many transcriptional targets, including some involved in triggering constriction. One is the secreted ligand Folded Gastrulation (Fog), and Concertina (Cta), a G protein $\alpha 12/13$ subunit, acts downstream of this ligand (Costa et al., 1994; Morize et al., 1998). However, the G-coupled receptor for Fog remains a mystery. RhoGEF2, a regulator of the Rho family GTPases, acts downstream of Fog–Cta signaling and links the signaling pathway with the cytoskeletal machinery (Barrett et al., 1997; Rogers et al., 2004). Expression of dominant-negative Rho1 results in ventral furrow defects similar to *fog* and *cta*, suggesting that RhoGEF2 mostly likely activates Rho1 to initiate cell shape change (Barrett et al., 1997).

The transmembrane protein T48, a RhoGEF2-binding partner regulated by Twi, appears to function in parallel to Fog–Cta signaling to recruit RhoGEF2 apically for activation (Kolsch et al., 2007). The existence of parallel pathways is supported by the fact that mutations in *fog, cta*, or *T48* lead to uncoordinated constriction of cells, but the ventral furrow eventually forms (Costa et al., 1994; Dawes-Hoang et al., 2005; Kolsch et al., 2007; Sweeton et al., 1991). In contrast, loss of RhoGEF2 severely disrupts apical constriction and the ventral furrow never forms (Barrett et al., 1997; Hacker and Perrimon, 1998). Interestingly, ectopic expression of Fog leads to ectopic apical constrictions in the embryo (Costa et al., 1994); however, in embryos lacking RhoGEF2, ectopic expression of Fog fails to produce ectopic constrictions (Barrett et al., 1997).

How do these proteins affect the contractile machinery? Further research suggests that pathway activation affects apical localization of RhoGEF2 and/or Myosin II. In wild-type ventral furrow cells, RhoGEF2 and Myosin II are first localized basally at the tips of cellularization furrows. At gastrulation onset they are relocalized apically and cells begin to constrict (Kolsch et al., 2007; Nikolaidou and Barrett, 2004). T48 and Cta each mildly affect the localization of RhoGEF2, but if both proteins are absent RhoGEF2 does not become apically localized (Kolsch et al., 2007). *cta* mutants have reduced and patchy accumulation of apical Myosin II, resulting in constriction of some, but not all, cells in the furrow (Nikolaidou and Barrett, 2004). Ventral furrow cells lacking RhoGEF2 fail to accumulate Myosin II apically in all cells and are therefore unable to constrict (Nikolaidou and Barrett, 2004), suggesting it is at the convergence of the constriction signals.

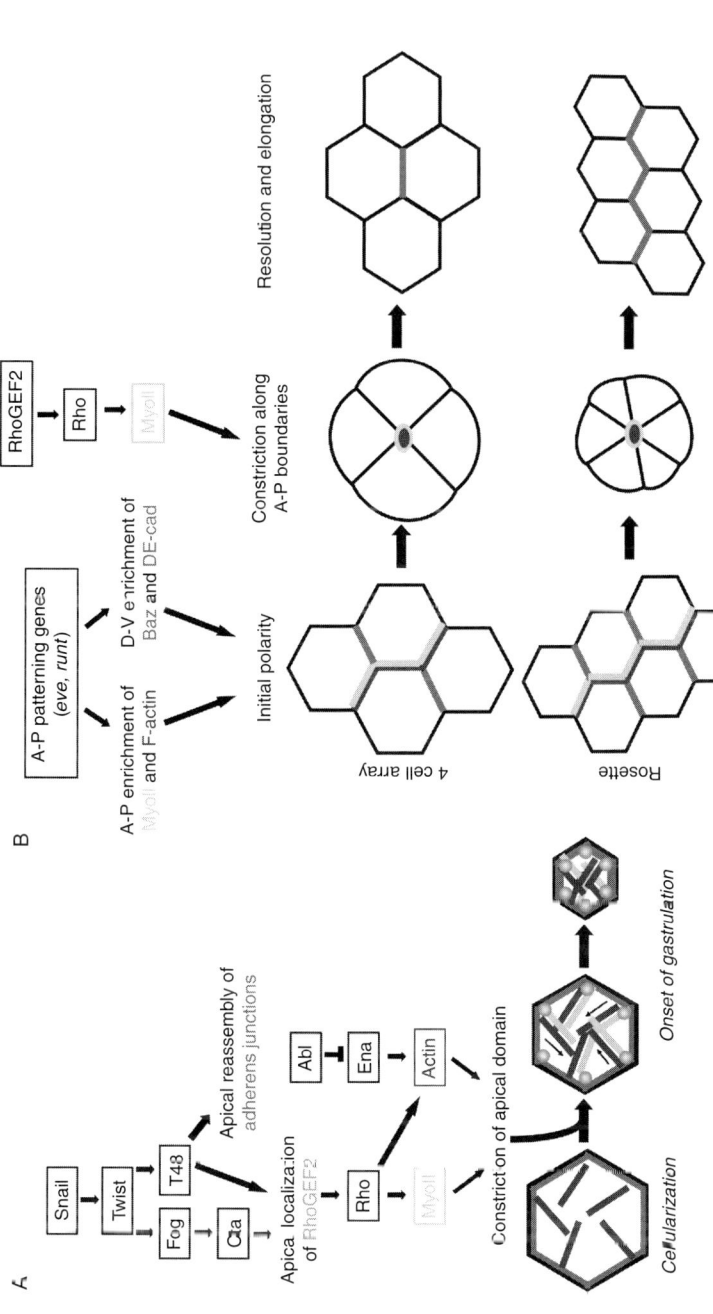

**Figure 3.3** Morphogenesis of the ventral furrow and germband. (A) Pathway to apical constriction in the ventral furrow. Transcription factors Twi and Snail specify the mesoderm and activate downstream effectors, T48, and Fog/Cta to apically localize RhoGEF2. RhoGEF2 signals through Rho to assemble Myosin II. RhoGEF2 also works in concert with Abl to establish an organized actin cytoskeleton. This leads to coordinated apical constriction in the ventral furrow. (B) Pathway to germband elongation. A-P patterning genes lead to the differential enrichment of actin/Myosin II and Baz (PAR-3)/DE-cad. This leads to unequal cortical tension. RhoGEF2 signals through Rho and Myosin II to constrict the A-P boundary. This allows the A-P neighbors to exchange positions with the D-V neighbors and leads to axis elongation. Colors in the model indicate differential enrichment of proteins, not total protein localization. (See Color Insert.)

Constriction involves assembly of both actin and Myosin II. Apical actin organization appears to be regulated by Abelson (Abl) kinase, a nonreceptor tyrosine kinase, and also affects the formation of the ventral furrow (Fox and Peifer, 2007). In *abl* mutants, furrow constriction is uncoordinated, but cells are eventually internalized, like *fog* and *cta* mutants. Interestingly, the localization of actin is disrupted in *abl* mutants, while it is not in *cta* mutants. However, in *RhoGEF2* mutants, the localization of F-actin is disrupted. This suggests that RhoGEF2 and Abl work in parallel to regulate actin localization in the ventral furrow. Enabled (Ena), an actin regulator, is a known target for Abl in other processes. Ena localization to AJs is normally downregulated in the ventral furrow cells. In *abl* mutants, Ena is not downregulated, resulting in ectopic, disorganized apical actin. Consistent with Ena downregulation being critical, reduction of *ena* in an *abl* mutant background suppresses the *abl* ventral furrow phenotype (Fox and Peifer, 2007). This suggests that through Ena, Abl helps to regulate actin so coordinated constriction can take place.

Once in place, how does a contractile cytoskeleton create cell shape change? Cells must both create and resist forces for coordinated cell shape changes to occur within a tissue. One hypothesis for apical constriction was the purse string model, where actin filaments localized in rings at cell junctions slide together with the help of Myosin II to reduce the apical area of all the cells in concert (Costa *et al.*, 1993; Young *et al.*, 1991). However, apical actin and Myosin II are not restricted to rings at AJs, but instead cover the entire apical surface. Recent work provided detailed insights into the process, revealing that apical constriction in the ventral furrow is created by pulsed contractions of this actomyosin apical network (Martin *et al.*, 2009). High-resolution live microscopy of ventral furrow cells revealed cyclic formation of Myosin II coalescences over the apical surface of the cells. These coalescences occur within a larger Myosin II network that appears to shrink with each pulsed coalescence, and then remain in this smaller state, suggesting a ratchet model for apical constriction. Additionally, these coalescences are attached to AJs at discrete sites and bend the plasma membrane inward, resulting in coordinated apical constriction across the epithelial sheet (Martin *et al.*, 2009). Indeed, if AJs are disrupted, Myosin II coalescences form, but no shape change occurs (Dawes-Hoang *et al.*, 2005).

The ratchet model is further supported by analysis of Myosin II localization in *twist* and *snail* mutant embryos. As mentioned above, apical constriction fails to occur in these mutants. Interestingly, in *twist* and *snail* mutants the localization of Myosin II becomes more concentrated at cell junctions (Martin *et al.*, 2009). If the purse string model were correct, cells should still be able to constrict with Myosin II localized at cell junctions, but they do not. However, it is possible that Myosin II is not properly activated at the junctions in *twist* and *snail* mutants. *twist* and *snail* differentially affected the formation of Myosin II coalescences. In *twist* mutants, Myosin

II coalescences were reduced with few pulsed constrictions (Martin et al., 2009), while in *snail* mutants both the Myosin II coalescences and pulsed constriction were lost. In double-mutant embryos, both Myosin II coalescences and pulsed constriction were also absent, suggesting that Snail is required to initiate apical constriction.

Actomyosin contractility also drives formation of the posterior midgut invagination (PMGI), which internalizes the endoderm. This cup-like structure forms as the ventral furrow seals and germband extension begins. The cells that will invaginate surround the pole cells at the posterior end of the embryo, with 10 cells extending dorsally and ventrally from the pole cells and 5 cells on each lateral side (Costa et al., 1993). PMGI is surprisingly similar to ventral furrow formation and is governed by many of the same players. Again, the process begins with apical flattening, proceeds with coordinated constriction, lengthening of cells in the apical–basal axis, downward shift of nuclei, and basal expansion, and as apical constriction proceeds, the cells begin to shorten, deepening the overall cup structure. Loss of *fog* or *cta* function completely blocks posterior midgut invagination, while it only delays the invagination of the ventral furrow (Parks and Wieschaus, 1991; Sweeton et al., 1991). Posterior midgut formation is also disrupted in embryos lacking RhoGEF2 or expressing a dominant-negative form of Rho1 (Barrett et al., 1997). Posterior midgut cells also relocalize Myosin II to their apical ends (Young et al., 1991). Together, these results suggest actomyosin contractility plays an important role in the apical constriction of the posterior midgut cells, as it does in the ventral furrow. However, other influences impact the PGMI. For example, the invagination has polarity, with more dorsal cells constricting before the more ventral cells (Sweeton et al., 1991). Internalization is also aided by the contraction of the dorsal side of the embryo, which pulls the posterior midgut cells, and extension of the germband, which pushes them and eventually seals the invagination into the interior of the embryo (Costa et al., 1993). It is also unclear how the contractile cytoskeleton behaves during the apical constriction of posterior midgut cells. It will be interesting to determine whether the purse string or ratchet model best fits the apical constriction of these cells.

## 3.2. Germband extension: Actomyosin contractility driving cell intercalation

The third event in gastrulation is germband extension (GBE), in which the ectoderm narrows in the dorsal–ventral (D–V) axis and lengthens in the anterior–posterior (A–P) axis. Because of the constraints of the eggshell, this pushes the posterior end of the embryo up and over the anterior end. Cell intercalation drives extension of the germband through a convergence and extension process (reviewed in Lecuit and Lenne, 2007; Zallen, 2007). As mentioned above, the germband begins to elongate as the posterior midgut invaginates, and eventually seals it. The posterior two thirds of the embryo contains the cells that will become the germband. In about 2 h the germband elongates along the A–P axis, doubling its length, and shortens in the D–V axis, halving its width (Costa et al., 1993). Most studies of GBE have focused on the anterior part of the germband because the cells move more slowly and are thus easier to image. Moreover, most of the actual cell intercalation occurs in this region.

Early studies of cell behavior in the germband revealed cells shift their positions relative to one another (Irvine and Wieschaus, 1994). Cells intercalate primarily between dorsal and ventral neighbors, and rarely between anterior and posterior neighbors. Before the onset of GBE, cells are arranged in a hexagonal array, resembling a honeycomb. As GBE proceeds cells become disordered, resulting in four-cell arrays (Bertet et al., 2004) and multicellular rosettes (Blankenship et al., 2006; Zallen and Wieschaus, 2004). Polarized cell junction remodeling accompanies and may drive this polarized cell behavior. The simple four-cell arrays begin with a long cell–cell contact between A–P neighbors, while D–V neighbors are not in contact (this is referred to as a Type I junction). Next the contact between A–P neighbors shrinks, so that all cells in the group are touching (a Type 2, X-shaped junction). Type 2 junctions then resolve so that D–V neighbors form a long cell–cell contact and A–P neighbors are separated (a Type 3 junction), completing a cell–cell intercalation event (Bertet et al., 2004; Fig. 3.3B). In addition to four-cell arrays, the germband also assembles multicellular rosettes, where the vertices of 5–11 cells meet. However, the behavior of these structures is similar to four-cell arrays. Contacts between A–P neighbors shorten until multiple cells meet to form a structure resembling a cut pie. Then, contacts between D–V neighbors grow displacing A–P neighbors away from each other along the A–P axis (Blankenship et al., 2006; Zallen and Wieschaus, 2004; Zallen and Zallen, 2004; Fig. 3.3B).

Planar polarized actomyosin activity is at work in the germband (Fig. 3.3B). Cells in the germband differentially localize proteins along their A–P and D–V boundaries. F-actin becomes enriched at the A–P boundary first, and then nonmuscle Myosin II follows (Blankenship et al., 2006; Zallen and Wieschaus, 2004). Significantly, Myosin II localizes at these contacts as they constrict during cell intercalation (Bertet et al., 2004). Moreover, partial loss of Myosin II activity in Myosin II zygotic mutants produces slight defects in GBE; while inhibition of Rho kinase (Rok), which normally phosphorylates and activates Myosin II, severely affects GBE (Bertet et al., 2004). Thus, actomyosin activity appears to constrict cell–cell contacts between A–P neighbors to drive cell intercalation. In fact, computer models suggest constriction of cell borders in a polarized direction is sufficient to result in the elongation of a group of cells (Honda et al., 2008; Rauzi et al., 2008). Before GBE, the cells are not polarized and therefore all cell boundaries probably experience similar tension on the junctions. However, when Myosin II becomes enriched along the A–P boundary, this increases tension and shrinks that boundary. This change irreversibly changes the tension in the system, so cells resolve into Type 3 to return to a more static state. Nanodissection experiments that disrupted the actomyosin cytoskeleton, but maintained cell integrity, confirmed there is tension along the A–P boundary (Rauzi et al., 2008). Interestingly, the apical polarity protein Baz (PAR-3) is enriched at the D–V boundary along with the AJ proteins E-cadherin and Arm/$\beta$-catenin (Blankenship et al., 2006; Zallen and Wieschaus, 2004). Zygotic *baz* mutants do not completely elongate their germband (Zallen and Wieschaus, 2004), but the mechanism involved is unclear. Further, in mutants disrupting Baz (PAR-3) localization, A–P localization of Myosin II and F-actin is unaffected (Zallen and Wieschaus, 2004), suggesting these proteins do not depend on each other for their polarized localization. However, disruption of the actin cytoskeleton enhances planar polarization of Baz (PAR-3) and AJ proteins (Harris and Peifer, 2007), suggesting that the actin cytoskeleton plays a role in preventing hyperpolarization of AJs, which might disrupt adhesion.

What directs the planar polarity of germband cells? Early studies revealed that A–P patterning, but not D–V patterning is essential for GBE (Irvine and Wieschaus, 1994). The A–P body axis in *Drosophila* is determined by sequentially restricted patterns of gene expression. Spatially restricted maternally contributed proteins provide positional cues to activate zygotic genes. The first zygotic genes activated are the gap genes, which provide regional information. In turn, the pair-rule genes are activated and define parasegments. Segment polarity genes further refine the anterior–posterior pattern within segments, and then finally, the homeotic genes define the identity of each segment. Interestingly, the pair-rule genes *even-skipped* (*eve*) and *runt* both are important for GBE. Loss of function or misexpression of Eve, Runt, or upstream A–P patterning genes disrupts GBE, suggesting these genes are required for polarizing cells so they are able to intercalate (Bertet et al., 2004; Irvine and Wieschaus, 1994; Zallen and Wieschaus, 2004). Moreover, mutants affecting A–P patterning disrupt the planar polarized localization of cytoskeletal and junctional proteins. These experiments suggest A–P patterning is the primary cue to set up polarity in germband cells to allow efficient cell intercalation. However, it is unclear how A–P patterning triggers planar polarized cell architecture and directed cell rearrangement. In vertebrates, planar cell polarity (PCP) genes are required for convergent extension, an analogous process (Goto and Keller, 2002). However, in *Drosophila*, the PCP genes *frizzled* and *disheveled* do not appear to play roles in GBE (Zallen and Wieschaus, 2004).

## 3.3. Microtubule arrays contributing to gastrulation

The examples above highlight the central role for the actin cytoskeleton in morphogenesis, but roles for MTs are also becoming apparent. After cellularization, the columnar epithelial cells of the *Drosophila* embryo develop a

MT network typical of animal epithelial cells (Bartolini and Gundersen, 2006). This network is comprised of lateral MT bundles running along the apical–basal axis plus a meshwork of MTs over the apical surface. Centrosomal MTs are downregulated at the end of cellularization by aPKC (Harris and Peifer, 2007) and at later stages centrosomes are stripped of $\gamma$-tubulin and become inactive (Rogers et al., 2008). Only mitotic epithelial cells build robust centrosomal-nucleated MT arrays (Rogers et al., 2008). These mitotic spindles are oriented in the plane of the epithelial sheet, allowing both daughter cells to remain in the epithelium. Recent papers have shown how both mitotic and noncentrosomal interphase MT arrays can contribute to cell and tissue shape change.

Oriented cell division was found to contribute to GBE during gastrulation (da Silva and Vincent, 2007). Specifically, a group of mitotic cells at the posterior tip of the germband orient their spindles both in the plane of the epithelium and in the same direction along the axis of tissue extension. In *string* (Cdc25) mutants, in which all cell division is blocked, the rate of GBE was reduced. Thus, oriented cell division in the posterior germband appears to promote full GBE, complementing the cell intercalation events in the more anterior germband described above. Interestingly, the divisions are oriented by the embryo's A–P patterning system, as are the cell intercalation events. Thus, this embryo-wide patterning system somehow impacts both the actin and MT cytoskeletons.

Microtubules have also been implicated in the initial morphogenesis of the amnioserosa (AS). This squamous epithelial sheet forms on the dorsal side of the embryo from columnar epithelial cells that form at cellularization. It was recently found that a novel mechanism, rotary cell elongation, converts this columnar epithelial sheet into the flattened, squamous epithelial sheet (Pope and Harris, 2008). As AS cells begin to flatten, lateral noncentrosomal MT bundles protrude into the apical domain and bend perpendicularly. MT inhibitors blocked this early step in the cell shape change, indicating MTs are needed to initiate cell flattening. This initial MT protrusion is followed by a full 90° rotation of cellular contents. The full MT network along with the centrosomes, ER and nuclei rotate as a unit in the direction of cell flattening. This 90° rotation of the cells, together with AJ remodeling, appears to generate the flattened squamous AS from the columnar cells formed at cellularization.

## 3.4. Dorsal closure: Integrating signaling, the cytoskeleton, and adhesion to drive epithelial sheet migration and sealing

One of the most studied models for how signaling pathways mold cell shape and behavior through effects on adhesion and the cytoskeleton is dorsal closure (reviewed in Harden, 2002; Jacinto et al., 2002; Martin and

Parkhurst, 2004; Xia and Karin, 2004). Dorsal closure encloses the embryo in epidermis. Prior to dorsal closure, the ventral and lateral sides of the embryo are enclosed in a columnar epidermal epithelium (Fig. 3.4A and B), while, as discussed above, the dorsal surface is covered by the AS. This squamous epithelium undergoes apoptosis at the end of dorsal closure, and thus does not contribute to the hatched larva. Dramatic events cover the dorsal surface with epidermis. It begins during germband retraction, when epidermal cells that border the AS, the "leading edge" (LE) cells, change shape, elongating along the dorsal–ventral axis and align in a row. They also begin to assemble an actomyosin cable at the leading edge (Fig. 3.4B), where they abut the AS. After germband retraction LE cells elongate further and begin extending lamellipodial and filopodial protrusions over the AS (Fig. 3.4D), while their more ventral neighbors also begin to elongate (Fig. 3.4B). At the same time, AS cells begin to apically constrict. As the epidermal sheets meet, filopodial protrusions reach out toward the opposing sheet. The cells match precisely with like neighbors on the other side in a zipping process (Fig. 3.4A) which establishes adhesion between the sheets and thus epithelial continuity.

Elegant experimental manipulations by Kiehart and colleagues revealed the many forces that are at play (Hutson et al., 2003; Kiehart et al., 2000). Closure is powered by both AS apical constriction and the contractile actomyosin cable, which acts like a supracellular purse string to close the opening (Fig. 3.4B). Remarkably, epidermal cells actually resist elongation rather than providing a motive force. Moreover, their data suggest that the driving forces are balanced in such a way that if one is compromised, the others can compensate, producing a robust system.

Genetic analysis revealed that several signaling pathways contribute to regulate the behavior of LE cells and their epidermal and amnioserosal neighbors. Mutations in the Jun N-terminal kinase (JNK) pathway have defects in dorsal closure, and JNK target genes are activated in LE cells (reviewed in Harden, 2002; Xia and Karin, 2004). One target gene is the TFG$\beta$ family ligand Decapentaplegic (Dpp), which is then thought to signal to more ventral epidermal cells. Consistent with this, mutations in Dpp receptors disrupt dorsal closure. Finally, Wnt signaling also regulates behavior of epidermal cells during closure.

## 3.5. Adding complexity to the JNK and Dpp pathways

Significant advances have been made in the past several years in our understanding of the mechanisms of signal transduction, and the role these signaling pathways play in dorsal closure. One key question is how JNK signaling is activated in LE cells. Mutational analysis (reviewed in Harden, 2002) revealed roles for the transcriptional activators Fos (*kayak*) and Jun, as well as JNK (*basket*), which phosphorylates Jun, and its upstream activator

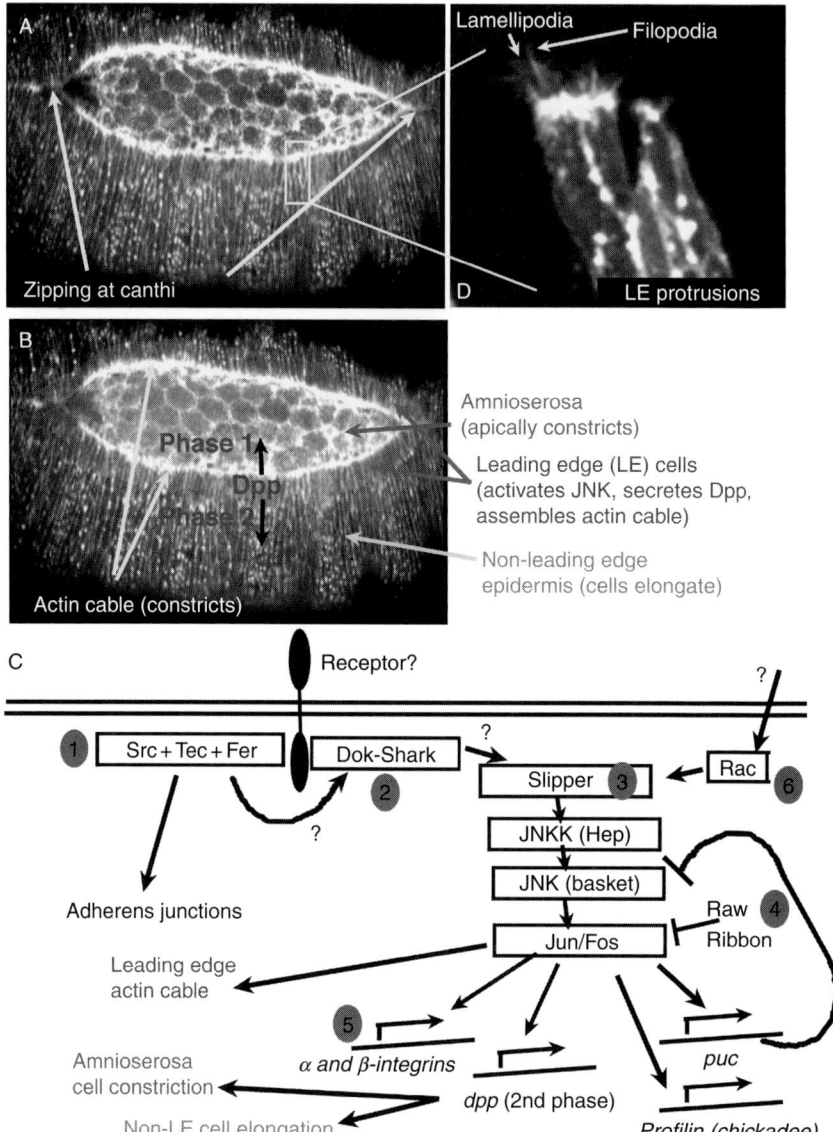

**Figure 3.4** Dorsal closure. (A, B) Still images of an embryo undergoing dorsal closure, with cells outlined using moesin–GFP. Key cell types and their behaviors are highlighted, as are the contractile actin cable, the source and direction of the two phases of Dpp signaling, and the location at which zipping occurs. (C) Portions of the signaling network that drives the cell behaviors of dorsal closure are illustrated, with suggested outcomes in terms of gene expression and cell behavior noted. Numbers refer to parts of this signaling network discussed in different sections of the text. (D) Close-up of LE cells, showing their lamellipodial and filopodial protrusions as visualized with GFP–actin. Region illustrated is similar to that boxed in (A). (See Color Insert.)

Hemipterous (Hep; homolog of the JNKK MKK7). The next upstream step in the kinase cascade remained unclear—both flies and mammals have several mixed lineage kinases proposed to be JNKK activators. Polaski et al. (2006) explored the role of one of these, Slipper (Slpr), in dorsal closure. Their data suggest that Slpr regulates the JNK pathway in embryonic and pupal morphogenesis and oogenesis (Fig. 3.4C3). Further, their data suggest that Tak, another JNKKK, does not regulate dorsal closure, but instead regulates the JNK pathway in response to Tumor necrosis factor (TNF)/Eiger signaling, a stress-signaling pathway. These data help clarify how different upstream inputs differentially activate JNK. However, they still leave open the question of what extracellular cues provide the ultimate trigger of JNK signaling.

Like many signal transduction pathways, the JNK pathway is kept in an OFF state by active repression outside its domain of activity. Previous work revealed that the JNK phosphatase Puckered acts as a negative feedback regulator to restrain ectopic JNK activity and also keep activation under control (Fig. 3.4C4; reviewed in Harden, 2002). Recent work revealed two other inputs that keep the JNK pathway off in nonleading edge epidermal cells (Bates et al., 2008; Fig. 3.4C4). Bates et al. explored the role of Raw. This novel protein localizes to the cytoplasm and restrains JNK activity outside of the LE cells by repressing the activity of the Fos/Jun transcription factor by an unknown mechanism. They also present evidence that the BTB/POZ domain DNA-binding protein Ribbon acts at a similar place in the JNK pathway. It will be interesting to explore their biochemical mechanisms of action.

Another key question involves the identity of Jun target genes that contribute to cell behavior during dorsal closure. Dpp was identified early as a target gene and plays an important role in dorsal closure (reviewed in Martin Blanco, 1997). A more direct cytoskeletal JNK target gene is *chickadee*, encoding profilin (Jasper et al., 2001). Homsy et al. (2006) looked for additional JNK target genes that might explain the severe delays in zipping seen in JNK mutants, leading ultimately to rupture of the connection between AS and LE cells. They observed that similar defects are seen in embryos mutant for the integrin α-subunit *scab*. Interestingly, they earlier identified an integrin β-subunit (*myospheroid*) as a JNK target gene by microarray analysis of embryos with activated JNK (Jasper et al., 2001). Both integrin subunits are elevated at the leading edge and in AS cells, and JNK signaling is both necessary and sufficient for integrin accumulation. Thus, integrins are also important target genes of JNK (Fig. 3.4C5); many other targets likely remain to be identified. Wada et al. (2007) expanded upon this, revealing that different populations of AS cells, presumably exposed to different levels of Dpp signal, play distinct roles in dorsal closure. In particular, the most ventral AS cells, which underlie the leading edge, play a key role. They make an actin purse string that abuts that made by the

LE cells. They also form complex connections to LE cells, which require the integrin α-chain Scab, and they exhibit novel protrusive behavior. All of these special behaviors require JNK-dependent Dpp signaling.

Others explored mechanisms of Dpp signaling. Fernandez et al. (2007) provide strong evidence that Dpp signaling occurs in two phases. The first phase of Dpp signaling, which occurs during germband retraction, is JNK-independent, with Dpp expression at the LE presumably activated by earlier dorsal–ventral patterning cues. This phase of Dpp signaling promotes AS apical constriction. Its effects and the activation of Dpp target genes are graded across the AS, with cells nearest the epidermis affected differentially. The second, later phase of Dpp action involves JNK-regulated expression of Dpp in LE cells. This regulates elongation of more ventral epidermal cells and also maintains proper adhesion between the AS and epidermis. Perhaps, the most striking observation supporting this two-phase model is that activating Dpp signaling in *either* the AS or epidermis is sufficient to rescue mutants lacking Dpp receptors in both tissues.

One Dpp target gene is Myosin II heavy chain (*zipper*). Zahedi et al. (2008) explored its regulation further, reporting that Dpp signaling is necessary but not sufficient for *zipper* expression. They suggest that cdc42 signaling, acting redundantly through two related kinase effectors, DACK and Fak-like Tyrosine Kinase (PR2), acts as a second input regulating *zipper* expression. Surprisingly, their data suggest that these kinases act in the AS, perhaps stimulating production some uncharacterized signal to the epidermis. The upstream inputs into cdc42, and DACK/PR2's downstream outputs remain to be characterized.

## 3.6. An embarrassment of riches: Five nonreceptor tyrosine kinases regulate dorsal closure

Another striking development in the past several years was the realization that several nonreceptor tyrosine kinases regulate dorsal closure. Among the first to be implicated were Src and Tec family kinases (Fig. 3.4C1). Flies have two Src family members, as well as a Tec kinase. Earlier work implicated Src42 and Tec29 in dorsal closure, and suggested they act upstream of the JNK pathway (Tateno et al., 2000). Takahashi et al. (2005) extended this analysis. They found that Src64 and Src42 act partially redundantly during germband retraction and dorsal closure. Double mutants have striking defects in cell behavior, with little cell elongation in the epidermis and a severely reduced actin cable. Surprisingly, however, expression of the JNK target gene *dpp* was expanded rather than lost, calling into question whether Src kinases act upstream of JNK. Instead, these authors suggest Src signaling regulates AJs, a hypothesis supported by strong genetic interactions, changes in AJ protein accumulation, and a physical interaction between Src42 and Armadillo (fly $\beta$-catenin; Takahashi et al.,

2005). These data reveal further complexity in signal integration, and suggest that AJs also play an important role in dorsal closure. Roles for both DE-cadherin and Armadillo in dorsal closure were subsequently identified and characterized, suggesting they strengthen adhesion between the AS and epidermis and regulate assembly of the actin cable (Gorfinkiel and Arias, 2007). Early relocalization of AJ proteins from the leading edge to "actin-nucleating centers," dots at the ends of the actin cables in each leading edge cell, suggest there might be temporally separate roles.

Fer kinase may also help regulate adhesion together with Src and Tec (Murray et al., 2006). Fer is strikingly enriched in LE cells. Zygotic loss of Fer slows but does not block dorsal closure. However, loss of both Fer and Src42 has more severe effects on dorsal closure, reducing assembly of the actin cable and accumulation of phosphotyrosine at the leading edge. Combining this data with analysis of a gain-of-function Fer mutant, Murray et al. suggested Fer regulates Dpp signaling and AJ regulation.

In addition to Src, Tec, and Fer, data also suggest a role for Shark (SH2 domain ankyrin repeat) kinase (Fernandez et al., 2000), which is essential for activation of JNK signaling (Fig. 3.4C2). In an effort to identify regulators/partners of Shark, Biswas et al. (2006) identified the adaptor protein Dok as a Shark interactor. Loss of maternal and zygotic Dok leads to dorsal closure failure, with strong defects in actin assembly and in activating *dpp* expression. Genetic tests suggest it acts upstream of Shark, JNK, and Jun. Based on this they propose activation of Src leads to Dok phosphorylation, recruiting Shark to the plasma membrane where it can be activated. It will be important to identify Shark targets to clarify its place in the JNK pathway.

## 3.7. Roles for novel regulators of adhesion and the cytoskeleton

In addition to the core AJ proteins, recent data implicate other potential adhesion regulators. Mammalian nectins are immunoglobulin-superfamily proteins that can regulate cell adhesion and junctional assembly. Flies have at least one nectin called Echinoid (Ed; Wei et al., 2005). It plays multiple roles in development, and recent data revealed regulating dorsal closure to be among these. Maternal/zygotic *ed* mutants have substantial defects in dorsal closure (Laplante and Nilson, 2006; Lin et al., 2007). The LE actin cable is reduced and the LE becomes very irregular. However, Ed does not appear to play a role in JNK activation. Two groups suggested different potential roles for Ed in dorsal closure. Juxtaposition of Ed wild-type and mutant cells in the ovarian follicle cell epithelium triggers assembly of a contractile actin structure at the interface. This, and the fact that Ed is expressed in the epidermis and absent from the AS, led Laplante and Nilson (2006) to speculate that Ed regulates assembly of the actin cable. Lin et al. (2007) report defects in assembly of other proteins at the leading edge in *ed* mutants, including Myosin II, the actin regulator Enabled, and Src42. Further, they find strong effects on the localization of unconventional myosin VI (Jaguar), and also strong genetic interactions between *ed* and *jaguar*. They support a model in which Ed regulates Jaguar's ability to stimulate actin contractility and thus cell shape change. Further work is needed to explore these two models.

Other potential integrators of adhesion, signaling and the cytoskeleton are the small GTPase Rap1 and its putative effector, the adapter protein Canoe. Both play important roles in dorsal closure (Boettner et al., 2003; Takahashi et al., 1998), though their mechanisms of action remain a bit mysterious. One new regulator of Rap1 was recently assigned its own role in dorsal closure. Mutants lacking the Rap1 Guanine nucleotide exchange factor PDZ–GEF have severe defects in elongation of epidermal cells during closure (Boettner and Van Aelst, 2007). Epistasis tests place it upstream of both Rap1 and Canoe, consistent with its known biochemical role. It will be important to identify upstream inputs into this pathway, as well as downstream molecular consequences.

## 3.8. Powering closure by old and new mechanisms

Earlier work from the Kiehart and Edwards labs established a quantitative model of the forces driving closure, which include contractile forces of AS cells and the LE actin cable, resistance of the epidermis, and zipping at the canthi—the anterior and posterior corners of the eye-shaped dorsal hole (Fig. 3.4A; Hutson et al., 2003). It has been assumed that contractility mediated by Myosin II is the driving force in both the AS and actin cable, but this had never been directly examined. Franke et al. (2005) tested this assumption, examining embryos with reduced levels of Myosin II heavy chain. This slows late stages of dorsal closure as LE actin becomes disorganized; ultimately the AS often rips from the epidermis. Restoring Myosin II function in either AS or LE cells rescues these defects, emphasizing the dual forces driving closure. Surprisingly, Myosin II also plays a role in zipping, and reducing its function leads to cell misalignment at the midline. This sets the stage for future examination of how Myosin II is regulated in each tissue to drive coordinated cell shape change.

Another key regulator of actomyosin activity at the LE is the small GTPase Rac1. Loss of all three *Drosophila* Racs severely disrupts morphogenesis, even before the onset of dorsal closure (Woolner et al., 2005). However, analysis of mutants with reduced Rac activity suggests that Rac is critical for assembling the LE actin cable and that it acts upstream of the JNK pathway in this process (Fig. 3.4C6). In embryos with reduced Rac activity closure continues, apparently driven by AS constriction, even though zipping is substantially impaired, further reinforcing the robust nature of the process. It will be interesting to explore how Rac activity is regulated during closure.

The Kiehart and Edwards labs also provided striking evidence for a surprising additional "force" driving closure—apoptosis (Toyama et al., 2008). Others had noted that a subset of AS cells constrict ahead of the others. Toyama et al. find that these cells constrict early because they undergo apoptosis earlier than their neighbors—this appears to be similar to the apical constriction that drives apoptosing MDCK cells to drop out of a cell monolayer. Blocking apoptosis reduces the force exerted by the AS and slows the rate of closure, while inducing early apoptosis increases force and speeds closure, suggesting a functional role for the normal apoptosis observed. Strikingly, early apoptosis is patterned, and their analysis suggests the pattern of cell death regulates the detailed timetable of apical constriction in different regions of the embryo. The mechanisms by which apoptosis stimulates apical constriction remains to be determined.

## 3.9. Reaching out to close the gap—Key roles for cell protrusions in dorsal closure

Another striking aspect of dorsal closure are the lamellipodial and filopodial protrusions produced by LE cells (Jacinto et al., 2000; Fig. 3.4D). One proposed function of these protrusions is to sense the environment. Epidermal cells from one side of the embryo precisely match their exact counterparts on the opposing side, making a perfect seam at the dorsal midline. It has been speculated that filopodia play a role in this. Consistent with this, Enabled, which plays a role in elongating actin filaments, is critical for precise matching at the dorsal midline (Gates et al., 2007). Reducing Ena function substantially reduces filopodia without eliminating lamellipodia. Under these circumstances zipping is substantially slowed, and segmental matching is impaired. This activity of Ena is regulated by the nonreceptor tyrosine kinase Abl (Stevens et al., 2008). Elevating Abl activity turns off Ena, disrupts filopodia in both the epidermis and the AS, and impairs zipping. Interestingly, reduced filopodial activity in the AS is accompanied by a striking increase in lamellipodia—the reasons for this remain unclear. Another player in regulating protrusive behavior is Rac (Woolner et al., 2005). When Rac is reduced all protrusive activity is diminished, while constitutively activating Rac leads to a dramatic increase in the size of protrusions. The mechanisms by which these different regulators of protrusion are controlled and coordinated remain important topics for future work.

Striking direct evidence for precise cell matching mediated by filopodia came from an elegant set of experiments by Millard and Martin (2008). They used GFP and RFP to mark actin in cells of the anterior and posterior compartments, respectively, and benefited from a consistent artifact in which a single anterior compartment cell was mispositioned in the midst of posterior cells. They then could watch anterior and posterior cells from each side of the embryo as the sheets approached one another, and examine their interactions with like and unlike cells. This revealed that there is preferential filopodial contact between cells of the same compartment, and subsequent formation of "filopodial tethers" that drew correct cells from the two sides together. These data graphically demonstrate that there must be adhesive molecules specific to different cell types—the identification of these remains a subject of great interest.

Another key issue involves identification of the machinery required to build filopodia and transport proteins to and from their tips. Liu *et al.* (2008) explored whether the unconventional myosin, Myosin XV or Sisyphus, plays a role in this process. Sisyphus is a member of the Myth–FERM domain class of myosins, which can move bidirectionally in filopodia. Sisyphus is enriched in LE cells, and RNAi knockdown leads to defects in zipping, disrupting the ability of cells to match at the dorsal midline. They searched for potential cargo proteins by a yeast two-hybrid screen, identifying several potentially interesting cytoskeletal, adhesion, and polarity proteins. They go on to demonstrate that one of these, DE-cadherin can co-IP with Sisyphus, and both colocalize in filopodia of cultured cells. They suggest Sisyphus traffics cargos in filopodia, helping shape the last steps of dorsal closure—it will be important to test this hypothesis.

MTs also function during dorsal closure (Jankovics and Brunner, 2006). Noncentrosomal MTs reorganize by 90° as LE cells flatten and elongate. The role of MTs in this initial cell shape change has not been tested directly. However, MTs are not essential for the tissue to move dorsally. Nonetheless, MT inhibitor studies and tissue specific expression of a MT severing protein showed a specific function for MTs in the zippering the two sides of the epidermis together as they meet at the dorsal midline. This appears to involve MTs protruding from the front of leading edge cells to support the structure of lamellipodia and filopodia that mediate zippering of the two epithelial sheets. It will be important to determine how MTs attain this organization, and how they regulate protrusive behavior.

## 4. Coordinating Cytoskeletal Machinery in Epithelial Cells

The spatial coordination of actin and MT cytoskeletal networks is apparent in epithelial cells as they form during cellularization, as they divide, and as they undergo morphogenesis. Embryonic patterning systems presumably regulate these coordinated actin and MT cytoskeletal networks for orderly morphogenesis. Indeed, as discussed above, the A–P patterning system can polarize both cortical actin and MT spindles in the germband. However, we are just beginning to dissect how such coordinated regulation occurs and how cytoskeletal assemblies affect one another in *Drosophila* embryonic epithelia.

One positive link between MTs and actomyosin contractility appears to occur during apical constriction. *Drosophila* RhoGEF2 is required for apical constriction of cells in the ventral furrow by regulating Rho1 (Barrett *et al.*, 1997; Hacker and Perrimon, 1998; Padash Barmchi *et al.*, 2005). The process of apical constriction can be mimicked in *Drosophila* S2 cell culture (Rogers *et al.*, 2004). Here, RhoGEF2 also induces constriction by activating Myosin II via Rho1, and remarkably RhoGEF2 was observed to travel to cell cortex on the plus ends of growing MTs in association with the +TIP protein EB1. This suggested a model in which MT associations allow RhoGEF2 to scan the cortex for cues signaling apical constriction (Rogers *et al.*, 2004). It will be interesting to understand exactly how this MT–actin crosstalk functions in the developing embryo. Interestingly, recent studies have also implicated MT-based delivery of a RhoGEF to the contractile ring of dividing cells (Glover *et al.*, 2008).

One negative link between actin and MTs is apparent in germband cells. During gastrulation (stage 7–8), actin disruption enhances AJ planar polarity at dorsal–ventral contacts (Harris and Peifer, 2007). MT disruption suppresses this effect, suggesting actin inhibits MT-based AJ positioning in these cells. In stage 9–10 germband cells, the same actin disruption causes a dramatic change in cell morphology (Pope and Harris, 2008). The apical domains of the cells elongate and their lateral regions rotate perpendicularly to become exposed to the embryo surface. Implicating actin–MT crosstalk, lateral MT bundles run into the extended apical domains of these actin-disrupted cells, and simultaneous MT disruption suppresses the cell shape change. These examples show how actin can antagonize MT effects on AJ positioning and epithelial cell shape in the *Drosophila* embryo. Notably, actin also antagonizes cortical MTs in other systems. For example, MT-based primary axons form where cortical actin is weakest (Bradke and Dotti, 1999). Actin also inhibits cortical MT protrusion in neutrophils (Eddy *et al.*, 2002) and Myosin IIA inhibits the accumulation of cortical MTs in mammalian cell culture (Even-Ram *et al.*, 2007). The mechanistic bases for these interactions remain to be defined.

## 5. Concluding Statement

The *Drosophila* embryo provides excellent models for diverse types of epithelial morphogenesis. Integrating these with work from other experimental systems will help define the general principles of the cell biology of development.

# REFERENCES

Adam, J. C., Pringle, J. R., and Peifer, M. (2000). Evidence for functional differentiation among *Drosophila* septins in cytokinesis and cellularization. *Mol. Biol. Cell* **11**, 3123–3135.

Afshar, K., Stuart, B., and Wasserman, S. A. (2000). Functional analysis of the *Drosophila* diaphanous FH protein in early embryonic development. *Development* **127**, 1887–1897.

Barrett, K., Leptin, M., and Settleman, J. (1997). The Rho GTPase and a putative RhoGEF mediate a signaling pathway for the cell shape changes in *Drosophila* gastrulation. *Cell* **91**, 905–915.

Bartolini, F., and Gundersen, G. G. (2006). Generation of noncentrosomal microtubule arrays. *J. Cell Sci.* **119**, 4155–4163.

Bates, K. L., Higley, M., and Letsou, A. (2008). Raw mediates antagonism of AP-1 activity in *Drosophila*. *Genetics* **178**, 1989–2002.

Bertet, C., Sulak, L., and Lecuit, T. (2004). Myosin-dependent junction remodelling controls planar cell intercalation and axis elongation. *Nature* **429**, 667–671.

Bilder, D., Schober, M., and Perrimon, N. (2003). Integrated activity of PDZ protein complexes regulates epithelial polarity. *Nat. Cell Biol.* **5**, 53–58.

Biswas, R., Stein, D., and Stanley, E. R. (2006). *Drosophila* Dok is required for embryonic dorsal closure. *Development* **133**, 217–227.

Blankenship, J. T., Backovic, S. T., Sanny, J. S., Weitz, O., and Zallen, J. A. (2006). Multicellular rosette formation links planar cell polarity to tissue morphogenesis. *Dev. Cell* **11**, 459–470.

Boettner, B., and Van Aelst, L. (2007). The Rap GTPase activator *Drosophila* PDZ-GEF regulates cell shape in epithelial migration and morphogenesis. *Mol. Cell. Biol.* **27**, 7966–7980.

Boettner, B., Harjes, P., Ishimaru, S., Heke, M., Fan, H. Q., Qin, Y., Van Aelst, L., and Gaul, U. (2003). The AF-6 homolog canoe acts as a Rap1 effector during dorsal closure of the *Drosophila* embryo. *Genetics* **165**, 159–169.

Bradke, F., and Dotti, C. G. (1999). The role of local actin instability in axon formation. *Science* **283**, 1931–1934.

Cavey, M., Rauzi, M., Lenne, P. F., and Lecuit, T. (2008). A two-tiered mechanism for stabilization and immobilization of E-cadherin. *Nature* **453**, 751–756.

Costa, M., Sweeton, D., and Wieschaus, E. (1993). Gastrulation in *Drosophila*: Cellular mechanisms of morphogenetic movements. *In* The Development of *Drosophila melanogaster*, (M. Bate and A. Martinez Arias, eds.), vol. I, pp. 425–466. Cold Spring Harbor Laboratory Press, New York.

Costa, M., Wilson, E. T., and Wieschaus, E. (1994). A putative cell signal encoded by the folded gastrulation gene coordinates cell shape changes during *Drosophila* gastrulation. *Cell* **76**, 1075–1089.

Cox, R. T., Kirkpatrick, C., and Peifer, M. (1996). Armadillo is required for adherens junction assembly, cell polarity, and morphogenesis during *Drosophila* embryogenesis. *J. Cell Biol.* **134**, 133–148.

Crawford, J. M., Harden, N., Leung, T., Lim, L., and Kiehart, D. P. (1998). Cellularization in *Drosophila* melanogaster is disrupted by the inhibition of rho activity and the activation of Cdc42 function. *Dev. Biol.* **204**, 151–164.

da Silva, S. M., and Vincent, J. P. (2007). Oriented cell divisions in the extending germband of *Drosophila*. *Development* **134**, 3049–3054.

Dawes-Hoang, R. E., Parmar, K. M., Christiansen, A. E., Phelps, C. B., Brand, A. H., and Wieschaus, E. F. (2005). folded gastrulation, cell shape change and the control of myosin localization. *Development* **132**, 4165–4178.

Eddy, R. J., Pierini, L. M., and Maxfield, F. R. (2002). Microtubule asymmetry during neutrophil polarization and migration. *Mol. Biol. Cell* **13,** 4470–4483.

Even-Ram, S., Doyle, A. D., Conti, M. A., Matsumoto, K., Adelstein, R. S., and Yamada, K. M. (2007). Myosin IIA regulates cell motility and actomyosin–microtubule crosstalk. *Nat. Cell Biol.* **9,** 299–309.

Fernandez, R., Takahashi, F., Liu, Z., Steward, R., Stein, D., and Stanley, E. R. (2000). The *Drosophila* shark tyrosine kinase is required for embryonic dorsal closure. *Genes Dev.* **14,** 604–614.

Fernandez, B. G., Arias, A. M., and Jacinto, A. (2007). Dpp signalling orchestrates dorsal closure by regulating cell shape changes both in the amnioserosa and in the epidermis. *Mech. Dev.* **124,** 884–897.

Fox, D. T., and Peifer, M. (2007). Abelson kinase (Abl) and RhoGEF2 regulate actin organization during cell constriction in *Drosophila*. *Development* **134,** 567–578.

Franke, J. D., Montague, R. A., and Kiehart, D. P. (2005). Nonmuscle myosin II generates forces that transmit tension and drive contraction in multiple tissues during dorsal closure. *Curr. Biol.* **15,** 2208–2221.

Frescas, D., Mavrakis, M., Lorenz, H., Delotto, R., and Lippincott-Schwartz, J. (2006). The secretory membrane system in the *Drosophila* syncytial blastoderm embryo exists as functionally compartmentalized units around individual nuclei. *J. Cell Biol.* **173,** 219–230.

Gates, J., Mahaffey, J. P., Rogers, S. L., Emerson, M., Rogers, E. M., Sottile, S. L., Van Vactor, D., Gertler, F. B., and Peifer, M. (2007). Enabled plays key roles in embryonic epithelial morphogenesis in *Drosophila*. *Development* **134,** 2027–2039.

Glover, D. M., Capalbo, L., D'Avino, P. P., Gatt, M. K., Savoian, M. S., and Takeda, T. (2008). Girds 'n' cleeks o' cytokinesis: Microtubule sticks and contractile hoops in cell division. *Biochem. Soc. Trans.* **36,** 400–404.

Gorfinkiel, N., and Arias, A. M. (2007). Requirements for adherens junction components in the interaction between epithelial tissues during dorsal closure in *Drosophila*. *J. Cell Sci.* **120,** 3289–3298.

Goto, T., and Keller, R. (2002). The planar cell polarity gene strabismus regulates convergence and extension and neural fold closure in *Xenopus*. *Dev. Biol.* **247,** 165–181.

Grevengoed, E. E., Fox, D. T., Gates, J., and Peifer, M. (2003). Balancing different types of actin polymerization at distinct sites: Roles for Abelson kinase and Enabled. *J. Cell Biol.* **163,** 1267–1279.

Grosshans, J., Wenzl, C., Herz, H. M., Bartoszewski, S., Schnorrer, F., Vogt, N., Schwarz, H., and Muller, H. A. (2005). RhoGEF2 and the formin Dia control the formation of the furrow canal by directed actin assembly during *Drosophila* cellularisation. *Development* **132,** 1009–1020.

Gumbiner, B. M. (2005). Regulation of cadherin-mediated adhesion in morphogenesis. *Nat. Rev. Mol. Cell Biol.* **6,** 622–634.

Hacker, U., and Perrimon, N. (1998). DRhoGEF2 encodes a member of the Dbl family of oncogenes and controls cell shape changes during gastrulation in *Drosophila*. *Genes Dev.* **12,** 274–284.

Harden, N. (2002). Signaling pathways directing the movement and fusion of epithelial sheets: Lessons from dorsal closure in *Drosophila*. *Differentiation* **70,** 181–203.

Harris, T. J., and Peifer, M. (2004). Adherens junction-dependent and -independent steps in the establishment of epithelial cell polarity in *Drosophila*. *J. Cell Biol.* **167,** 135–147.

Harris, T. J., and Peifer, M. (2005). The positioning and segregation of apical cues during epithelial polarity establishment in *Drosophila*. *J. Cell Biol.* **170,** 813–823.

Harris, T. J., and Peifer, M. (2007). aPKC controls microtubule organization to balance adherens junction symmetry and planar polarity during development. *Dev. Cell* **12,** 727–738.

Homsy, J. G., Jasper, H., Peralta, X. G., Wu, H., Kiehart, D. P., and Bohmann, D. (2006). JNK signaling coordinates integrin and actin functions during *Drosophila* embryogenesis. *Dev. Dyn.* **235,** 427–434.

Honda, H., Nagai, T., and Tanemura, M. (2008). Two different mechanisms of planar cell intercalation leading to tissue elongation. *Dev. Dyn.* **237,** 1826–1836.

Hutson, M. S., Tokutake, Y., Chang, M. S., Bloor, J. W., Venakides, S., Kiehart, D. P., and Edwards, G. S. (2003). Forces for morphogenesis investigated with laser microsurgery and quantitative modeling. *Science* **300,** 145–149.

Irvine, K. D., and Wieschaus, E. (1994). Cell intercalation during *Drosophila* germband extension and its regulation by pair-rule segmentation genes. *Development* **120,** 827–841.

Jacinto, A., Wood, W., Balayo, T., Turmaine, M., Martinez-Arias, A., and Martin, P. (2000). Dynamic actin-based epithelial adhesion and cell matching during *Drosophila* dorsal closure. *Curr. Biol.* **10,** 1420–1426.

Jacinto, A., Woolner, S., and Martin, P. (2002). Dynamic analysis of dorsal closure in *Drosophila*: From genetics to cell biology. *Dev. Cell* **3,** 9–19.

Jankovics, F., and Brunner, D. (2006). Transiently reorganized microtubules are essential for zippering during dorsal closure in *Drosophila melanogaster*. *Dev. Cell* **11,** 375–385.

Jasper, H., Benes, V., Schwager, C., Sauer, S., Clauder-Munster, S., Ansorge, W., and Bohmann, D. (2001). The genomic response of the *Drosophila* embryo to JNK signaling. *Dev. Cell* **1,** 579–586.

Kiehart, D. P., Galbraith, C. G., Edwards, K. A., Rickoll, W. L., and Montague, R. A. (2000). Multiple forces contribute to cell sheet morphogenesis for dorsal closure in *Drosophila*. *J. Cell Biol.* **149,** 471–490.

Knust, E., and Bossinger, O. (2002). Composition and formation of intercellular junctions in epithelial cells. *Science* **298,** 1955–1959.

Kolsch, V., Seher, T., Fernandez-Ballester, G. J., Serrano, L., and Leptin, M. (2007). Control of *Drosophila* gastrulation by apical localization of adherens junctions and RhoGEF2. *Science* **315,** 384–386.

Laplante, C., and Nilson, L. A. (2006). Differential expression of the adhesion molecule Echinoid drives epithelial morphogenesis in *Drosophila*. *Development* **133,** 3255–3264.

Lecuit, T. (2004). Junctions and vesicular trafficking during *Drosophila* cellularization. *J. Cell Sci.* **117,** 3427–3433.

Lecuit, T., and Lenne, P. F. (2007). Cell surface mechanics and the control of cell shape, tissue patterns and morphogenesis. *Nat. Rev. Mol. Cell Biol.* **8,** 633–644.

Lecuit, T., and Wieschaus, E. (2000). Polarized insertion of new membrane from a cytoplasmic reservoir during cleavage of the *Drosophila* embryo. *J. Cell Biol.* **150,** 849–860.

Lee, O. K., Frese, K. K., James, J. S., Chadda, D., Chen, Z. H., Javier, R. T., and Cho, K. O. (2003). Discs-Large and Strabismus are functionally linked to plasma membrane formation. *Nat. Cell Biol.* **5,** 987–993.

Leptin, M. (1999). Gastrulation in *Drosophila*: The logic and the cellular mechanisms. *EMBO J.* **18,** 3187–3192.

Leptin, M., and Grunewald, B. (1990). Cell shape changes during gastrulation in *Drosophila*. *Development* **110,** 73–84.

Lin, H. P., Chen, H. M., Wei, S. Y., Chen, L. Y., Chang, L. H., Sun, Y. J., Huang, S. Y., and Hsu, J. C. (2007). Cell adhesion molecule Echinoid associates with unconventional myosin VI/Jaguar motor to regulate cell morphology during dorsal closure in *Drosophila*. *Dev. Biol.* **311,** 423–433.

Liu, R., Woolner, S., Johndrow, J. E., Metzger, D., Flores, A., and Parkhurst, S. M. (2008). Sisyphus, the *Drosophila* myosin XV homolog, traffics within filopodia transporting key sensory and adhesion cargos. *Development* **135,** 53–63.

Macara, I. G. (2004). Par proteins: Partners in polarization. *Curr. Biol.* **14,** R160–R162.

Martin, P., and Parkhurst, S. M. (2004). Parallels between tissue repair and embryo morphogenesis. *Development* **131**, 3021–3034.
Martin, A. C., Kaschube, M., and Wieschaus, E. F. (2009). Pulsed contractions of an actin–myosin network drive apical constriction. *Nature* **457**, 495–499.
Martin-Blanco, E. (1997). Regulation of cell differentiation by the *Drosophila* Jun kinase cascade. *Curr. Opin. Genet. Dev.* **7**, 666–671.
Matsumura, F. (2005). Regulation of myosin II during cytokinesis in higher eukaryotes. *Trends Cell Biol.* **15**, 371–377.
Mazumdar, A., and Mazumdar, M. (2002). How one becomes many: Blastoderm cellularization in *Drosophila melanogaster*. *Bioessays* **24**, 1012–1022.
Medina, E., Williams, J., Klipfell, E., Zarnescu, D., Thomas, G., and Le Bivic, A. (2002). Crumbs interacts with moesin and beta(Heavy)-spectrin in the apical membrane skeleton of *Drosophila*. *J. Cell Biol.* **158**, 941–951.
Millard, T. H., and Martin, P. (2008). Dynamic analysis of filopodial interactions during the zippering phase of *Drosophila* dorsal closure. *Development* **135**, 621–626.
Morize, P., Christiansen, A. E., Costa, M., Parks, S., and Wieschaus, E. (1998). Hyperactivation of the folded gastrulation pathway induces specific cell shape changes. *Development* **125**, 589–597.
Murray, M. J., Davidson, C. M., Hayward, N. M., and Brand, A. H. (2006). The Fes/Fer non-receptor tyrosine kinase cooperates with Src42A to regulate dorsal closure in *Drosophila*. *Development* **133**, 3063–3073.
Nance, J. (2005). PAR proteins and the establishment of cell polarity during C. elegans development. *Bioessays* **27**, 126–135.
Nelson, W. J. (2003). Adaptation of core mechanisms to generate cell polarity. *Nature* **422**, 766–774.
Nikolaidou, K. K., and Barrett, K. (2004). A Rho GTPase signaling pathway is used reiteratively in epithelial folding and potentially selects the outcome of Rho activation. *Curr. Biol.* **14**, 1822–1826.
Ohno, S. (2001). Intercellular junctions and cellular polarity: The PAR-aPKC complex, a conserved core cassette playing fundamental roles in cell polarity. *Curr. Opin. Cell Biol.* **13**, 641–648.
Padash Barmchi, M., Rogers, S., and Hacker, U. (2005). DRhoGEF2 regulates actin organization and contractility in the *Drosophila* blastoderm embryo. *J. Cell Biol.* **168**, 575–585.
Papoulas, O., Hays, T. S., and Sisson, J. C. (2005). The golgin Lava lamp mediates dynein-based Golgi movements during *Drosophila* cellularization. *Nat. Cell Biol.* **7**, 612–618.
Parks, S., and Wieschaus, E. (1991). The *Drosophila* gastrulation gene concertina encodes a G alpha-like protein. *Cell* **64**, 447–458.
Pelissier, A., Chauvin, J. P., and Lecuit, T. (2003). Trafficking through Rab11 endosomes is required for cellularization during *Drosophila* embryogenesis. *Curr. Biol.* **13**, 1848–1857.
Pilot, F., Philippe, J. M., Lemmers, C., and Lecuit, T. (2006). Spatial control of actin organization at adherens junctions by a synaptotagmin-like protein Btsz. *Nature* **442**, 580–584.
Polaski, S., Whitney, L., Barker, B. W., and Stronach, B. (2006). Genetic analysis of slipper/mixed lineage kinase reveals requirements in multiple Jun-N-terminal kinase-dependent morphogenetic events during *Drosophila* development. *Genetics* **174**, 719–733.
Pope, K. L., and Harris, T. J. (2008). Control of cell flattening and junctional remodeling during squamous epithelial morphogenesis in *Drosophila* Development **135**, 2227–2238.
Rauzi, M., Verant, P., Lecuit, T., and Lenne, P. F. (2008). Nature and anisotropy of cortical forces orienting *Drosophila* tissue morphogenesis. *Nat. Cell Biol.* **10**, 1401–1410.
Rickoll, W. L., and Counce, S. J. (1980). Morphogenesis in the embryo of *Drosophila melanogaster*—Germ band extension. *Wilhelm Roux's Arch.* **188**, 163–177.

Riggs, B., Rothwell, W., Mische, S., Hickson, G. R., Matheson, J., Hays, T. S., Gould, G. W., and Sullivan, W. (2003). Actin cytoskeleton remodeling during early *Drosophila* furrow formation requires recycling endosomal components Nuclear-fallout and Rab11. *J. Cell Biol.* **163,** 143–154.

Rogers, S. L., Wiedemann, U., Hacker, U., Turck, C., and Vale, R. D. (2004). *Drosophila* RhoGEF2 associates with microtubule plus ends in an EB1-dependent manner. *Curr. Biol.* **14,** 1827–1833.

Rogers, G. C., Rusan, N. M., Peifer, M., and Rogers, S. L. (2008). A multicomponent assembly pathway contributes to the formation of acentrosomal microtubule arrays in interphase *Drosophila* cells. *Mol. Biol. Cell* **19,** 3163–3178.

Royou, A., Field, C., Sisson, J. C., Sullivan, W., and Karess, R. (2004). Reassessing the role and dynamics of nonmuscle myosin II during furrow formation in early *Drosophila* embryos. *Mol. Biol. Cell* **15,** 838–850.

Schejter, E. D., and Wieschaus, E. (1993). bottleneck acts as a regulator of the microfilament network governing cellularization of the *Drosophila* embryo. *Cell* **75,** 373–385.

Simpson, L., and Wieschaus, E. (1990). Zygotic activity of the nullo locus is required to stabilize the actin–myosin network during cellularization in *Drosophila*. *Development* **110,** 851–863.

Sisson, J. C., Field, C., Ventura, R., Royou, A., and Sullivan, W. (2000). Lava lamp, a novel peripheral golgi protein, is required for *Drosophila melanogaster* cellularization. *J. Cell Biol.* **151,** 905–918.

Sokac, A. M., and Wieschaus, E. (2008a). Local actin-dependent endocytosis is zygotically controlled to initiate *Drosophila* cellularization. *Dev. Cell* **14,** 775–786.

Sokac, A. M., and Wieschaus, E. (2008b). Zygotically controlled F-actin establishes cortical compartments to stabilize furrows during *Drosophila* cellularization. *J. Cell Sci.* **121,** 1815–1824.

Sotillos, S., Diaz-Meco, M. T., Caminero, E., Moscat, J., and Campuzano, S. (2004). DaPKC-dependent phosphorylation of Crumbs is required for epithelial cell polarity in *Drosophila*. *J. Cell Biol.* **166,** 549–557.

Stevens, T. L., Rogers, E. M., Koontz, L. M., Fox, D. T., Homem, C. C., Nowotarski, S. H., Artabazon, N. B., and Peifer, M. (2008). Using Bcr–Abl to examine mechanisms by which Abl kinase regulates morphogenesis in *Drosophila*. *Mol. Biol. Cell* **19,** 378–393.

Sweeton, D., Parks, S., Costa, M., and Wieschaus, E. (1991). Gastrulation in *Drosophila*: The formation of the ventral furrow and posterior midgut invaginations. *Development* **112,** 775–789.

Takahashi, K., Matsuo, T., Katsube, T., Ueda, R., and Yamamoto, D. (1998). Direct binding between two PDZ domain proteins Canoe and ZO-1 and their roles in regulation of the jun N-terminal kinase pathway in *Drosophila* morphogenesis. *Mech. Dev.* **78,** 97–111.

Takahashi, M., Takahashi, F., Ui-Tei, K., Kojima, T., and Saigo, K. (2005). Requirements of genetic interactions between Src42A, armadillo and shotgun, a gene encoding E-cadherin, for normal development in *Drosophila*. *Development* **132,** 2547–2559.

Tateno, M., Nishida, Y., and Adachi-Yamada, T. (2000). Regulation of JNK by Src during *Drosophila* development. *Science* **287,** 324–327.

Tepass, U. (1996). Crumbs, a component of the apical membrane, is required for zonula adherens formation in primary epithelia of *Drosophila*. *Dev. Biol.* **177,** 217–225.

Tepass, U., and Hartenstein, V. (1994). The development of cellular junctions in the *Drosophila* embryo. *Dev. Biol.* **161,** 563–596.

Tepass, U., Tanentzapf, G., Ward, R., and Fehon, R. (2001). Epithelial cell polarity and cell junctions in *Drosophila*. *Annu. Rev. Genet.* **35,** 747–784.

Thomas, J. H., and Wieschaus, E. (2004). src64 and tec29 are required for microfilament contraction during *Drosophila* cellularization. *Development* **131**, 863–871.

Toyama, Y., Peralta, X. G., Wells, A. R., Kiehart, D. P., and Edwards, G. S. (2008). Apoptotic force and tissue dynamics during *Drosophila* embryogenesis. *Science* **321**, 1683–1686.

Turner, F. R., and Mahowald, A. P. (1976). Scanning electron microscopy of *Drosophila* embryogenesis. 1. The structure of the egg envelopes and the formation of the cellular blastoderm. *Dev. Biol.* **50**, 95–108.

Wada, A., Kato, K., Uwo, M. F., Yonemura, S., and Hayashi, S. (2007). Specialized extraembryonic cells connect embryonic and extraembryonic epidermis in response to Dpp during dorsal closure in *Drosophila*. *Dev. Biol.* **301**, 340–349.

Warn, R. M., and Magrath, R. (1983). F-actin distribution during the cellularization of the *Drosophila* embryo visualized with FL-phalloidin. *Exp. Cell Res.* **143**, 103–114.

Warn, R. M., and Warn, A. (1986). Microtubule arrays present during the syncytial and cellular blastoderm stages of the early *Drosophila* embryo. *Exp. Cell Res.* **163**, 201–210.

Warn, R. M., Bullard, B., and Magrath, R. (1980). Changes in the distribution of cortical myosin during the cellularization of the *Drosophila* embryo. *J. Embryol. Exp. Morphol.* **57**, 167–176.

Wei, S. Y., Escudero, L. M., Yu, F., Chang, L. H., Chen, L. Y., Ho, Y. H., Lin, C. M., Chou, C. S., Chia, W., Modolell, J., and Hsu, J. C. (2005). Echinoid is a component of adherens junctions that cooperates with DE-cadherin to mediate cell adhesion. *Dev. Cell* **8**, 493–504.

Woolner, S., Jacinto, A., and Martin, P. (2005). The small GTPase Rac plays multiple roles in epithelial sheet fusion dynamic studies of *Drosophila* dorsal closure. *Dev. Biol.* **282**, 163–173.

Xia, Y., and Karin, M. (2004). The control of cell motility and epithelial morphogenesis by Jun kinases. *Trends Cell Biol.* **14**, 94–101.

Yamada, S., Pokutta, S., Drees, F., Weis, W. I., and Nelson, W. J. (2005). Deconstructing the cadherin–catenin–actin complex. *Cell* **123**, 889–901.

Young, P. E., Pesacreta, T. C., and Kiehart, D. P. (1991). Dynamic changes in the distribution of cytoplasmic myosin during *Drosophila* embryogenesis. *Development* **111**, 1–14.

Zahedi, B., Shen, W., Xu, X., Chen, X., Mahey, M., and Harden, N. (2008). Leading edge-secreted Dpp cooperates with ACK-dependent signaling from the amnioserosa to regulate myosin levels during dorsal closure. *Dev. Dyn.* **237**, 2936–2946.

Zallen, J. A. (2007). Planar polarity and tissue morphogenesis. *Cell* **129**, 1051–1063.

Zallen, J. A., and Wieschaus, E. (2004). Patterned gene expression directs bipolar planar polarity in *Drosophila*. *Dev. Cell* **6**, 343–355.

Zallen, J. A., and Zallen, R. (2004). Cell-pattern disordering during convergent extension in *Drosophila*. *J. Phys.: Condens. Matter* **16**, S5073–S5080.

CHAPTER FOUR

# Cell Topology, Geometry, and Morphogenesis in Proliferating Epithelia

William T. Gibson* *and* Matthew C. Gibson[†]

## Contents

1. Introduction                                                                88
2. Conservation of Epithelial Architecture                                     89
3. Introduction to Cellular Topology                                           92
4. Conservation of Topological Structure in Proliferating Epithelia            92
5. Topological Inference in Epithelia: Maximum Entropy Methods                 94
6. Topological Models: The Simplest Models of Epithelia                        95
7. The Smallest Geometrical Model: Cleavage Plane Orientation
   in a Single Cell                                                            97
8. Nongeometric Mechanisms of Division Orientation: A Larger
   Morphogenetic Space                                                         99
9. Scaling Up: Geometrical Models and Cellular Mechanics
   in Proliferating Epithelia                                                 100
   9.1. Dirichlet models                                                      100
   9.2. Cellular Potts models                                                 102
   9.3. Subcellular element models                                            103
   9.4. Finite-element models                                                 103
10. Putting It All Together: Genetics, Geometry, and Biophysics               104
11. Future Directions                                                         108
12. Conclusion                                                                109
References                                                                    110

## Abstract

Epithelia are sheets of tightly adherent cells that line both internal and external surfaces in a vast array of metazoans. During development, an intrinsic consequence of coupling tight adhesion with cellular proliferation is the emergence of an epithelial form characterized by a stereotyped distribution of polygonal cell shapes. Despite the near universality of this constraint on cell shape and tissue

[*] Program in Biophysics, Harvard University, Cambridge, Massachusetts, USA
[†] The Stowers Institute for Medical Research, Kansas City, Missouri, USA

organization, very little is known about the possible implications of cell pattern geometry for mechanical properties of tissues or key biological processes, such as planar polarization, tissue remodeling, and cell division. In this chapter, through an examination of increasingly complex models, we highlight what is known about the role of mitotic proliferation in the emergence of epithelial cell geometry, and examine some possible implications for tissue morphogenesis. Ideally, continued progress in this area will address a major conceptual challenge in biology, which is to understand aspects of morphogenesis that are not explicitly directed by genetic control, but instead emerge from the complex interactions between geometric and biomechanical properties of epithelial tissues.

## 1. INTRODUCTION

From the simplest to the most complex metazoans, epithelial morphogenesis is a fundamental component of development, organogenesis, and even disease progression. A staggering variety of organisms depend on epithelia and their derivatives for development and homeostasis. Across widely divergent evolutionary clades, epithelial architecture retains certain essential structural features. These include apical/basal cell polarization, formation of cell–cell junctions, and the constitution of a paracellular diffusion barrier, all of which enable epithelia to serve an even greater diversity of biological functions. Still, while the advantages of epithelial organization are clear, the corresponding constraints on morphogenesis are often poorly understood.

The paper-folding art of origami is a simple but powerful conceptual model for epithelial morphogenesis. Using a finite number of simple folding rules, an infinite number of scale-invariant morphologies can be achieved, which derive entirely from a simple, planar sheet (Huzita and Scimemi, 1989, as cited in Nagpal, 2001, 2002). This analogy is intended to emphasize the role of macroscopic architectural changes that occur seemingly independently of the microscopic and two-dimensional structure of the epithelium. In real epithelial sheets, however, an additional layer of constraint and complexity is introduced by the requirements for growth, proliferation, and the control of neighbor cell relationships. Consider a flat, monolayer sheet of proliferating columnar cells. Intuitively, cell division and cell rearrangement in the plane of the epithelium change the polygonal cell pattern geometry on the microscopic level. A critical and unresolved problem is thus to define the implications of the polygonal cell pattern geometry for macroscopic morphogenesis.

There is considerable evidence that morphological transformations at the cellular level are relevant for tissue-level morphogenesis. Three examples include coordinated apical constriction, oriented cell division, and localized, differential control of cellular proliferation, among numerous others (Baena-Lopez *et al.*, 2005; Gong *et al.*, 2004; Lecuit and Lenne, 2007; O'Brochta and Bryant, 1985; Saburi *et al.*, 2008). However, these cases all involve some sort of global pattern control to orchestrate the cellular changes. An open question is whether uncoordinated aspects of the planar pattern geometry could also be of macroscopic relevance, for example, in setting the spatial or temporal noise in morphogenesis. Alternatively, do some microscale geometries make epithelial sheets more or less structurally stiff or strong? Could understanding the polygonal cell-packing pattern be relevant for understanding cell–cell signaling, for example, in planar cell polarity? These are speculative subjects, but they underscore the importance of understanding the cellular geometry of proliferating epithelia.

In this chapter, we highlight what is known about cell proliferation-dependent cell shape dynamics, with an emphasis on its broader relevance for higher-order aspects of morphogenesis. We first review epithelial structure, both molecular and cellular, to establish both the intrinsic and the emergent properties of epithelial architecture. Next, in order of increasing complexity, we consider models explaining the emergence of epithelial planar pattern geometry. We start with the simplest possible models, working up through complex geometric and biophysical simulators. Throughout, we consider the role of planar pattern geometry in epithelial morphogenesis, and conclude with an examination of the interactions among epithelial packing, biophysical processes, and tissue morphogenesis.

## 2. Conservation of Epithelial Architecture

The different types of epithelia are commonly classified by thickness, cellular morphology, and cellular connectivity. Simple epithelia are a single layer thick; stratified epithelia have two or more layers. Simple epithelia are typically classified as one of four types based on morphology of the component cells: squamous, cuboidal, columnar, or pseudostratified. Squamous cells, for example, are shaped like flattened, interlocking polygonal plates or scales, whereas cuboidal cells are isometric in vertical section (Gray, 1995). Columnar cells have height to width ratios significantly greater than one, and like cuboidal cells, are polygonal when sectioned horizontally (Gray, 1995). There is one additional epithelial category (considered to be a simple epithelium), the pseudostratified type, where elongate, spindle-shaped cells interdigitate their nuclei within the plane of the epithelium but nonetheless remain a monolayer (Wright and Alison, 1984). By analogy with the simple epithelia, the stratified epithelia also contain squamous, cuboidal, and columnar varieties. The critical difference between simple and stratified epithelia is that at least one layer of the latter category has lost contact with the basal lamina, and differentiated (Wright and Alison, 1984). For simplicity, in this chapter we focus exclusively on simple columnar epithelia.

While the essential features of epithelial construction are conserved among metazoa, there are clear differences in the architecture among different evolutionary clades (Knust and Bossinger, 2002; Tepass et al., 2001). The scope of animal epithelia and plant epidermis covered here is sufficiently expansive that a full enumeration of the comparative structural differences is not possible. For purposes of illustration in discussing epithelial architecture, we place emphasis on *Drosophila* simple epithelia, which are particularly well characterized, in terms of both macroscopic and molecular structure.

A primary feature of epithelia, in *Drosophila* and throughout the animal phyla, is cell polarization. Polarization, in turn, facilitates formation of a paracellular diffusion barrier, specialization of plasma membrane proteins, and directional transport in the form of secretion and absorption. The plasma membrane of each epithelial cell is divided into immiscible apical and basolateral domains (Tepass et al., 2001). Importantly, both the apical and the basal domains of the neighboring cells align with each other, endowing the epithelium with a globally faithful, local polarity. Separating the apical and basolateral domains is the zonula adherens (ZA), an adhesive belt encircling the cell (Knust and Bossinger, 2002). The apical zone is subdivided into the apical surface and the marginal zone, where cell-cell contact occurs apical to the Zonula Adherens (Tepass et al., 2001). In *Drosophila*, septate junctions (SJs) lie basal to the ZA and constitute a paracellular permeability barrier, functionally analogous to the vertebrate tight junction (Bilder, 2001; Genova and Fehon, 2003; Gibson and Perrimon, 2003).

The molecular architecture of epithelia has been studied extensively in animal tissues, partially owing to the prominent role they play in human cancers. ZA proteins in *Drosophila* include DE-cadherin, and the scaffold proteins Armadillo (the *Drosophila* orthologue of $\beta$-catenin), $\alpha$-catenin, Canoe (homologue of mammalian Afadin). SJs include the transmembrane proteins Neurexin IV and Fasciclin III, and the scaffold proteins Scribble, Disks Large, Coracle, and Lethal giant larvae (Lgl) (Bilder, 2001; Gibson and Perrimon, 2003; Knust and Bossinger, 2002; Tepass et al., 2001). Additional SJ proteins are encoded by such genes as *contactin, neuroglian, gliotactin, sinuous, $Na^+/K^+$ ATPase, lachesin,* and *megatrachea,* as well as *varicose* (Banerjee et al., 2006; Wu et al., 2007). Interestingly, several of the SJ proteins are critical for apical/basal polarity. Scribble, Disks Large, and Lethal giant larvae are also neoplastic tumor suppressors, thus linking epithelial morphology with control of epithelial proliferation (Hariharan and Bilder, 2006).

Currently, cellular geometry within simple epithelia is best understood in cases when it can be modeled as a planar network, such as at the apical junctions, where the mechanics are constrained (Farhadifar et al., 2007). When sectioned apically, monolayer epithelial cells form ordered polygonal arrays, resembling a froth of soap bubbles (Fig. 4.1). However, in the pseudostratified *Drosophila* wing imaginal disk epithelium, the three-dimensional cellular geometry is considerably more complex below the level of the SJs where cells no longer tightly adhere. This cellular disorder can be attributed to the fact that the relatively large cell nuclei cyclically migrate along the apical–basal axis in concert with the phase of the cell cycle. During cell division, the nuclei are just beneath the apical surface, and the morphology of the dividing cell is almost spherical. The dividing cell thus deforms the apical geometries of its neighbors. Nevertheless, the contacts between neighboring cells tightly adhere and do not rearrange, in spite of the stretching and compression induced by their mitotic neighbor (Gibson et al., 2006). As a result, the "interkinetic" mode of cell division, reliant on cell cycle phase-coupled nuclear movements, has little effect on the polygonal geometry of the apical epithelial surface. By comparison with the deformable cell contacts of animal epithelia, the geometry of plant epidermis appears to be simple, stiff, and regular, and without the complication of nuclear migration along the apical/basal axis. Cucumber epidermal cells, for example, have a slight apical curvature, and are either flat or have a shallow pyramidal point at the basal level. Overall, they are close to being simple, stiff, polygonal prisms (Lewis, 1928). In light of these fundamental structural differences, one might expect animal epithelia and plant epidermis to have very different cellular geometries. In fact, their cellular geometries resemble one another to an unexpected degree, at least in apical cross section.

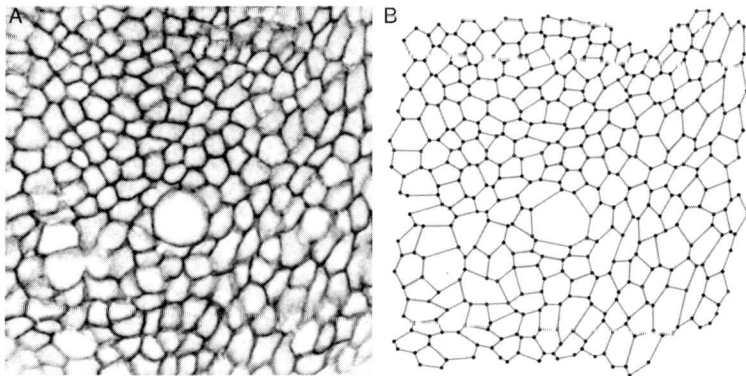

**Figure 4.1** Epithelial topology at the level of cell–cell junctions. (A) An apical cross section through the pseudostratified *Drosophila* wing disk epithelium, stained with antibodies against the SJ component Disks Large to outline cell boundaries. (B) A polygonal approximation to the apical section geometry. Note the presence of non-hexagonal cells.

## 3. Introduction to Cellular Topology

In contrast to cellular geometry, which specifies cell shape, *cellular topology* refers to the connectivity among cells in a tissue. Intuitively, one can imagine stretching or deforming a sheet of cells in such a way that the cells' respective shapes change, but all neighbor relationships are preserved. For such deformations, the sheet's topology remains unchanged. By contrast, processes such as perforation or tearing, in which cell contacts are broken, or convergent extension, in which cell contacts are both made and broken, can change the sheet's topology significantly (Zallen and Zallen, 2004). In epithelia, various elementary processes, such as cell division, cell rearrangement, and cell disappearance, can be shown to modify cell sheet topology in stereotyped ways (Dubertret and Rivier, 1997). Moreover, in many biological systems, cell topology is expected to correlate with geometric variables, such as cell area (Rivier and Lissowski, 1982). Therefore, as a first approximation to geometrically complex morphogenetic processes, topological descriptions can provide fundamental insight into how tissue-level connectivity emerges from elementary cellular transformations.

## 4. Conservation of Topological Structure in Proliferating Epithelia

The columnar cells of both animal epithelia and plant epidermis, which differ in both cellular morphology and molecular architecture, nevertheless look quite similar when viewed in apical section. Quantitative analysis of the topological distributions in both types of monolayers reveals unexpected similarities that distinguish epithelial cell packings from other cellular structures such as soap bubble foams. This conservation of topological structure raises questions about why a given pattern geometry might be preferable to another, or whether the stereotypical polygonal pattern is simply an inevitable consequence of cell division.

The observation that apical sections of proliferating epidermal sheets have constant distributions of polygon types was first made by Lewis (1928) in the cucumber. The distributions of cellular polygons have since been measured in a wide range of divergent organisms, both animal and plant (Gibson *et al.*, 2006; Korn and Spalding, 1973; Zallen and Zallen, 2004). The polygon distributions are remarkably similar within select metazoan epithelia (differing by only a few percent), and are also similar between certain metazoans and some plant epidermis (Fig. 4.2). For example, the cucumber epidermis and the *Drosophila* larval wing disk epithelium have an almost identical distribution of polygon types, with a peak of approximately

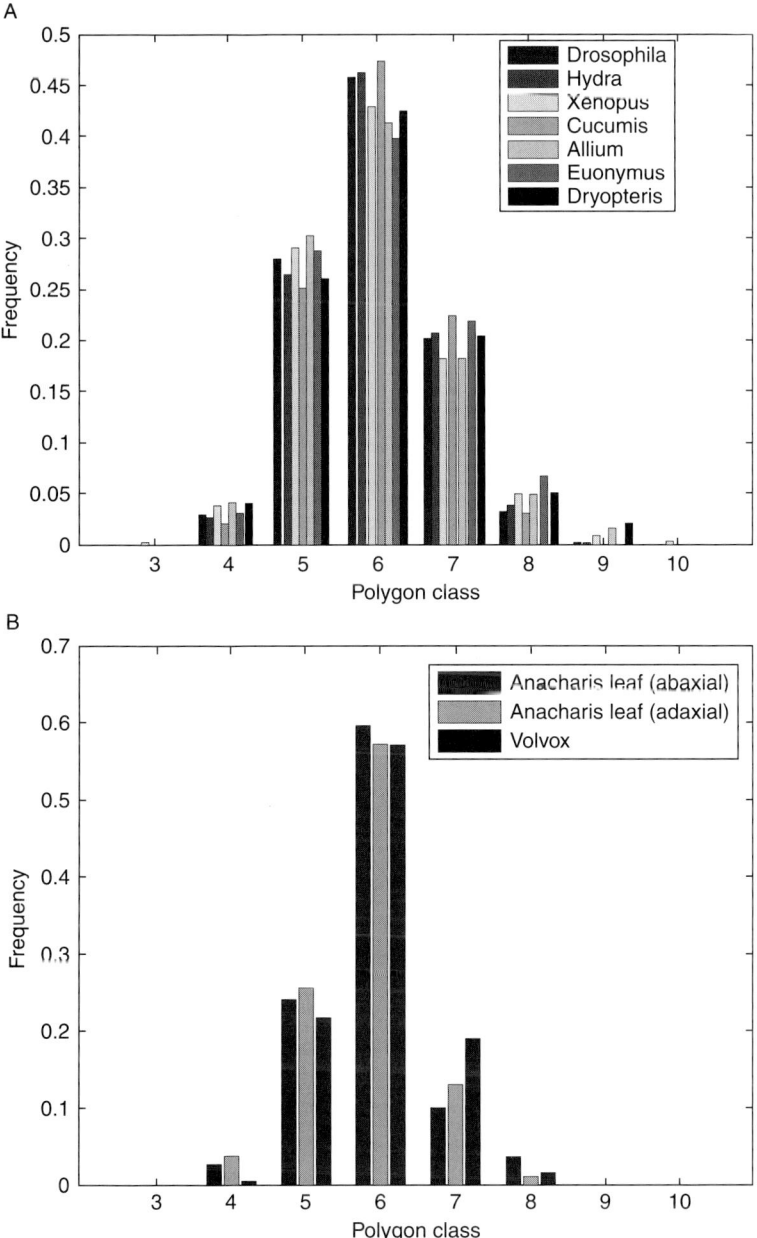

**Figure 4.2** Distributions of cellular polygons in epithelia and plant epidermis. (A) The distribution of polygonal cell types in diverse animal epithelia and plant epidermis. Note the mode of hexagons, and the conservation of the general form in both plants and animals. (B) Two distributions of polygonal cell types that differ from the widely observed distribution seen in (A). Sources of data: *Drosophila*, *Hydra*, *Xenopus* (Gibson et al., 2006); *Cucumis* (Lewis, 1928); and *Allium*, *Dryopteris*, *Euonymus*, *Anacharis*, *Volvox* (Korn and Spalding, 1973). (See Color Insert.)

45% hexagons (Gibson et al., 2006; Lewis, 1928). Moreover, a similar distribution of polygons is also observed in onion epidermis, fern epidermis, and the epithelium of the simple cnidarian, *Hydra* (Gibson et al., 2006; Korn and Spalding, 1973). On the other hand, some species, such as the plant *Anacharis*, with a hexagonal frequency near 60%, have a significantly different distribution of polygon types (Fig. 4.2B; Gibson et al., 2006; Korn and Spalding, 1973; Lewis, 1928). This indicates while there may be significant topological conservation in "default state" cell layers (which exhibit uniform proliferation and little cell rearrangement), biological mechanisms can clearly produce variant distributions under a range of circumstances in both animals and plants.

The distributions of cellular polygons observed in proliferating epithelia seem to share two prominent features. First, without exception, the mode of the distribution is at six-sided cells. Second, the form of the distribution is unimodal, with a rapidly decreasing right tail (Fig. 4.2B). Four-sided cells are rare (between 1% and 5% of the population), whereas three-sided cells are either ultra-rare ($<1/10^6$) or nonexistent (*Xenopus* epithelia present an exception to this rule). The reason for the absence of three-sided cells is unknown, but is probably due (at least in part) to highly symmetric mitoses in epithelial cells having very regular geometries. Cells most likely to give rise to three-sided daughters via division, such as four-sided cells, may also have extremely low division probabilities.

The widely observed similarities in epithelial cell topology naturally lead to the question of whether these similarities arise from conserved division mechanisms. While this question is unsolved theoretically, experimental evidence consistent with this hypothesis has been previously reported (Korn and Spalding, 1973). Quantitatively different planar pattern geometries are seen in plant tissues having qualitatively different division mechanisms (Fig. 4.2B). While the proper controls have not been done, the study suggests that quantitatively different division mechanisms could generate quantitatively different distributions of polygon types. For a first look at theoretical treatment of such questions, see (Cowan and Morris, 1988).

## 5. Topological Inference in Epithelia: Maximum Entropy Methods

The simplest models of epithelial geometry are actually direct statistical inferences about packed sheets of polygons known as *maximum entropy methods*. In epithelia, such methods algebraically compute the most likely configuration of polygons based on a small number of basic geometric assumptions (Rivier et al., 1995). Maximum entropy calculations have yielded excellent local predictions about the neighbor relationships among the different polygon classes (Dubertret and Rivier, 1997; Peshkin *et al.*, 1991). The basic prediction is that many sided cells and few-sided cells are more likely to neighbor one another than would be expected by chance. Such correlations are not only relevant for modeling epithelial topology, they also provide fundamental insight into tissue architecture. Many-sided cells are expected to be larger than fewer-sided cells, both based on experimental observation and based on statistical inference (Lewis, 1928; Rivier and Lissowski, 1982). By anticorrelating the many-sided and the few-sided cells, the tissue reduces the frequency at which multiple large cells are crowded together, or at which multiple small cells are stretched to remain neighbors. Because these tissues are built using division mechanisms, the inference suggests an indirect link between proliferation and epithelial biophysics, and by extension, morphogenesis.

## 6. Topological Models: The Simplest Models of Epithelia

To complement statistical inference methods, simple topological models incorporate biologically plausible mechanisms to generate the empirically observed polygonal cell shape distributions. For a proliferating epithelial sheet, the three most likely candidate mechanisms include cell division, cell rearrangement, and cell disappearance. Empirical studies indicate that the polygon frequencies are in equilibrium or nearly so (Gibson *et al.*, 2006; Korn and Spalding, 1973; Lewis, 1928). Therefore, the simplest possible model to describe the polygon frequencies will specify the rates at which each polygon type is created and destroyed in an attempt to match the distributions observed empirically. During cell division, the mother polygon cell is destroyed, whereas two new daughter polygon cells are created. In addition, the two polygon cells that abut the division plane gain one side each (Fig. 4.3). The myriad ways in which cells can divide and gain sides, when combined with neighbor correlations between different polygon classes, makes predicting topological dynamics nontrivial. At least three groups have independently built mathematical models that closely approximate steady-state topological dynamics (Dubertret and Rivier, 1997; Gibson *et al.*, 2006; Korn and Spalding, 1973). Analytical models are essential for understanding the dynamics of dividing cell sheets. Nevertheless, such models have limitations. In particular, they predict global average dynamics in terms of local average dynamics, which in turn depend on neighbor correlations. Currently, such neighbor correlations are inferred based on equilibrium assumptions and maximum entropy (Dubertret and Rivier, 1997; Miri and Rivier, 2006; Peshkin *et al.*, 1991). Thus, only division rules having a stable equilibrium, and which are well captured

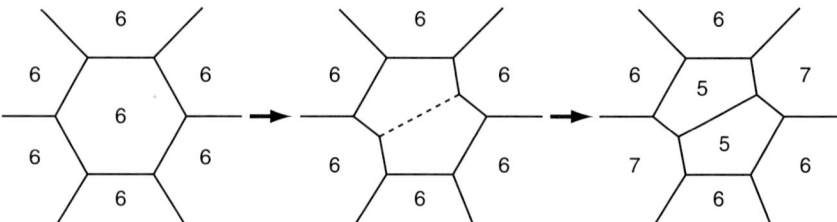

**Figure 4.3** Mitosis in polygonal cells. Following division of the central cell, the two neighboring cells abutting the cleavage plane effectively gain one side each, transitioning from hexagons to heptagons. Daughter cells of the division tend to have a lesser number of sides than the mother polygon. These simple transformations drive the epithelial topology to a heterogeneous, yet predictable, equilibrium.

in terms of local mean-field behavior, can be represented by such analytical models. Topological simulators, which explicitly represent and store neighbor relationships between cells in computer memory, make no such approximations and permit a more general space of division mechanisms to be explored.

In the future, we anticipate that topological simulators will permit answering general questions about topological dynamics. For example, simulators may be useful for determining whether a particular cell division mechanism uniquely corresponds to a specific steady-state distribution of polygon types. When uniqueness holds, the division mechanism can automatically be inferred from the polygon distribution, which is useful for interpreting empirical polygon distributions. For example, the experiments of Korn and Spalding (1973) found several epithelia having hexagonal frequencies close to 60%, a significant deviation from the other studied epithelia, where the frequency is closer to 45%. Uniqueness would imply that these epithelia must also be using quantitatively different division mechanisms or cell rearrangements, which is consistent with the study's qualitative observations (Korn and Spalding, 1973).

A related question concerns the full range of possible polygon distributions that can be reached by a system using all possible division mechanisms. This second question has direct biological relevance for *Drosophila* wing development. For example, it is not currently known why cells undergo extensive neighbor exchanges to achieve hexagonal repacking during pupal wing morphogenesis. During this process, hexagonal frequency in the wing changes from about 45% to nearly 80% (Classen *et al.*, 2005). During larval development, the 45% hexagonal frequency appears to be achieved as a direct consequence of cell division (Gibson *et al.*, 2006). This raises the question of whether the tissue must employ cellular rearrangement to elevate the frequency of hexagons, perhaps because achieving an 80% frequency of hexagons is impossible by division alone. In this case, rule-based modeling could be used to explore the full range of topologies achievable through cell division, and provide new hypotheses about when and why epithelial tissues utilize directed cell rearrangements during morphogenesis.

In spite of the promise that topological models hold for answering fundamental questions about epithelial proliferation and morphogenesis, they have important limitations. First, by definition, topological models do not incorporate space into their dynamics. Therefore, no matter how well epithelial dynamics are captured, topological models cannot predict large-scale macroscopic changes. Second, because any two polygons of the same class are assumed to have the same properties irrespective of their spatial location, processes that involve spatially correlated variables or directionality cannot be properly captured. For example, although topological relationships are a useful conceptual framework for understanding complex cell dynamics such as rosette formation and convergent extension during *Drosophila* germ-band elongation, meaningful simulations of these processes must be geometrical in nature (Bertet *et al.*, 2004; Blankenship *et al.*, 2006; Rauzi *et al.*, 2008). An even simpler example concerns oriented cell divisions, in which case the topological dynamics may look identical for oriented versus nonoriented divisions despite the fact that the geometrical dynamics are fundamentally different (Baena-Lopez *et al.*, 2005; Gong *et al.*, 2004). These limitations of topological models lead us to next consider methods for modeling geometrical aspects of epithelial organization.

## 7. THE SMALLEST GEOMETRICAL MODEL: CLEAVAGE PLANE ORIENTATION IN A SINGLE CELL

The geometric complexity within the plane of an epithelium emerges from interactions between cell shape, cell adhesion, and cell division. Very simple geometric models of proliferating epithelia can be built by assuming that cell division orientation is determined by a cell's local geometry, which emerges from mechanical interactions. Importantly, there is substantial evidence, in both plant and animal cells, that local cell geometry is the default mechanism for determining the cleavage plane.

As early as the nineteenth century, several geometrical rules for cell division were observed in plants. Hofmeister observed that cells tend to divide orthogonal to their long axes. Errera formulated a rule holding that cleavage planes tend to find the shortest path that will divide a cell into two equal halves, which has been confirmed in trichome cells of the Venus flytrap (Dumais, 2007; Smith, 2001). Sach's rule holds that dividing plant cells orient their newly formed cell walls perpendicular to previously formed walls (Lynch and Lintilhac, 1997). Still, the mechanistic puzzle of how the geometry of a dividing cell influences its cleavage plane orientation is not fully understood. Using laser microsurgery in *Nautilocalyx* cells, Goodbody *et al.* (1991) demonstrated that the strands connecting the premitotic nucleus to the cellular cortex are under tension. Arguing that the cortical connection points for the tensile strands are able to move along the cortex (as demonstrated in a simple analogue model), the work supports a minimal distance configuration of tensile strands (Flanders *et al.*, 1990; Goodbody *et al.*, 1991). Under this model, the spatial distribution of tensile strands, and therefore the spatial distribution of internal tension, depends on the geometry of the cell cortex (Flanders *et al.*, 1990).

In the context of plant cells, the distribution of internal tension is important because tension has been implicated as a regulator of cleavage plane orientation. Dividing tobacco cells imbedded in agarose gel blocks under compression have been reported to orient their cleavage planes either parallel to, or perpendicular to, the direction of the principle stress tensor. As a more direct link to cell geometry, the same study suggests that the short axis of the dividing cell is strongly correlated with the compressive stress tensor (Lynch and Lintilhac, 1997). Looking more globally at plant tissue, compression has been shown to induce coplanar cleavage orientations in otherwise disorganized tissue (Lintilhac and Vesecky, 1984). Therefore, it is tempting to speculate that, at least in some cases, dividing cells sense the direction of stress in a tissue based on their geometric strain, and then respond by orienting their cleavage planes to relieve the stress. Recently, Hamant *et al.* (2008) demonstrated a strong correlation between the direction of maximum stress and the orientation of microtubules in the *Arabidopsis* meristem. These experiments suggest that in *Arabidopsis*, cell-autonomous, stress-guided, microtubule alignment-based processes feed back on morphogenetic processes, including tissue folding and cell division. The mechanisms guiding such feedbacks are unknown, but may involve mechanotransduction (Ingber, 2006; Wang *et al.*, 1993).

The geometric biophysics of division plane orientation in animal cells is arguably less well understood than it is in plants. Orienting the division plane so as to divide the long axis is thought to be a default orientation mechanism, although sufficient data to make a general statement are lacking (Strauss *et al.*, 2006). Following a cell cycle-dependent time lag, dissociated *Xenopus* blastula cells with experimentally induced long axes divided perpendicular to the long axis up to 100% of the time (Strauss *et al.*, 2006). Similar division orientation preferences have been shown for the first cleavage of compressed *Xenopus* eggs and mouse zygotes, and also in the blastular wall of starfish embryos (Black and Vincent, 1988; Gray *et al.*, 2004; Honda, 1983). Mitotic spindle orientation is a valuable, if imperfect, predictor of eventual division plane orientation in some systems, and thus revealing about how division planes are determined in animal cells. Spindles in frog blastulas have been shown to orient according to the long axis the majority of the time (Strauss *et al.*, 2006). Additionally, cultured normal rat kidney (NRK) cells reorient the spindle in a dynein-dependent manner so as to divide the long axis when the cellular cortex is deformed (O'Connell and Wang, 2000). Thus, without any additional information, a line perpendicular to a cell's long axis appears to be the best estimate of division plane orientation for an animal cell.

By analogy with plant cells, geometric correlates of division orientation in animal cells are likely due to biophysical mechanisms. The work of Thery *et al.* (2005) shows that placing HeLa cells onto micropatterns printed with fibronectin, which interacts with integrins, is able to strongly bias their spindle orientation. The work also provides evidence that internal actin-binding protein distributions are correlated with these external ECM patterns, thus suggesting a partial mechanism (Thery *et al.*, 2005). Proof of principle was provided in a simple, torque-based model with impressive predictive power (Thery *et al.*, 2007). Such mechanisms may help explain how cells may use extracellular matrix proteins to biophysically "read" their geometry.

## 8. Nongeometric Mechanisms of Division Orientation: A Larger Morphogenetic Space

Genetically directed mechanisms of cleavage plane orientation not solely driven by geometry or biophysics make possible a substantially larger space of morphogenetic transformations, and thus morphologies. An important class of nongeometric division orientation mechanisms includes the molecular control of mitotic spindle orientation, which is partially correlated with division plane orientation. In plants, molecular mechanisms are well known to be involved in orienting cleavage orientation (Jurgens, 2005). For example, the preprophase band (PPB), a ring of microtubules and F-actin, designates the future site of the cleavage plane on the cell cortex (Jurgens, 2005; Smith, 2001). It was recently shown that the *Arabidopsis* protein *tangled* colocalizes with the PPB and predicts the future cleavage sites throughout mitosis and cytokinesis (Walker *et al.*, 2007). Moreover, in *tangled* mutants, cleavage plane orientation is aberrant (Smith *et al.*, 1996). Thus, at least some plant cells appear to decide on their division orientation long before cytokinesis begins. Such fine control over division orientation might be expected to be essential for the control of organ shape. However, in *tangled* mutants, organ shape is normal, suggesting that division-plane independent mechanisms are operating (Smith *et al.*, 1996).

Molecular mechanisms guiding division plane and/or mitotic spindle orientation are well studied in animal cells, but the cellular "decision" concerning division orientation is less well understood. In *Drosophila*, a number cellular junction components have been implicated in spindle orientation mechanisms, including the adherens junction components E-cadherin and Canoe (Le Borgne et al., 2002; Speicher et al., 2008). In mammalian cells, α-catenin is an additional example (Lechler and Fuchs, 2005). Planar cell polarity (PCP) is also implicated in spindle orientation. In developing zebrafish, the dorsal epiblast divides the short axes (not the long axes) of the cells in a PCP pathway-dependent manner (Gong et al., 2004). Additional players include integrins in both *Drosophila* and mammals (Fernandez-Minan et al., 2007; Lechler and Fuchs, 2005). Also, the microtubule plus-end tracking proteins APC and EB1 are implicated in *Drosophila* and also in human cell culture (Draviam et al., 2006; Green et al., 2005; Lu et al., 2001; Rogers et al., 2002; Yamashita et al., 2003). Thus, both intrinsic and extrinsic mechanisms are involved in spindle orientation, and by extension, division plane orientation. These findings suggest that attempting to model morphogenesis using only geometric rules for guiding the division plane can be a vast oversimplification.

##  9. SCALING UP: GEOMETRICAL MODELS AND CELLULAR MECHANICS IN PROLIFERATING EPITHELIA

By comparison with geometrical division orientation mechanisms, nongeometric mechanisms make tissue growth relatively more complex. Without question, such mechanisms are an essential part of proliferation and morphogenesis. Nevertheless, for the most basic understanding of tissue growth, nongeometrical complications are negligible, because over a range of tissue types and species, we expect nongeometrical biases to average out. In other words, it is reasonable to consider geometric mechanisms driving cleavage plane orientation as a default system that can be overridden in instances of direct molecular control. To consider the emergence of epithelial structure in the default geometrical frame, here we consider four types of models: Dirichlet models, cellular Potts models, subcellular element models, and finite-element models, as well as their implications for proliferation and morphogenesis.

### 9.1. Dirichlet models

Dirichlet models are remarkable both for their simplicity and for their accuracy in predicting epithelial geometries. A Dirichlet domain is best visualized in the 2D plane. Suppose many different random dots lie in the plane. The Dirichlet domain for a particular dot is the region of space that is closer to that dot than to all other dots (Honda, 1978). When such a domain is computed for every dot in the plane, and the borders separating the individual domains are drawn, one is left with a polygonal tiling that looks strikingly like an epithelial sheet (Fig. 4.4A). Starting with an image of an epithelium, the Dirichlet approximation to the epithelial geometry is constructed by placing a dot at the center of mass of each cell, and then constructing the Dirichlet domains. Not all cellular structures match the Dirichlet domains, and the degree to which a structure deviates from the approximation can be quantified (Honda, 1978). Nevertheless, for a first-order approximation, it looks quite realistic, and is an illustration of how strong the space constraints are in epithelial sheets.

To probe the underlying forces specifying the geometry of an actual cellular sheet, one might consider the regularity of the cells. Using a "boundary shortening" procedure, one can, in an iterative fashion, shorten the

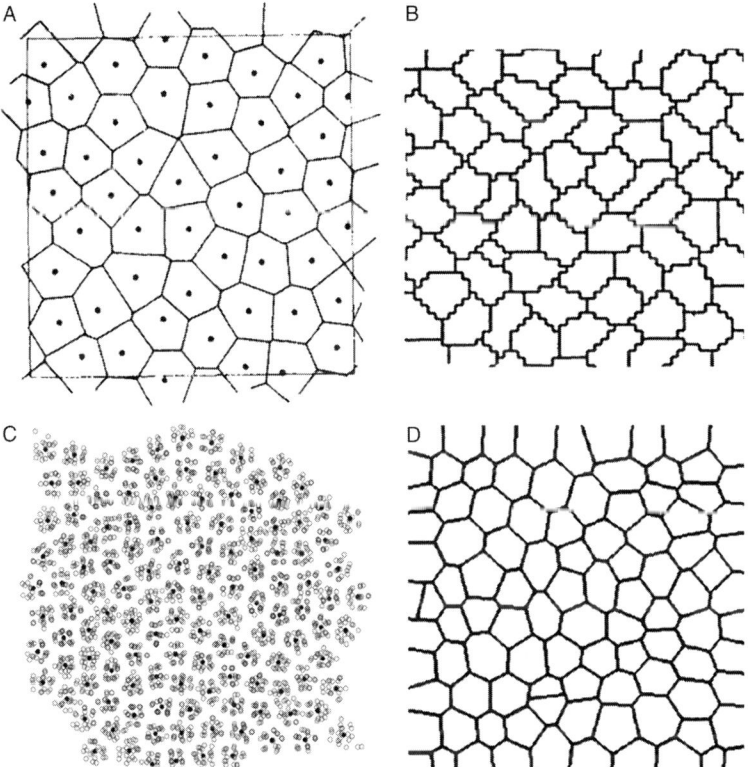

**Figure 4.4** Visual output from models for epithelial geometry. (A) A Dirichlet model of space partitioning (Honda, 1978). Note the resemblance to natural epithelia. (B) Output from a cellular Potts model (Glazier and Graner, 1993). The cellular geometries are free to assume nonpolygonal shapes. (C) Visual output from a subcellular element model (Newman, 2005). Each cell is represented by a cloud of physically interacting points, visible in black. (D) Visual output from a finite-element model (Brodland and Veldhuis, 2002).

perimeters of the cells' boundaries without changing their areas (Honda and Eguchi, 1980). Interestingly, different cellular structures are able to shorten their boundaries to differing extents, depending on the regularity of the cellular shapes to begin with (Honda and Eguchi, 1980). Especially regular structures, such as soap bubbles, are already close to the minimum perimeter, whereas more irregularly packed cell sheets may be able to shorten their boundaries substantially. The differing abilities to shorten are thought to reflect a property of contractility in the cell boundaries, which was recently included in a model of epithelial geometry by Farhadifar *et al.* (Farhadifar *et al.*, 2007; Honda and Eguchi, 1980). Contractility of cells is a fundamental property that can influence morphogenesis in three dimensions, and may impact the dynamics of spindle orientation for planar representations.

The Dirichlet approximation has several applications to epithelia. First, it can be used to model how a sheet of cells responds to cellular disappearance (Honda, 1978). Second, in combination with an iterative, center-of-gravity-based relaxation procedure, it can be used to model cell division (Honda, 1983; Honda *et al.*, 1984). Thus, using only a few simple assumptions, a coarse-grained approximation of temporal froth evolution can be achieved for epithelial systems. These predictions demonstrate the degree to which epithelial proliferation, and morphogenesis by extension, is constrained by the geometry of the starting material.

## 9.2. Cellular Potts models

A second standard technique, the cellular Potts model (CPM) is a general method for simulating cellular dynamics and is useful for studying proliferating epithelia. Such models are based on Hamiltonians, or effective energy functions, in order to determine the probability that one mechanical configuration will transition to another. The art of designing these models is in specifying the Hamiltonian. Experimentally, the challenge is justifying the Hamiltonian being used. Practically, these models' utility is in simulating complex cellular shapes that could not be studied in closed form.

An extremely simple CPM for simulating proliferating epithelia can be found in the work of Mombach *et al.* (1993). Here, the energy function is based on interfacial energy. Mitosis (through the cell's approximate center of mass) occurs when the area to perimeter ratio exceeds a threshold. The significance of the model is that it demonstrates how the dynamics and geometry of epithelial sheets depend on fundamental mechanical quantities, such as interfacial energy. An ambitious and interesting model of morphogenesis can be found in the work of Graner and Glazier (1992, 1993), which simulates cell sorting for two classes of cells based on differential adhesion. Here, the energy function is based both on surface energy and on the difference between a cell's actual size and its "preferred" size (Graner and Glazier, 1992). More recently, Farhadifar *et al.* (2007) used the basic ideas of the CPM to construct a more sophisticated incarnation incorporating line tensions to simulate proliferation and rearrangement in the *Drosophila* wing disk. Like the model of Graner and Glazier (1992), this model uses a preferred size in the energy function, but now the Hamiltonian also includes a contractility term and a preferred perimeter (Farhadifar *et al.*, 2007). The power of these models is that they are completely general—any cellular system can be simulated, with varying degrees of complexity, provided a proper energy function is specified. Because the energy functions are usually based on realistic physical principles, the framework allows large regions of developmental space to be explored by changing a few fundamental parameters. Developmentally, different parameter regimes may correspond to very different types of morphogenesis.

## 9.3. Subcellular element models

A new and exciting alternative to CPMs can be found in the subcellular element models developed by Newman (2005). These models simulate cellular interactions by representing each cell as a cloud of points. Each point belongs to exactly one cell, and has over damped, elastic interactions with the other points in the cell, which depend on absolute distances between particles. In addition to these interactions, there is also a random noise component. An additional term models interactions between points in neighboring cells when those points are in close proximity. Thus, cellular geometry and tissue morphology emerge as the collective interactions of these point masses (Fig. 4.4C). When division is introduced into such models, the cellular structures produced look qualitatively realistic (Newman, 2008). The true power of such models likely lies in the future, when three-dimensional biological complexities such as cell polarity and cell–cell signaling can be introduced. Nonplanar, dynamical rearrangements of epithelial sheets, such as folding or buckling, could be simulated in this manner. Such models also make it possible to compute the local stresses and strains that are operating in a cell at a particular instant, which might be useful for testing hypotheses about mechanotransduction (Ingber, 2006).

## 9.4. Finite-element models

A fourth class of models, termed finite-element models (Fig. 4.4D), simulates cellular mechanics in terms of deformable connections between point masses, with additional physical constraints. The most basic finite-element models essentially represent cells as idealized, polygonal rubber balloons. In such models, the tricellular junctions are small point masses, which are connected to the other tricellular junctions by idealized, straight springs. Keeping with the balloon analogy, the internal pressure is taken to be inversely proportional to the 2D cellular area (Prusinkiewicz and

Lindenmayer, 1990). On the other end of the complexity spectrum, Brodland and colleagues (Brodland and Wiebe, 2004; Chen and Brodland, 2000) have developed rigorous finite-element simulators of epithelial growth and morphogenesis based on mechanical properties including viscosities and interfacial tensions. Such models have been used to answer fundamental questions about the macroscopic mechanical effects of cellular anisotropy, and about whether lamellipodia might be sufficient to drive convergent extension (Brodland, 2006; Brodland and Veldhuis, 2002; Brodland and Wiebe, 2004). They have also been used as counterevidence against the differential adhesion hypothesis of cell sorting (Brodland, 2002). Thus, finite-element models are ideal for answering fundamentally mechanical questions. However, they are also among the most complex models to implement.

Ultimately, whatever the underlying model of the cellular mechanics, the goal of a geometrical model is to asymptotically capture the local geometry of cells. This is not just a modeling issue; dynamic control of cell geometry is a fundamental problem that a developing tissue must solve to grow and pattern itself in a robust, reproducible way. To consider just one of the challenges, when the new side is placed between the new daughter cells, how long should the side be? After many iterations of a division algorithm, if the lengths of the inserted sides are not realistically implemented, then the global statistics of cell geometry will be affected. Also, how much tension should a newly inserted side be under? This is a free parameter in most simulations, but could be estimated from the ablation methods of Farhadifar *et al.* (2007). Depending on the system, steady-state cell geometry may also feed back on tissue growth by influencing spindle orientation. In the future, we anticipate that robust geometrical models will serve as dynamic lattices upon which complex models of cell–cell signaling and morphogenesis can be layered. The intricate feedback between genetics, geometry, and biophysics will likely bring us one step closer to realistic simulations of development.

## 10. Putting It All Together: Genetics, Geometry, and Biophysics

Geometrical models of epithelial organization are alone insufficient to provide insight into how the simplest tissues grow and pattern themselves. Building on the geometric and biophysical machinery discussed above, it is necessary to integrate active cellular behavior into the picture, which will eventually be understood at the intersection of genetics, geometry, and biophysics. Below we discuss four systems in which all three components interact strongly, and consider the implications for morphogenesis.

One of the most basic aspects of organ development concerns how cells in a growing tissue know when to stop proliferating, or when they are proliferating too quickly. In the *Drosophila* wing disk, it has been observed that the proliferation rate is on average roughly uniform across the epithelium, with some heterogeneity (Dubatolova and Omelyanchuk, 2004; Milan *et al.*, 1996; Shraiman, 2005). In this system, cells that divide more slowly than their neighbors are eliminated by cell death, while those that divide too rapidly are able to target slower-dividing neighbors for elimination (Li and Baker, 2007). This phenomenon, termed cell competition, suggests that cells know how quickly they are proliferating relative to their neighbors (de la Cova *et al.*, 2004; Morata and Ripoll, 1975; Simpson, 1979; Simpson and Morata, 1981). How this information is computed is not known. Recently, Shraiman (2005) analyzed the continuous mechanical implications of differential rates of growth, and has proposed a mechanical basis for a negative feedback mechanism regulating cellular growth and apoptosis, which is implicated in cell competition. More recently, Hufnagel *et al.* (2007) demonstrated, using a geometrical, energy based polygonal model not unlike the model of Farhadifar *et al.* (2007), that mechanical feedback is a potential mechanism for controlling organ size. This model achieves a uniform rate of cellular proliferation and uses a plausible mechanism for an organ to sense when it has reached a critical size (Hufnagel *et al.*, 2007). Such models demonstrate the sufficiency of mechanical stress as a negative regulator of growth to reproduce empirically observed growth trends. However, there are plausible alternatives. Recently, Senoo-Matsuda and Johnston (2007) demonstrated experimental *in vitro* evidence of bidirectional diffusible signaling molecules to explain how cells know their growth rates relative to their neighbors during cell competition. It is not clear whether such a mechanism could operate *in vivo*, or between wild-type cells. Perhaps a simple diffusion model based on these experiments could be extended in a geometrical framework as an alternative to the stress-based models of Hufnagel *et al.* (2007) and Shraiman (2005).

To consider a second system in which biophysics and cell geometry are both essential in the context of genetics, we look at the problem of hexagonal repacking that occurs during pupal wing morphogenesis, also in *Drosophila*. During this stage of wing development, the frequency of hexagons increases from approximately 45% to nearly 80%, and appears to result from iterative T1 transitions (Classen *et al.*, 2005). The work of Classen *et al.* (2005) suggests that cells recycle cadherin during junctional remodeling. Additionally, mutants for PCP proteins and dynamin are defective in hexagonal repacking (Classen *et al.*, 2005). Based on this evidence, it is tempting to speculate that the cells actively regulate the localization of cadherin so as to enable a T1 transition to a lower energy configuration at each junctional remodeling step, and thus bring their packing configuration closer to a hexagonal lattice. One might further speculate that the PCP pathway biases each step in the sequence of iterations. Geometrical modeling would be a highly appropriate approach with which to study the repacking. An open question is why an iterative T1 procedure is not sufficient to make the lattice perfectly hexagonal. Is it simply a matter of structural noise, or is this the lowest energy state achievable?

A related question concerns why the pupal wing disk undergoes repacking in the first place. One possibility concerns the directional uniformity of the tiny hairs that point distally in wild-type wings, but can have different orientations in PCP mutants. Recently, Amonlirdviman et al. (2005) modeled PCP as a reaction–diffusion system utilizing directionally biased positive feedback on a perfectly regular hexagonal lattice. Importantly, however, the hexagonally packed pupal wing is not perfectly hexagonal and regular. A natural question is whether the proposed mechanism would function just as well on a more realistic irregular lattice. Recently, Ma et al. (2008) tested these assumptions both experimentally and computationally. Their analysis strongly suggests that planar packing geometry is a critical parameter for the proper functioning of the PCP-based distal hair alignment mechanism. The earlier work of Classen et al. (2005) did not uncover such a correlation, although their topological metric may not have been sensitive enough to detect such differences. A second possible reason for hexagonal repacking is simply structural. Would a more densely packed wing be stiffer or stronger, or result in structure with more homogenous mechanical properties? Might two hexagonally packed wings have better mirror-image symmetry than two irregularly packed wings? By simulating very regular versus very irregular packings, a finite-element model could yield insight into these questions. The *Drosophila rho-associated kinase* (*Drok*) gene has been shown to link PCP signaling to the cytoskeleton (Winter et al., 2001). Therefore, it is also possible that PCP programs are able to change the packing geometry, which would bring the feedback full circle.

A third system in which biophysics strongly interacts with cellular geometry in a genetic context is in the *Drosophila* embryonic epithelium during germ-band extension. There are two very different analyses of this process, which have significantly different implications biophysically. The first analysis describes germ-band extension as an active process involving iterative, directional T1 transitions (Bertet et al., 2004). The second describes complex rearrangements of cellular neighbor relationships, termed "rosettes," which may involve local tissue-level coordination (Blankenship et al., 2006). In terms of biophysical, geometric simulation, a very important issue concerns whether rosette-like structures are actively controlled tissue movements, or whether they are expected to emerge by chance from multiple, locally aligned T1 processes.

A recent study by Rauzi et al. (2008) provides evidence for the latter hypothesis based on cell and tissue-level biophysical simulations of germ-band extension. The argument rests on two observations. First, using only T1 processes, and a physical model and parameter regime consistent with empirical measurements of germ-band elongation as a function of the number of T1 iterations, the authors observe similar frequencies of rosette like structures *in silico* and *in vivo*, which in both cases are rare. Second, it is shown experimentally that rosette structures can be decomposed into multiple T1-like processes using laser ablation. Moreover, such ablations can be used to infer similar levels of tension for the two processes (Rauzi et al., 2008). Importantly, this analysis shows that chance alignments of three-way vertices, in combination with T1 processes, may be sufficient to produce rosettes at realistic frequencies. However, it does not preclude the existence of active biological programs which align and coordinate rosette formation. Analysis of the distribution of $n$-way vertices (where $n$ is an integer), may be a useful way to distinguish between the two cases. Additional simulations, which consider different models of how elastic tension varies with junction orientation, are needed to ensure that the frequencies generated are not an artifact of the modeling assumptions (Rauzi et al., 2008). Such methods might be complemented by geometric analysis of rosette formation, as well as genetic screens for rosette-formation defects. Thus, whether rosette formation is a process distinct from T1 transitions in the germ band is currently unresolved.

A fourth biophysically and genetically complex morphogenetic process is convergent extension in the notochord of the Ascidian *Ciona Savignyi*. Here, 40 cells in an initially rounded packing shape intercalate to produce a single column of flattened cells. The intercalation process involves penetration of cellular projections between neighboring cells (Miyamoto and Crowther, 1985). The system is especially well suited to address convergent extension in a biophysical context because cells (1) do not divide and (2) depend less strongly on neighboring tissues for their movement than they might in a vertebrate system (Veeman et al., 2008). It was recently shown that normal convergent extension in *Ciona* requires the Prickled protein (encoded by *aim*), which biases the direction of lamellipodial extensions that are believed to drive convergent extension (Jiang et al., 2005). A subsequent analysis of the mutant *chongmague* (*chm*) suggests that another reason that convergent extension fails in the *prickled* mutant is because PCP signaling is required for maintenance of the polarized distribution of a laminin protein encoded by *chm*. Moreover, even in the absence of PCP, *chm* mutants are able to partially complete convergent extension (Veeman et al., 2008). Therefore, in this organism, there is a complex interaction between planar polarity signaling, laminin maintenance/polarization, and the biophysics of convergent extension. Mutation of the PCP player *disheveled* is also known to disrupt convergent extension in *Ciona intestinalis* (Keys et al., 2002). In *Xenopus*, expressing a mutant form of *Disheveled* causes defects both in cellular polarization and in convergent extension (Wallingford et al., 2000). The complexity of convergent extension highlights why traditional genetic analysis alone is insufficient to describe these PCP and convergent extension phenotypes.

In terms of a biophysical understanding, the extended Potts model has been used to successfully capture the cellular rearrangements seen in convergent extension in *Xenopus* based on anisotropic differential adhesion. Interestingly, such models are also able to mimic Ascidian convergent extension (Zajac *et al.*, 2003). However, energy-based models are not mechanistic, and additional work remains to be done to determine how the separate forces indirectly associated with *chm* and *prickled* contribute to convergent extension. Previous models were also successful in reproducing convergent extension but require additional constraints (Weliky *et al.*, 1991). One possible next step might be to layer simulations of PCP components on top of the physical models to test how these rearrangements might be controlled by PCP.

## 11. Future Directions

Some of the primary challenges in understanding both cellular topology and cellular geometry are not theoretical, but instead empirical and technological. We suggest two key future directions. First, both statistical inference and mathematical modeling depend heavily on empirical constraints. Yet currently, there is no high-throughput means with which to gather empirical statistics on cellular geometry. Classen *et al.* (2005) employed an image processing software package to infer cellular topology, which we have independently tested (W. T. Gibson, unpublished data). Such programs represent a first step in the high-throughput transition, but are currently quite sensitive to experimental noise, scale, and imaging conditions. We therefore argue that the field is currently limited most by statistical power, and image analysis methods. For the same reason, most of the progress in the field has been made in studying static images. Solving problems in image processing will also give the field a substantial boost in studying dynamics. Currently, data are plentifully available, but the available image processing methods are the limiting factor.

One should not be left with the impression that the "lower" levels of the complexity hierarchy—topology and geometry—are in any way completely understood. Even at the most basic level, topology, there are areas of statistical inference that have not been attempted. For example, to our knowledge no study has yet attempted to predict the relative frequencies of the different classes of tricellular junctions (the number of such junctions bordering cells having $i$, $j$, and $k$ sides, where $i$, $j$, and $k$ are arbitrary), probably because such inferences would be difficult to verify empirically. As a consequence, there are limits to our ability to mathematically model the processes of hexagonal repacking or convergent extension in terms of topology, because both depend on the relative frequencies of the different classes of tricellular junctions. It is also unclear whether dynamical topological models of epithelial proliferation will need to be revisited, because their topological kinetics have never been measured empirically.

Space partitioning plays a prominent role in setting up the geometric structure of epithelia, as can clearly be seen from Dirichlet constructions (Honda, 1983). Nevertheless, such constraints are not sufficient to fully specify geometry. A second major hurdle, as image processing improves, is to understand cell geometry as an emergent property of tissue mechanics and cellular rearrangements. The geometric parameters of rearrangement can be measured using live imaging. However, the underlying biophysical forces will have to be inferred, using a combination of statistical inference, modeling, and experimental methods. The work of Farhadifar et al. (2007) offers an example of how such properties might be tested experimentally. The final step is to construct geometrical simulators of cell and tissue mechanics, and then to test how closely such models are able to mimic the geometric parameters (including the statistical moments of angle measurements, side lengths, etc.) of actual tissues.

While this chapter has primarily focused on very simple animal tissues, the field has much to learn from the highly realistic models of tissue morphogenesis being developed by the plant community. In some respects, plant tissues are a more natural choice for a model system, due to their ease of culture, the viability of their genetic knockouts, their structural integrity, and their simple, elegant architecture. Recently, Grieneisen et al. (2007) considered the transport of Auxin in a growing root using both computational modeling and genetic perturbations. Importantly, the computational model incorporates realistic cellular geometry, and is therefore able to treat diffusion and permeability separately. The work considers the influence of cell geometry on Auxin distribution and transport, as well as the influence of Auxin transport on cell geometry and tissue patterning. Similarly complex models of plant phyllotaxis based on Auxin transport have been developed in the work of Smith et al. (2006). If even these complex plant tissues are amenable to computational modeling and experimental validation, then plant epidermis may provide an ideal model system for studying epithelial proliferation and morphogenesis in the future, as originally suggested by Lewis (1928).

## 12. CONCLUSION

Development is often considered in terms of gene networks and deterministic decisions, yet the emergent, biophysical properties of a developing tissue are essential for its morphogenesis. These properties emerge stochastically and macroscopically, and cannot be explicitly encoded into the developmental-genetic program, even if the genetic program is tuned to exploit them. Consequently, these properties are difficult to understand through the traditional logic of molecular-genetic analysis, requiring the creative deployment of new modeling and simulation-based methodologies. Epithelial proliferation as it relates to morphogenesis is perhaps the simplest such relationship in development. The emergence of planar packing geometry is beginning to be understood, and it has certainly been shown to correlate with morphogenetic events. Still, much work remains to be done in order to understand the dynamic relationship between proliferation and epithelial cell packing, and to establish whether packing geometry plays an essential role in morphogenesis. Future work will likely expand the repertoire of quantitative models for tissue architecture and thereby extend our understanding of epithelial morphogenesis beyond the limits of traditional genetic analysis.

## REFERENCES

Amonlirdviman, K., Khare, N. A., Tree, D. R., Chen, W. S., Axelrod, J. D., and Tomlin, C. J. (2005). Mathematical modeling of planar cell polarity to understand domineering nonautonomy. *Science* **307,** 423–426.

Baena-Lopez, L. A., Baonza, A., and Garcia-Bellido, A. (2005). The orientation of cell divisions determines the shape of *Drosophila* organs. *Curr. Biol.* **15,** 1640–1644.

Banerjee, S., Sousa, A. D., and Bhat, M. A. (2006). Organization and function of septate junctions: An evolutionary perspective. *Cell Biochem. Biophys.* **46,** 65–77.

Bertet, C., Sulak, L., and Lecuit, T. (2004). Myosin-dependent junction remodelling controls planar cell intercalation and axis elongation. *Nature* **429,** 667–671.

Bilder, D. (2001). PDZ proteins and polarity: Functions from the fly. *Trends Genet.* **17,** 511–519.

Black, S. D., and Vincent, J. P. (1988). The first cleavage plane and the embryonic axis are determined by separate mechanisms in *Xenopus laevis*. II. Experimental dissociation by lateral compression of the egg. *Dev. Biol.* **128,** 65–71.

Blankenship, J. T., Backovic, S. T., Sanny, J. S., Weitz, O., and Zallen, J. A. (2006). Multicellular rosette formation links planar cell polarity to tissue morphogenesis. *Dev. Cell* **11,** 459–470.

Brodland, G. W. (2002). The differential interfacial tension hypothesis (DITH): A comprehensive theory for the self-rearrangement of embryonic cells and tissues. *J. Biomech. Eng.* **124,** 188–197.

Brodland, G. W. (2006). Do lamellipodia have the mechanical capacity to drive convergent extension? *Int. J. Dev. Biol.* **50,** 151–155.

Brodland, G. W., and Veldhuis, J. H. (2002). Computer simulations of mitosis and interdependencies between mitosis orientation, cell shape and epithelia reshaping. *J. Biomech.* **35,** 673–681.

Brodland, G. W., and Wiebe, C. J. (2004). Mechanical effects of cell anisotropy on epithelia. *Comput. Methods Biomech. Biomed. Eng.* **7,** 91–99.

Chen, H. H., and Brodland, G. W. (2000). Cell-level finite element studies of viscous cells in planar aggregates. *J. Biomech. Eng.* **122,** 394–401.

Classen, A. K., Anderson, K. I., Marois, E., and Eaton, S. (2005). Hexagonal packing of *Drosophila* wing epithelial cells by the planar cell polarity pathway. *Dev. Cell* **9**, 805–817.

Cowan, R., and Morris, V. B. (1988). Division rules for polygonal cells. *J. Theor. Biol.* **131**, 33–42.

de la Cova, C., Abril, M., Bellosta, P., Gallant, P., and Johnston, L. A. (2004). *Drosophila* myc regulates organ size by inducing cell competition. *Cell* **117**, 107–116.

Draviam, V. M., Shapiro, I., Aldridge, B., and Sorger, P. K. (2006). Misorientation and reduced stretching of aligned sister kinetochores promote chromosome missegregation in EB1- or APC-depleted cells. *EMBO J.* **25**, 2814–2827.

Dubatolova, T., and Omelyanchuk, L. (2004). Analysis of cell proliferation in *Drosophila* wing imaginal discs using mosaic clones. *Heredity* **92**, 299–305.

Dubertret, B., and Rivier, N. (1997). The renewal of the epidermis: A topological mechanism. *Biophys. J.* **73**, 38–44.

Dumais, J. (2007). Can mechanics control pattern formation in plants? *Curr. Opin. Plant Biol.* **10**, 58–62.

Farhadifar, R., Roper, J. C., Aigouy, B., Eaton, S., and Julicher, F. (2007). The influence of cell mechanics, cell–cell interactions, and proliferation on epithelial packing. *Curr. Biol.* **17**, 2095–2104.

Fernandez-Minan, A., Martin-Bermudo, M. D., and Gonzalez-Reyes, A. (2007). Integrin signaling regulates spindle orientation in *Drosophila* to preserve the follicular-epithelium monolayer. *Curr. Biol.* **17**, 683–688.

Flanders, D. J., Rawlins, D. J., Shaw, P. J., and Lloyd, C. W. (1990). Nucleus-associated microtubules help determine the division plane of plant epidermal cells: Avoidance of four-way junctions and the role of cell geometry. *J. Cell Biol.* **110**, 1111–1122.

Genova, J. L., and Fehon, R. G. (2003). Neuroglian, Gliotactin, and the $Na^+/K^+$ ATPase are essential for septate junction function in *Drosophila*. *J. Cell Biol.* **161**, 979–989.

Gibson, M. C., and Perrimon, N. (2003). Apicobasal polarization: Epithelial form and function. *Curr. Opin. Cell Biol.* **15**, 747–752.

Gibson, M. C., Patel, A. B., Nagpal, R., and Perrimon, N. (2006). The emergence of geometric order in proliferating metazoan epithelia. *Nature* **442**, 1038–1041.

Glazier, J. A., and Graner, F. (1993). Simulation of the differential adhesion driven rearrangement of biological cells. *Phys. Rev. E Stat. Phys. Plasmas Fluids Relat. Interdiscip. Topics* **47**, 2128–2154.

Gong, Y., Mo, C., and Fraser, S. E. (2004). Planar cell polarity signalling controls cell division orientation during zebrafish gastrulation. *Nature* **430**, 689–693.

Goodbody, K. C., Venverloo, C. J., and Lloyd, C. W. (1991). Laser microsurgery demonstrates that cytoplasmic strands anchoring the nucleus across the vacuole of premitotic plant cells are under tension. Implications for division plane alignment. *Development* **113**, 931–939.

Graner, F., and Glazier, J. A. (1992). Simulation of biological cell sorting using a two-dimensional extended Potts model. *Phys. Rev. Lett.* **69**, 2013–2016.

Gray, H. (1995). Gray's Anatomy: The Anatomical Basis of Medicine and Surgery. Churchill Livingstone, New York.

Gray, D., Plusa, B., Piotrowska, K., Na, J., Tom, B., Glover, D. M., and Zernicka-Goetz, M. (2004). First cleavage of the mouse embryo responds to change in egg shape at fertilization. *Curr. Biol.* **14**, 397–405.

Green, R. A., Wollman, R., and Kaplan, K. B. (2005). APC and EB1 function together in mitosis to regulate spindle dynamics and chromosome alignment. *Mol. Biol. Cell* **16**, 4609–4622.

Grieneisen, V. A., Xu, J., Maree, A. F., Hogeweg, P., and Scheres, B. (2007). Auxin transport is sufficient to generate a maximum and gradient guiding root growth. *Nature* **449**, 1008–1013.

Hamant, O., Heisler, M. G., Jonsson, H., Krupinski, P., Uyttewaal, M., Bokov, P., Corson, F., Sahlin, P., Boudaoud, A., Meyerowitz, E. M., Couder, Y., and Traas, J. (2008). Developmental patterning by mechanical signals in *Arabidopsis*. *Science* **322**, 1650–1655.

Hariharan, I. K., and Bilder, D. (2006). Regulation of imaginal disc growth by tumor-suppressor genes in *Drosophila*. *Annu. Rev. Genet.* **40**, 335–361.

Honda, H. (1978). Description of cellular patterns by Dirichlet domains: The two-dimensional case. *J. Theor. Biol.* **72**, 523–543.

Honda, H. (1983). Geometrical models for cells in tissues. *Int. Rev. Cytol.* **81**, 191–248.

Honda, H., and Eguchi, G. (1980). How much does the cell boundary contract in a monolayered cell sheet? *J. Theor. Biol.* **84**, 575–588.

Honda, H., Yamanaka, H., and Dan-Sohkawa, M. (1984). A computer simulation of geometrical configurations during cell division. *J. Theor. Biol.* **106**, 423–435.

Hufnagel, L., Teleman, A. A., Rouault, H., Cohen, S. M., and Shraiman, B. I. (2007). On the mechanism of wing size determination in fly development. *Proc. Natl. Acad. Sci. USA* **104**, 3835–3840.

Huzita, H., and Scimemi, B. (1989). The algebra of paper folding (origami). *In Proceedings of the First International Meeting of Origami Science and Technology*, (H. Huzita, ed.), pp. 215–222.

Ingber, D. E. (2006). Cellular mechanotransduction: Putting all the pieces together again. *FASEB J.* **20**, 811–827.

Jiang, D., Munro, E. M., and Smith, W. C. (2005). Ascidian prickle regulates both mediolateral and anterior–posterior cell polarity of notochord cells. *Curr. Biol.* **15**, 79–85.

Jurgens, G. (2005). Cytokinesis in higher plants. *Annu. Rev. Plant Biol.* **56**, 281–299.

Keys, D. N., Levine, M., Harland, R. M., and Wallingford, J. B. (2002). Control of intercalation is cell-autonomous in the notochord of *Ciona intestinalis*. *Dev. Biol.* **246**, 329–340.

Knust, E., and Bossinger, O. (2002). Composition and formation of intercellular junctions in epithelial cells. *Science* **298**, 1955–1959.

Korn, R. W., and Spalding, R. M. (1973). The geometry of plant epidermal cells. *New Phytol.* **72**, 1357–1365.

Le Borgne, R., Bellaiche, Y., and Schweisguth, F. (2002). *Drosophila* E-cadherin regulates the orientation of asymmetric cell division in the sensory organ lineage. *Curr. Biol.* **12**, 95–104.

Lechler, T., and Fuchs, E. (2005). Asymmetric cell divisions promote stratification and differentiation of mammalian skin. *Nature* **437**, 275–280.

Lecuit, T., and Lenne, P. F. (2007). Cell surface mechanics and the control of cell shape, tissue patterns and morphogenesis. *Nat. Rev. Mol. Cell Biol.* **8**, 633–644.

Lewis, F. T. (1928). The correlation between cell division and the shapes and sizes of prismatic cells in the epidermis of cucumis. *Anatom. Rec.* **38**, 341–376.

Li, W., and Baker, N. E. (2007). Engulfment is required for cell competition. *Cell* **129**, 1215–1225.

Lintilhac, P. M., and Vesecky, T. B. (1984). Stress-induced alignment of division plane in plant tissues grown *in vitro*. *Nature* **307**, 363–364.

Lu, B., Roegiers, F., Jan, L. Y., and Jan, Y. N. (2001). Adherens junctions inhibit asymmetric division in the *Drosophila* epithelium. *Nature* **409**, 522–525.

Lynch, T. M., and Lintilhac, P. M. (1997). Mechanical signals in plant development: A new method for single cell studies. *Dev. Biol.* **181**, 246–256.

Ma, D., Amonlirdviman, K., Raffard, R. L., Abate, A., Tomlin, C. J., and Axelrod, J. D. (2008). Cell packing influences planar cell polarity signaling. *Proc. Natl. Acad. Sci. USA* **105**, 18800–18805.

Milan, M., Campuzano, S., and Garcia-Bellido, A. (1996). *Proc. Natl. Acad. Sci. USA* **93**, 640–645.

Miri, M., and Rivier, N. (2006). Universality in two-dimensional cellular structures evolving by cell division and disappearance. *Phys. Rev. E Stat. Nonlin. Soft Matter Phys.* **73**, 031101.

Miyamoto, D. M., and Crowther, R. J. (1985). Formation of the notochord in living ascidian embryos. *J. Embryol. Exp. Morphol.* **86**, 1–17.

Mombach, J. C., de Almeida, R. M., and Iglesias, J. R. (1993). Mitosis and growth in biological tissues. *Phys. Rev. E Stat. Phys. Plasmas Fluids Relat. Interdiscip. Topics* **48**, 598–602.

Morata, G., and Ripoll, P. (1975). Minutes: Mutants of *Drosophila* autonomously affecting cell division rate. *Dev. Biol.* **42**, 211–221.

Nagpal, R. (2001). *In Programmable Self-Assembly: Constructing Global Shape Using Biologically-Inspired Local Interactions and Origami Mathematics* p. 105. Ph.D. Thesis, Department of Electrical Engineering and Computer Science, Massachusetts Institute of Technology, Cambridge, MA.

Nagpal, R. (2002). Programmable pattern-formation and scale-independence. *Proceedings of the Fourth International Conference on Complex Systems (ICCS)*.

Newman, T. J. (2005). Modeling multi-cellular systems using sub-cellular elements. *Math. Biosci. Eng.* **2**, 611–622.

Newman, T. J. (2008). Grid-free models of multicellular systems, with an application to large-scale vortices accompanying primitive streak formation. *Curr. Top. Dev. Biol.* **81**, 157–182.

O'Brochta, D. A., and Bryant, P. J. (1985). A zone of non-proliferating cells at a lineage restriction boundary in *Drosophila*. *Nature* **313**, 138–141.

O'Connell, C. B., and Wang, Y. L. (2000). Mammalian spindle orientation and position respond to changes in cell shape in a dynein-dependent fashion. *Mol. Biol. Cell* **11**, 1765–1774.

Peshkin, M. A., Strandburg, K. J., and Rivier, N. (1991). Entropic predictions for cellular networks. *Phys. Rev. Lett.* **67**, 1803–1806.

Prusinkiewicz, P., and Lindenmayer, A. (1990). *The Algorithmic Beauty of Plants*. Springer-Verlag, New York.

Rauzi, M., Verant, P., Lecuit, T., and Lenne, P. F. (2008). Nature and anisotropy of cortical forces orienting *Drosophila* tissue morphogenesis. *Nat. Cell Biol.* **10**, 1401–1410.

Rivier, N., and Lissowski, A. (1982). On the correlation between sizes and shapes of cells in epithelial mosaics. *J. Phys. A: Math. Gen.* **15**, L143–L148.

Rivier, N., Schliecker, G., and Dubertret, B. (1995). The stationary state of epithelia. *Acta Biotheor.* **43**, 403–423.

Rogers, S. L., Rogers, G. C., Sharp, D. J., and Vale, R. D. (2002). *J. Cell Biol.* **158**, 873–884.

Saburi, S., Hester, I., Fischer, E., Pontoglio, M., Eremina, V., Gessler, M., Quaggin, S. E., Harrison, R., Mount, R., and McNeill, H. (2008). Loss of Fat4 disrupts PCP signaling and oriented cell division and leads to cystic kidney disease. *Nat. Genet.* **40**, 1010–1015.

Senoo-Matsuda, N., and Johnston, L. A. (2007). Soluble factors mediate competitive and cooperative interactions between cells expressing different levels of *Drosophila* Myc. *Proc. Natl. Acad. Sci. USA* **104**, 18543–18548.

Shraiman, B. I. (2005). Mechanical feedback as a possible regulator of tissue growth. *Proc. Natl. Acad. Sci. USA* **102**, 3318–3323.

Simpson, P. (1979). Parameters of cell competition in the compartments of the wing disc of *Drosophila*. *Dev. Biol.* **69**, 182–193.

Simpson, P., and Morata, G. (1981). Differential mitotic rates and patterns of growth in compartments in the *Drosophila* wing. *Dev. Biol.* **85**, 299–308.

Smith, L. G. (2001). Plant cell division: Building walls in the right places. *Nat. Rev. Mol. Cell Biol.* **2**, 33–39.

Smith, L. G., Hake, S., and Sylvester, A. W. (1996). The tangled-1 mutation alters cell division orientations throughout maize leaf development without altering leaf shape. *Development* **122**, 481–489.

Smith, R. S., Guyomarc'h, S., Mandel, T., Reinhardt, D., Kuhlemeier, C., and Prusinkiewicz, P. (2006). A plausible model of phyllotaxis. *Proc. Natl. Acad. Sci. USA* **103**, 1301–1306.

Speicher, S., Fischer, A., Knoblich, J., and Carmena, A. (2008). The PDZ protein Canoe regulates the asymmetric division of *Drosophila* neuroblasts and muscle progenitors. *Curr. Biol.* **18**, 831–837.

Strauss, B., Adams, R. J., and Papalopulu, N. (2006). A default mechanism of spindle orientation based on cell shape is sufficient to generate cell fate diversity in polarised *Xenopus* blastomeres. *Development* **133**, 3883–3893.

Tepass, U., Tanentzapf, G., Ward, R., and Fehon, R. (2001). Epithelial cell polarity and cell junctions in *Drosophila*. *Annu. Rev. Genet.* **35**, 747–784.

Thery, M., Racine, V., Pepin, A., Piel, M., Chen, Y., Sibarita, J. B., and Bornens, M. (2005). The extracellular matrix guides the orientation of the cell division axis. *Nat. Cell Biol.* **7**, 947–953.

Thery, M., Jimenez-Dalmaroni, A., Racine, V., Bornens, M., and Julicher, F. (2007). Experimental and theoretical study of mitotic spindle orientation. *Nature* **447**, 493–496.

Veeman, M. T., Nakatani, Y., Hendrickson, C., Ericson, V., Lin, C., and Smith, W. C. (2008). Chongmague reveals an essential role for laminin-mediated boundary formation in chordate convergence and extension movements. *Development* **135**, 33–41.

Walker, K. L., Muller, S., Moss, D., Ehrhardt, D. W., and Smith, L. G. (2007). *Arabidopsis* TANGLED identifies the division plane throughout mitosis and cytokinesis. *Curr. Biol.* **17**, 1827–1836.

Wallingford, J. B., Rowning, B. A., Vogeli, K. M., Rothbacher, U., Fraser, S. E., and Harland, R. M. (2000). Dishevelled controls cell polarity during *Xenopus* gastrulation. *Nature* **405**, 81–85.

Wang, N., Butler, J. P., and Ingber, D. E. (1993). Mechanotransduction across the cell surface and through the cytoskeleton. *Science* **260**, 1124–1127.

Weliky, M., Minsuk, S., Keller, R., and Oster, G. (1991). Notochord morphogenesis in *Xenopus laevis*: Simulation of cell behavior underlying tissue convergence and extension. *Development* **113**, 1231–1244.

Winter, C. G., Wang, B., Ballew, A., Royou, A., Karess, R., Axelrod, J. D., and Luo, L. (2001). *Drosophila* Rho-associated kinase (Drok) links Frizzled-mediated planar cell polarity signaling to the actin cytoskeleton. *Cell* **105**, 81–91.

Wright, N., and Alison, M. (1984). The Biology of Epithelial Cell Populations. Oxford University Press, New York.

Wu, V. M., Yu, M. H., Paik, R., Banerjee, S., Liang, Z., Paul, S. M., Bhat, M. A., and Beitel, G. J. (2007). *Drosophila* Varicose, a member of a new subgroup of basolateral MAGUKs, is required for septate junctions and tracheal morphogenesis. *Development* **134**, 999–1009.

Yamashita, Y. M., Jones, D. L., and Fuller, M. T. (2003). Orientation of asymmetric stem cell division by the APC tumor suppressor and centrosome. *Science* **301**, 1547–1550.

Zajac, M., Jones, G. L., and Glazier, J. A. (2003). Simulating convergent extension by way of anisotropic differential adhesion. *J. Theor. Biol.* **222**, 247–259.

Zallen, J. A., and Zallen, R. (2004). Cell-pattern disordering during convergent extension in *Drosophila*. *J. Phys.: Condens. Matter* **16**, S5073–S5080.

CHAPTER FIVE

# Principles of *Drosophila* Eye Differentiation

Ross Cagan

## Contents

| | |
|---|---|
| 1. Introduction | 116 |
| 2. The Developing *Drosophila* Eye | 116 |
|    2.1. The *Drosophila* adult eye | 117 |
|    2.2. Establishing the eye field | 117 |
|    2.3. The morphogenetic furrow and emergence of the ommatidial core | 119 |
|    2.4. Signaling from the peripodial membrane | 121 |
|    2.5. Planar cell polarity | 121 |
|    2.6. Fundamental principles I: *Drosophila* eye assembly | 123 |
| 3. Patterning the *Drosophila* Eye: Morphogenesis and Cell Movements | 124 |
|    3.1. Differential adhesion patterns the cone cell quartet | 125 |
|    3.2. Emergence of the interommatidial lattice | 126 |
|    3.3. Fundamental principles II: Long-range patterning | 127 |
| 4. Conclusion | 131 |
| Acknowledgments | 132 |
| References | 132 |

## Abstract

The *Drosophila* eye is one of nature's most beautiful structures and one of its most useful. It has emerged as a favored model for understanding the processes that direct cell fate specification, patterning, and morphogenesis. Though composed of thousands of cells, each fly eye is a simple repeating pattern of perhaps a dozen cell types arranged in a hexagonal array that optimizes coverage of the visual field. This simple structure combined with powerful genetic tools make the fly eye an ideal model to explore the relationships between local cell fate specification and global tissue patterning. In this chapter, I discuss the basic principles that have emerged from three decades of close study. We now understand at a useful level some of the basic principles of cell fate selection and the importance of local cell–cell communication.

Department of Developmental and Regenerative Biology, Mount Sinai School of Medicine, New York, NY, USA

We understand less of the processes by which signaling combines with morphogenesis and basic cell biology to create a correctly patterned neuroepithelium. Progress is being made on these fundamental issues, and in this chapter I discuss some of the principles that are beginning to emerge.

## 1. INTRODUCTION

The retina is a favorite tissue of developmental neurobiologists as it is a relatively simple neuroectoderm. Two major types of retinas have evolved. The *simple camera eye* is favored by larger animals including most vertebrates. It is composed of three components: a lens, a retina, and a pigment layer. The simple camera eye has the considerable virtue that it the more sensitive eye with higher resolving power. But it requires a minimum distance between lens and retina due to the refractive index of a biological lens. Smaller animals including most insects favor the *compound eye*. It also contains lens, retina, and pigment layer, but the three are compressed and compartmentalized to fit onto tiny heads. The two types of retinas share many of the molecular factors required to direct development as well as many of the channels, etc., required to permit function, as well as similar neuronal subtypes. They differ in their overall structure and the germ layers from which they emerge.

Understanding construction of epithelia—and neuroepithelia in particular—remains a central challenge. We will need to improve our understanding of how local features at the level of individual cells influences higher order tissue patterning. In this chapter, I examine the processes that shape the emerging *Drosophila* eye with an emphasis on tissue morphogenesis and patterning. I will outline basic findings and encourage the reader to follow the references provided for greater detail. Instead, I focus on providing an overview to place these issues in context, then discuss some basic principles that have emerged from these studies.

## 2. THE DEVELOPING *DROSOPHILA* EYE

The *Drosophila* eye has proven a powerful model for understanding how local cell–cell signaling activates signal transduction pathways to direct cell fate. The eye has provided several of the first examples of directed cell fate within an epithelium and, along the way, has helped us better understand how the Notch, RTK/Ras, Hedgehog, and Wnt pathways can work together as cell fate "switches." Recently, developmental biologists have sought to look deeper into the cell biology of emerging cell types to understand how cell intrinsic properties contribute to building an epithelium. Other systems such as *Drosophila* embryonic gastrulation are better suited to understanding large movements of cells across an epithelium. However, the retina remains a model of choice for understanding the smaller-scale movements of cells within an epithelium that are critical for its proper assembly. In this section, we will review work that looks (1) inward at the details of a cells' biology and how it affects overall eye field structure and (2) outward at the cell biology 'rules' that govern rearrangement of cells into useful patterns. This work represents a maturing of the fly eye field, further increasing its utility as one of the best understood developing epithelia.

## 2.1. The *Drosophila* adult eye

The *Drosophila* compound eye contains approximately 700 (male) to 750 (female) unit eyes known as 'ommatidia'; the adult structure is presented in Fig. 5.1. Each ommatidium consists of a core of eight photoreceptor neurons, capped by four non-neuronal cone cells and two primary pigment cells that together form an 'iris.' These ommatidial cores are optically insulated from neighboring ommatidia by an interweaving hexagonal lattice of secondary and tertiary pigment cells that prevent light from passing between ommatidia. This arrangement is astonishingly precise, aiming each ommatidium in an exacting outward angle to evenly cover the visual field. Assembling this highly derived structure presents a challenge: the acuity of the fly's vision is directly related to the precision by which it can construct and place each ommatidium. Insects have solved this challenge through an evolving series of patterning choices that coordinate cell signaling, cell proliferation, cell death, and cell movements. The result is one of nature's more stunning structures and, from a developmental biologist's standpoint, one of its most useful.

## 2.2. Establishing the eye field

Clonal analysis indicated that the eye field is derived from approximately six cells that are set aside early in embryogenesis (Wieschaus and Gehring, 1976). These six cells give rise to a structure with several basic strengths that make it in many ways an ideal model system.

As with most other adult *Drosophila* structures, each eye emerges from an 'imaginal disk' (from *imago*, meaning adult insect), specifically the eye portion of the eye/antennal disk ('eye disk'). This peninsula of tissue begins as an ectoderm-derived infolding that is originally established—and split into two eye fields—through the combined efforts of Decapentaplegic (Dpp) and

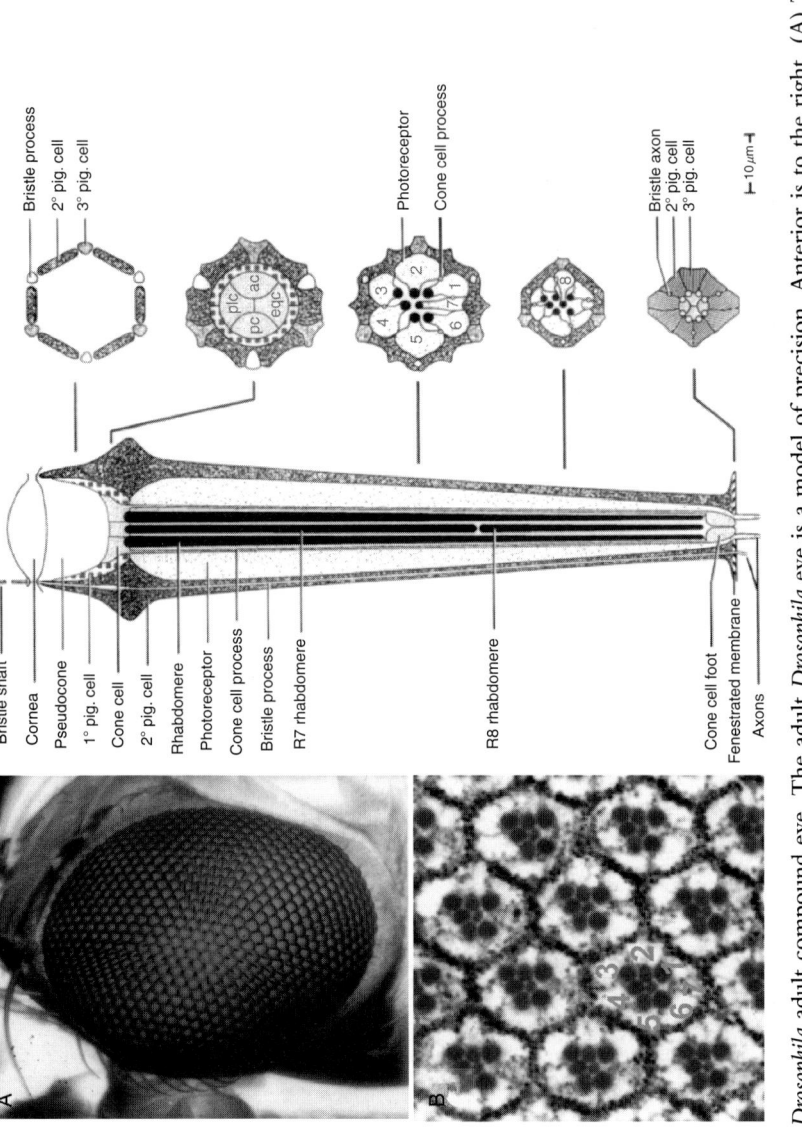

**Figure 5.1** The *Drosophila* adult compound eye. The adult compound eye is composed of more than 700 precisely arranged ommatidia. Anterior is to the right. (A) The adult compound eye combines with the eye's curvature to evenly cover the fly's visual field. (B) Section through an adult eye approximately 30 μm below the eye's surface, showing a field of 10 whole ommatidia. The seven apical photoreceptor neurons of one ommatidium are labeled within the cell bodies; their (blue) rhabdomeres are arranged in a trapezoid. Note how the precise hexagonal lattice of red pigmented 2°/3°s between ommatidia (darkened by methylene blue staining) optically insulate each ommatidium. (C) Schematic of a single ommatidium drawn to scale. The third "section" matches panel B (from Cagan and Ready, 1989b).

Hedgehog (Hh) pathways (Chang et al., 2001) in a manner analogous to signals that establish the vertebrate eye field (Yang, 2004). These signals then act with the Notch and EGF-receptor pathways to induce proliferation as well as a complex web of signals that establish and define the eye field (Fig. 5.2; Chang et al., 2003; Kenyon et al., 2003; Kumar and Moses, 2001). The emergence of the *Drosophila* eye from the ectoderm overlying the brain is more similar to the vertebrate lens than its retina; nevertheless, the two retinas show marked similarity in their molecular underpinnings (Chang et al., 2001; Pichaud et al., 2001). Infolding of the fly eye disk results in a two-layered tissue: the future eye field thickens into a pseudostratified epithelium that is covered by a thin squamous epithelium, the 'peripodial membrane'. This eye anlage is easily plucked out of the organism and imaged as a whole mount, a useful property.

Six core transcription factors (*eye, toy, optix, so, eya, dac*) interact amongst themselves and other noncore factors to establish the eye field (Fig. 5.2). Each of these six factors is active in establishing the mammalian eye, and mutations can lead to serious diseases of the retina and elsewhere. In addition to establishing the eye field (i.e., conferring competence to a set of cells for response to eye-specific signals), these six factors are active in promoting cell proliferation and cell fate by acting with numerous other factors in the nucleus and at the surface (reviewed in Kumar, 2008; Pappu and Mardon, 2004).

## 2.3. The morphogenetic furrow and emergence of the ommatidial core

The *Drosophila* life cycle includes three larval stages that together span four days. For most of larval life the eye disk proliferates, broadly establishing the eye field and dividing it into dorsal and ventral regions. Cell-type differentiation begins midway through the final larval stage. The first sign is emergence of the 'morphogenetic furrow', a physical indentation in the eye field that appears initially near the posterior edge of the eye/antennal disk. This furrow 'sweeps' anteriorly as progressive rows of cells utilize actin/myosin dynamics to alter their shapes and sink basally (Benlali et al., 2000; Escudero et al., 2007). Ahead (anterior to) the morphogenetic furrow, cells continue to proliferate and expand the eye field. The furrow itself is a point of cell cycle arrest as cells enter G1. Cell fate determination begins within the furrow and continues to progress behind (posterior) to it. Despite providing a striking demarcation in the emerging eye field—a sort of moving anterior/posterior boundary dependent on Hedgehog and Dpp signaling (Heberlein et al., 1993; Ma et al., 1993)—the function of the morphogenetic furrow is not known.

Our emerging understanding of the processes that direct cell fate determination represents a striking success of modern developmental biology, it has been well reviewed (e.g., Nagaraj and Banerjee, 2004; Voas and Rebay, 2004; Wernet and Desplan, 2004). In short, local signals are shared between

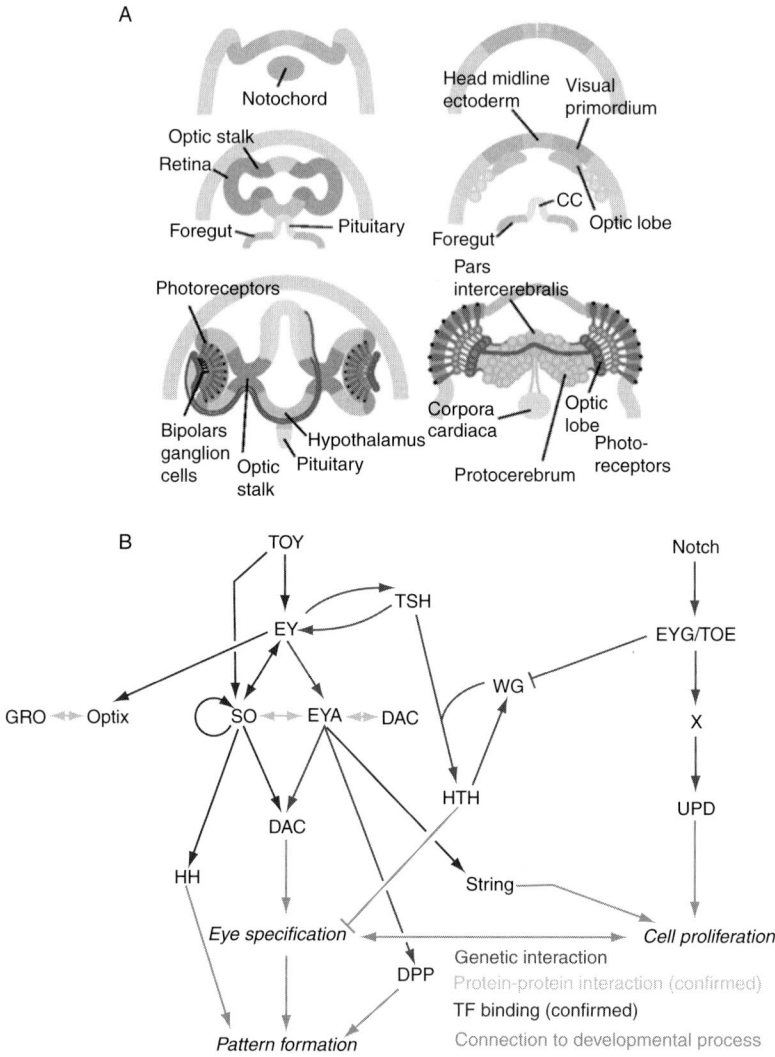

**Figure 5.2** Establishing the eye field. (A) Comparing emergence of vertebrate and *Drosophila* eyes during early neurogenesis. The *Drosophila* eye remains as part of the surface ectoderm (from Chang *et al.*, 2001). (B) The network of factors known to establish the eye field (adapted from Kumar, 2008). (See Color Insert.)

cells that are 'read' as cell fate information. Therefore, the position of a cell within the eye field determines its fate. This process leads to progressive induction of cell fates beginning with photoreceptor neurons (first neuron R8, then R2 and R5, R3/R4, R1/R6, and finally R7 within each ommatidium). Four glial-like Semper or 'cone cells' are recruited to each photoreceptor octet and, in the young pupa, two additional glial-like 'primary pigment cells' (1°s) complete the 14-cell ommatidial core (Fig. 5.3). Finally, an interweaving lattice of secondary and tertiary pigment cells (2°s, 3°s) and sensory bristle organules emerge in the pupa to organize the ommatidial array into a precise pattern. The cone and pigment cells serve to secrete the overlying lens, turn over rhabdomere membrane, and limit the pathways of light permitted to enter each ommatidium.

While we understand quite a bit regarding the signals that direct particular cell fates, we understand very little about the spatial organization by which these cells collect together. For example, all cell types exempting R8 and R7 emerge in symmetric pairs in which each cell mirrors its partner's position across the growing core (some insects exhibit pairwise R7-like neurons). The presumption is that this symmetry reflects symmetric cell signaling within the growing core (Freeman, 1997; Tomlinson and Ready, 1987), but the factors that place R1/R6 on a side (initially) posterior to R3/R4 is not understood. Nor do we understand the mechanisms that limit the number of photoreceptor neurons specifically to eight in *Drosophila*. The *Drosophila* eye presents a uniquely simple model for studying these small-scale issues of tissue patterning and remodeling yet these issues remain a challenge to the field.

## 2.4. Signaling from the peripodial membrane

One of the truly surprising findings that have emerged from studying eye development is the direct regulation of cell fate by the overlying peripodial membrane. Recall that the peripodial lies as a thin squamous cap over the eye field. Two different groups (Cho *et al.*, 2000; Gibson and Schubiger, 2000) identified a remarkable physical process that emerges and extends from individual cells within the peripodial membrane to contact individual cells within the underlying eye field. This microtubule-dependent extension brings with it signal transduction molecules including Hedgehog, Dpp (a BMP ortholog), and Wingless (Wnt) that help regulate cell fate. This is a remarkable and nearly unprecedented example of individual cells from one epithelium physically reaching across to another to direct development of individual target cells. If commonly utilized in development, the significance of this mechanism is large and holds the potential to endlessly complicate the study of cell fate induction.

## 2.5. Planar cell polarity

Another aspect of patterning and morphogenesis in the fly eye is planar cell polarity. Nearly all epithelia have an intrinsic, cell autonomous polarity. This polarity is reflected across the plane of the eye's surface as each ommatidium rotates, leaving photoreceptor R7 toward the dorsal/ventral equator at the eye disk's center (Fig. 5.3A). For each ommatidium to achieve its correct polarity the initially symmetric R3 and R4 pair must

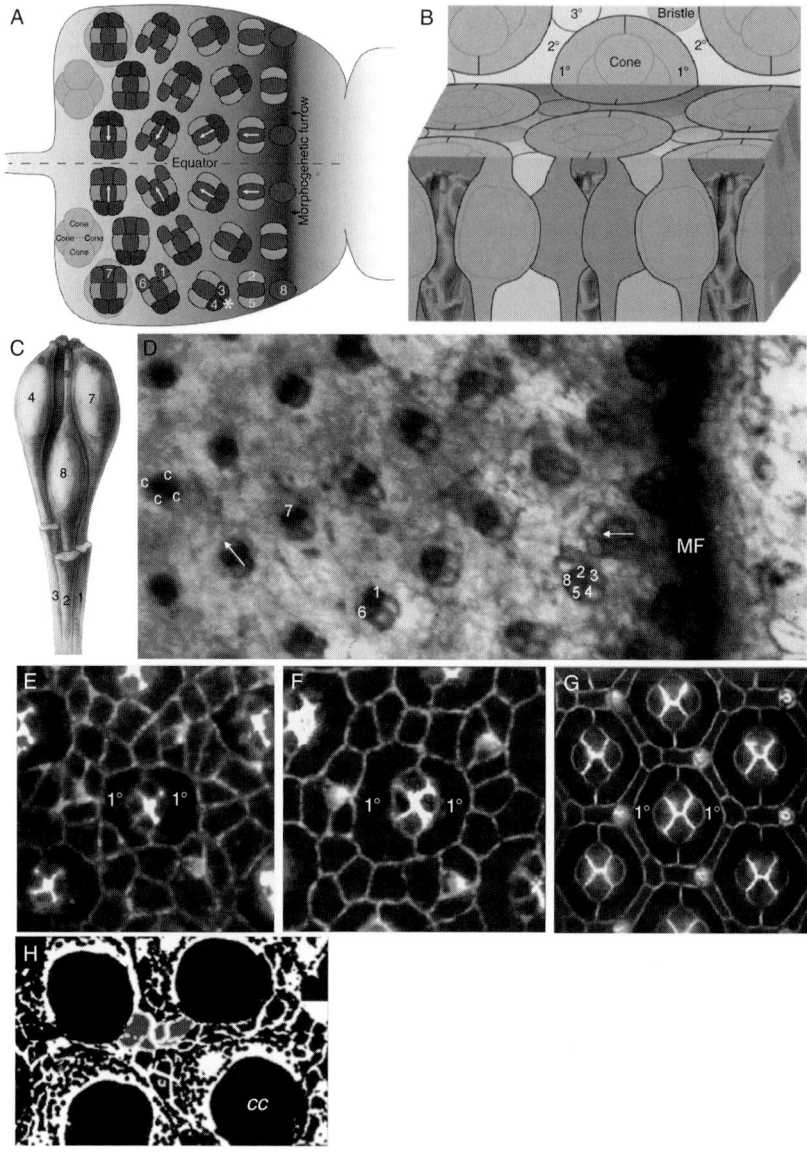

**Figure 5.3** Larval and pupal eye development. Ommatidial differentiation and eye pattern formation begin in the third instar larva (panels A–D) and continue through early pupal stages (panels E–G). Anterior is to the right. (A) Photoreceptors emerge progressively and in symmetric pairs: R8, R2/R5, R3/R4, R1/R6, and finally R7 emerge stepwise and symmetrically across each ommatidium's central axis. Planar cell polarity (PCP) is revealed by progressive rotation of each ommatidium (yellow arrows). Distinguishing photoreceptors R3 and R4 (★) within each ommatidium is a necessary step for establishing its polarity. (B) In the pupal eye, the apical portions of the cone cells and pigment cells (1°, 2°, 3°) push over the photoreceptors to dominate the

be distinguished, presumably to break the ommatidium's initial symmetry and provide it 'handedness'. Once established, each ommatidium then acts on this polarity by rotating 90° in a clockwise (if located within the dorsal half of the eye field) or counterclockwise (ventral) fashion. Several factors important in developing an R3/R4 distinction are known and, interestingly, these same factors mediate planar cell polarity in other animals (reviewed in Seifert and Mlodzik, 2007; Simons and Mlodzik, 2008). These include the antagonistic protein complexes Frizzled/Flamingo and Strabismus/Prickle that partition within a cell to provide autonomous polarity. In a mechanism that is not well understood, a signal involving noncanonical Wnt pathway activity provides spatial information to each R3/R4 pair across the eye field.

The precise mechanism that distinguishes R3 from R4 is not understood in part because no ligand has been identified capable of signaling across the eye field (e.g., none of the Wnt orthologs clearly play this role). Further looming, we understand very little about the cell biological processes that mediate the physical rotation of ommatidia that turn 90° in the span of several hours. Reducing the activity of one gene is known to affect specifically rotation: the serine–threonine kinase Nemo (Choi and Benzer, 1994; Fiehler and Wolff, 2008). Other factors mediating both surface adhesion and cytoskeletal rearrangement are likely to be centrally involved, but little data have surfaced for either of these components.

## 2.6. Fundamental principles I: *Drosophila* eye assembly

The eye is an elegantly simple structure. Its constituent ommatidia are organized into a precisely staggered pattern evolved to optimize coverage of the fly's visual field. The importance of this repeating pattern as a strength of the eye as a model system should not be underestimated: subtle changes

---

surface; the photoreceptor octet has been pushed below the surface to lie just below the cone cells and 1°s (★). The Z-axis is provided to indicate that, while cell movements initiate at the apical surface, each cell extends to the basement membrane. The area of each cell's apical profile is determined primarily by the proximity of its nucleus to the surface. (C) Breakaway schematic view of a larval photoreceptor cluster (from Kramer *et al.*, 1991). (D) A portion of a developing larval eye disk viewed at its apical surface. Cell membranes are highlighted with a cobalt sulfide stain. The morphogenetic furrow (MF) and the progression of photoreceptor neurons (1–8) and cone cells (c) are labeled. Arrows emphasize progression rotation of ommatidia. Compare with panel A. (E–G) Three views of a pupal eye field at 20, 24, and 42 h APF, respectively. IPCs are pseudocolored green. Initially unpatterned (panel D), IPCs rapidly re-arrange into single file as excess IPCs are removed by programmed cell death (panel E). Panel F shows the final pattern (compare to panel B). (H) Many compound eyes do not re-organize their interommatidial pigment cells to achieve tight patterning of the ommatidial array. Shown is a view of a mature cockroach eye with a region of six loosely organized 2°-like cells emphasized by pseudocoloring. Cone cells from an ommatidium (cc) are indicated. The adult ommatidial array remains poorly aligned (from Nowel, 1980). (See Color Insert.)

are more easily and confidently observed when a field contains hundreds of repetitions. Through the concerted efforts of many laboratories we have learned many of the fundamental principles utilized to assemble a fly eye:

*The eye matures in a stepwise fashion.* The eye emerges in just a few days and is easily accessible both physically and genetically, permitting us to readily observe in detail many of the basic processes that assemble the eye. The eye first enlarges through proliferation before specific cell types emerge. General axes—D/V, A/P—are also set early and independent of cell-type specification. Finally, the animal monitors its eye carefully. Expression of, for example, oncogenes targeted to specifically disrupt the size and integrity of the eye field causes the animal to delay pupariation (Pedraza et al., 2004; Read et al., 2004), presumably to give the eye time to right itself.

*Molecular pathways are utilized reiteratively.* The same core signaling pathways are utilized multiple times throughout eye development, particularly the EGFR/Ras, Notch, Dpp (BMP), Wg (Wnt), and Hedgehog pathways. This phenomenon has been observed in most developing tissues, and studies in the fly eye have been central to our understanding of how a single pathway can direct development of so many cell decisions from fate to morphogenesis. A common motif is the dynamic progression of transcription factors and other molecular mediators: general surface signals achieve specificity by altering (e.g., phosphorylating, degrading, etc.) evolving sets of targets. In the past, this changing cellular set of targets was somewhat mysteriously referred to as "competence." While we have a deepening understanding of how this works in terms of evolving transcription factors, our understanding of evolving morphogenesis-related factors (e.g., regulators of junction or cytoskeleton dynamics) is less mature.

*Position determines cell fate.* A cell's position and its competence determines how it will act on a 'generic' surface signal. For example, a cell next to the R8 photoreceptor cell that receives a *Notch*-plus-*Ras* signal in the mature larva will develop as an R7 photoreceptor cell due to expression of factors such as *prospero* (Xu et al., 2000). The same signal provided a day later by a cone cell can act on factors such as *dPax2* in its neighbor to direct a glial-like support cell fate (Flores et al., 2000). Same signal, different response.

## 3. Patterning the *Drosophila* Eye: Morphogenesis and Cell Movements

In this section, we examine a more recent focus of the fly eye field. The remarkable simplicity and precise pattern of the eye is also a powerful advantage for examining the details of how cell–cell contacts and cell movements mediate tissue patterning. This task has been helped recently with the establishment of live imaging. The eye disk migrates to the surface of the animal early in pupation; opening the cuticle allows use of higher resolution microscopy, and junction-associated green fluorescent protein (GFP) nicely outlines the apical surface of cells. In this section, we focus on events in the pupal eye with an emphasis on the morphogenesis of pattern formation.

### 3.1. Differential adhesion patterns the cone cell quartet

Beginning in the late larval stages, the four cone cells within each ommatidium push through the photoreceptor cluster to generate a tight cap at the top of the retina. This cap will eventually serve as the ommatidium's 'iris.' The apical profile of the four-cell cluster is approximately ellipsoid (Fig. 5.3). Hayashi and Carthew (2004) pointed out that this configuration is predicted by "soap bubble"-based models that focus on differences in relative adhesions. They demonstrated that this cone cell quartet has strongly selective preference for adherence amongst themselves based on their selective expression of N-cadherin. Neighboring 1°s do not express N-cadherin, effectively segregating the cone cells and encouraging them to collect into a predictably shaped cluster. Loss of N-cadherin led to a more elongate "cruciform" shape of the cone cell quartet; additional removal of E-cadherin led to still more dramatic shape changes (Hayashi and Carthew, 2004).

This work provides an excellent example of understanding local cell arrangements based on differential adhesion. Utilizing this concept allowed Hayashi and Carthew to predict the final patterns of various numbers of cone cell groups—altered through gene mutation—based on the principle of most efficient packing (Fig. 5.5); indeed, the precise angle of cell contacts was successfully modeled mathematically (Hilgenfeldt et al., 2008; Kafer et al., 2007). Applying the concept of minimization of free energy to cell packing was first proposed by Steinberg (Steinberg, 1970; Steinberg and Poole, 1981). In Steinberg's "Differential Adhesion" model, collections of cells have a natural drive to sort into configurations that minimize the system's free energy: for example, cells with high levels of the homophilic adhesive protein E-cadherin will naturally sort together at the core of a cell collection, pushing cells with lower "adhesiveness" to the periphery. More specifically, cone cells minimize the system's free energy by decreasing surface contacts with 1°s, a drive that—paired with "elastic tension" to promote stable forms—encourages rounding of the cone cell quartet's outer profile. Patterning of the cone cells presents an elegant example of this process in the context of two-dimensional sorting of their apical profiles. Next, I consider an alternative strategy, increasing cell contacts between different cells, that nevertheless heeds the same taskmaster of decreasing overall free energy.

## 3.2. Emergence of the interommatidial lattice

My laboratory has focused on the events that rearrange interommatidial cells into a remarkably precise hexagonal lattice of 2°s, 3°s, and bristle organules that weave through and organize the ommatidial array. This process is a useful model for long-range patterning and re-organization across an epithelium and has even been modeled computationally (D. Larson et al., unpublished data). Initially, 'interommatidial precursor cells' (IPCs)—precursors of the 2°s and 3°s—lie unpatterned between the ommatidial cores (Fig. 5.3E). Within a few hours, (1) the IPCs line up in single file while (2) some of the IPCs are removed to get to the final required cell number. When the process is complete, each 2° is stretched between two 1°s and each 3° (and each bristle, sort of) sits between three 1°s (Fig. 5.3G). This process aligns ommatidia into a staggered array that best covers the visual field. While organizing the ommatidial lattice should improve visual precision this organization is not the rule in compound eyes: most arthropoda leave their pigment cell lattice poorly organized (e.g., Fig. 5.3H), emphasizing the derived nature of this patterning process. Nevertheless, some familiar factors mediate its progression.

Current work on IPC patterning is beginning to link signal transduction with adhesion and cell biology. Four signaling pathways play a primary role in 2°/3° maturation and assembly: Notch, EGFR, Wg, and Dpp (BMP-like) activities combine to regulate cell fate, cell positioning, and cell death; disruption of any of these pathways leads to incorrect patterning (Cagan and Ready, 1989a; Cordero et al., 2004, 2007; Freeman, 1996; Reiter et al., 1996; Yu et al., 2002). Their roles are not fully understood but, roughly speaking, EGFR primarily regulates all cell fates, Wg regulates cone cell fate (which affects later patterning; J. Cordero and R. Cagan, unpublished data) and cell death, and Dpp regulates cell positioning and a correctly formed apical profile. Notch has the most severe effects and regulates several aspects including cell fate, cell death, and cell adhesion/assembly. Recent data suggest that Notch acts in part through regulating Hibris (Hbs) and Roughest (Rst) (S. Bao and R. Cagan, unpublished data). Hbs and Rst are members of the Nephrin superfamily (Nephrin and Neph1, respectively) of adhesion-like transmembrane proteins that show heterophilic adhesion to each other (Ramos et al., 1993). Loss of either leads to incorrect IPC patterning due to a failure of IPCs to correctly move into and establish their required positions (Bao and Cagan, 2005; Reiter et al., 1996). Two other Nephrin superfamily members, Sns and Kirre, play a somewhat redundant role in this process (S. Bao and R. Cagan, unpublished data).

This reliance on adhesion-like factors has led to a "Preferential Adhesion" model in which the epithelium has a drive toward minimizing the system's overall free energy by maximizing contacts between 2°/3°s and 1°s. This model is yet another twist on Steinberg's Differential Adhesion model: by utilizing two heterophilic adhesion proteins, potential pattern complexity greatly increases from simple soap-bubble-like collections of cells to complex patterns of ommatidia woven together by a separate hexagonal pigment cell lattice. However, more is to come on this process. Nephrin family members—including Hbs, Rst, Sns, and Kirre—can directly signal into cells and recent work suggests they provide a link between the surface and rearrangement of the cytoskeleton (Johnson et al., 2008; Seppa et al., 2008). This rearrangement would presumably mediate cell movements during the patterning process. Recent work suggests that control of the apical profile, for example, through nuclear movements (as the nucleus rises toward the surface the apical profile increases to accommodate its size) plays an important role in patterning (D. Larson, R. Johnson, M. Swat, J. Cordero, J. Glazier, and R. Cagan, unpublished data).

## 3.3. Fundamental principles II: Long-range patterning

Cells are placed in their correct positions using basic *patterning rules*. So what have we learned so far from work on patterning events in the pupal eye? A few points are worth mentioning, although clearly this is a work in progress:

*The eye epithelium steadily calibrates cell number.* To achieve a precise pattern, the eye field likely must keep the number of cells between ommatidial pattern elements within a certain range. Too many IPCs will be difficult to efficiently remove, and indeed the adult eye does contain a few ectopic 2°s (which share a single 2° niche and do not disrupt the overall pattern). Too few IPCs will result in direct contact between neighboring ommatidia and the potential for optical bleed-through.

To calibrate IPC number the eye uses a steady combination of cell cycle and cell death. Regarding the former, after establishment of the initial ommatidial 'precluster' (assembly of photoreceptor neurons R2–R5 and R8) the remaining cells undergo a final 'second wave' of proliferation to provide additional cells between nascent ommatidia (Ready et al., 1976). Levels of proliferation are calibrated through Notch and EGFR pathway activities that, in turn, spatially control proliferation through signaling from the ommatidia themselves (Baker, 2001; Baonza and Freeman, 2005; Firth and Baker, 2005).

The second method of regulating cell number is through programmed cell death. Excess cells are cued to die by classic apoptosis (Cagan and Ready, 1989b; Wolff and Ready, 1991). Two mechanisms are known to regulate the position and number of programmed cell deaths. In the larva, competition for ommatidial-provided EGFR ligand leads to the death of cells located far from ommatidia (Baker, 2001; Yang and Baker, 2003). In the pupal eye, IPC patterning requires removal of cells simultaneous to their rearrangement (Cagan and Ready, 1989b; Wolff and Ready, 1991). Death is preceded by activation of the classical *Drosophila* apoptotic pathway, which leads to activation of multiple caspases and subsequent apoptosis (reviewed in Brachmann and Cagan, 2003). Cell death in the pupal eye appears to be a competition between cells to maximize contact with the neighboring 1°s, but the precise nature of the mechanisms that determine which cell lives and which dies within a niche is not understood.

*Patterning begins at the apical surface.* Based on tissue reconstruction from electron micrographs, contacts between cells in the emerging eye occur first at the apical surface. This has been most clearly catalogued in the pupal eye (Cagan and Ready, 1989b), where cells make striking changes across the epithelium. Typically as a cell moves within the epithelial field, it (1) extends a small 'process' to contact a cell 1–2 cell diameters away, (2) moves its apical surface to establish a contact and junction with that cell, and (3) the contact is 'zippered down' basally to move the remainder of the cell into the new position (Fig. 5.4; Cagan and Ready, 1989b). This 'apical first' style of cell movement simplifies the patterning process by reducing it to a two-dimensional problem. Most important signaling and junction-related proteins are found at or apical to the apical adherens junctions. Again, collecting these factors at the surface aids in simplifying the signals and adhesions needed to correctly move and pattern cells.

Some questions still remain regarding the role of the adherens junctions themselves in the patterning process. As discussed below, cadherins are important for patterning cone cells and altering them leads to a poorly patterned quartet. Adherens junctions—as assessed by the presence of cadherins—do show some dynamic properties but for the most part there is not a clean correlation between absence of adherens junctions and cell movements. Movies that visualize cell movements in the live eye *in situ*

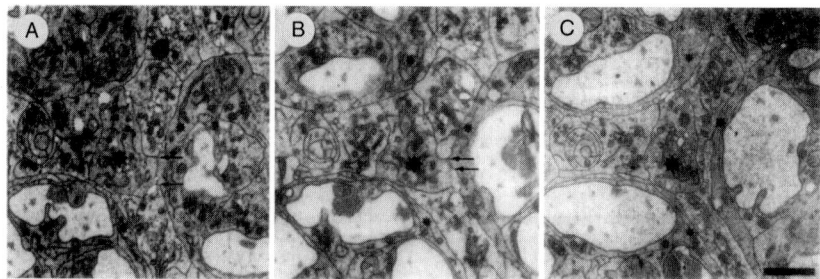

**Figure 5.4** Cells make new contacts initially at the apical surface. A recently established 3° is pseudocolored in green. Electron micrographs (from Cagan and Ready, 1989b). (A) At the apical surface of the emerging pupal eye, the 3° has contacted 1°s from three different ommatidia. Arrows indicate the extent of the 3°s contact with one of the 1°s, which is pseudocolored in brown. (B) One micron deep, the 3°/1° contact is smaller (arrows). (C) A deeper view demonstrates how the contact has yet to extend basally; two neighboring IPCs occlude the 3° and likely had originally 'competed' to become the niche's single 3°. The emergent 3° will rapidly extend its 1° contact basally through this region and to the retina's basal 'floor.' (See Color Insert.)

show that cells with apparently significant junctions still freely exchange contacts and move to other sites (Larson et al., 2008). Either junction strength is regulated in a manner other than by regulating the presence of Cadherin at the cells' surface (perhaps it is regulated by cytoplasmic proteins that link Cadherins to the cytoskeleton), adherens junctions are more easily broken than expected, or other factors override a cell's junction with its neighbor.

*Events are integrated.* Work on patterning of both the cone cells and the pigment cells have emphasized the importance of cell–cell adhesion. However, we know that both Cadherin and Nephrin family members also signal through interactions with cytoplasmic proteins. Mutations in some of these proteins—for example, orthologs of CD2AP (Cindr) and ZO-1 (Pyd)—show patterning defects (Fig. 5.5E; Johnson *et al.*, 2008; Seppa *et al.*, 2008), emphasizing this point. The most likely result of these sorts of intracellular signals is modification of the actin cytoskeleton. Indeed, morphogenetic events in the larval eye (Corrigall *et al.*, 2007; Escudero *et al.*, 2007; Schlichting and Dahmann, 2008) as well as the pupal eye (Johnson *et al.*, 2008; Seppa *et al.*, 2008) require precise actin remodeling as cells execute fine movements. We appreciate that dynamic regulation of cells' cytoskeleton is a central component to regulating morphogenetic furrow formation, ommatidial rotation, and selective adhesion. However, we only poorly understand how signals and adhesion are translated into re-organization of the cytoskeleton and subsequently back to the surface to affect cell movements. Nor do we understand how this process is regulated spatially to move cells into their proper niche or rotate ommatidia in the proper direction. The connection between long-range signaling, short-range cell–cell interactions, and actin remodeling holds perhaps the greatest potential for surprises over the next few years.

*More complex patterns require more surface factors.* Cone cells are an elegant and simple system that, at their core, appear to require a single predominant adhesive factor: N Cadherin. This simplicity is reflected in four-cell packing that cleanly reflects the dynamics of soap bubbles (Hayashi and Carthew, 2004). Introducing two factors that interact in heterophilic adhesion dramatically increases the potential complexity of the system, and IPC patterning provides an example. In addition to adding an extra adhesion factor, IPC patterning also requires precise, dynamic expression of each of the factors: Hbs (and Sns) must be expressed in the cone cells and 1°s at precisely the correct stage, and Rst (and Kirre) must be expressed in a complementary fashion in the IPCs. To accomplish this, regulation comes at the level of both controlled expression and protein turnover (S. Bao and R. Cagan, unpublished data). This is turn opens several points of potential regulation to achieve still greater complexity and nuance. Therefore, the progression of patterning in the pupal eye permits us to study increasingly complex epithelial patterning mechanisms, a useful training exercise for grasping the still greater complexity of the maturing mammalian nervous system.

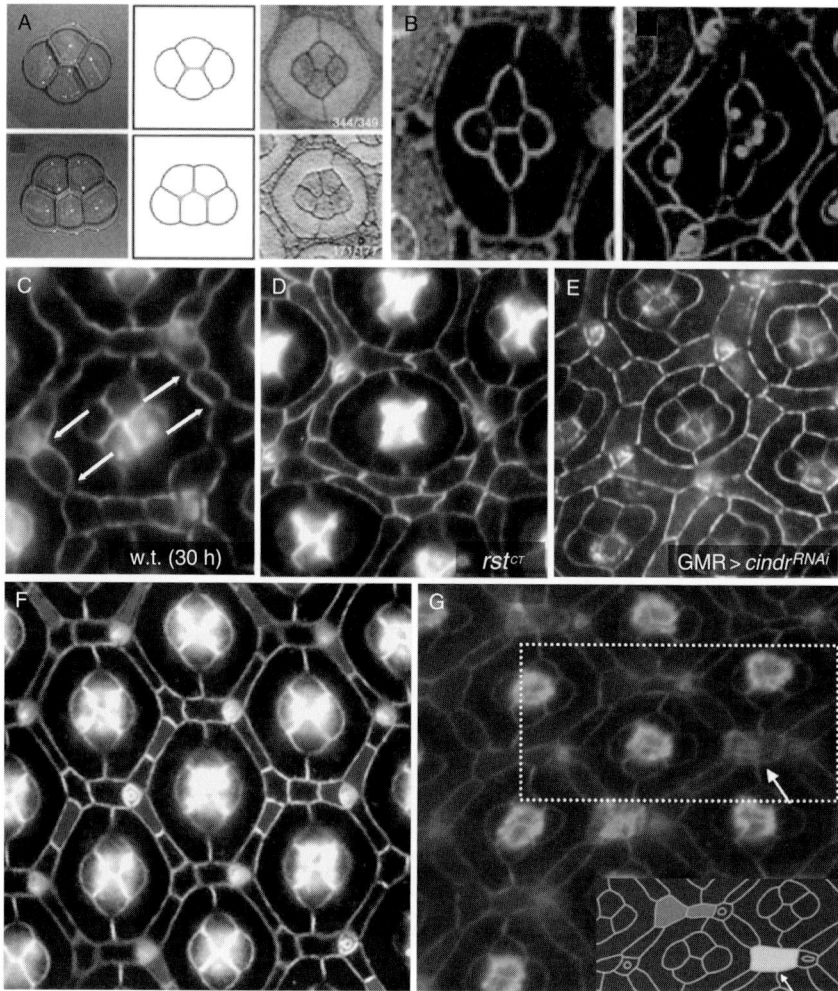

**Figure 5.5** Adhesion directs assembly of cone cells and pigment cells. (A) Cone cell assemblies reflect the adhesive properties observed in adherent soap bubbles (from Hayashi and Carthew, 2004). (B) Loss of N-cadherin led to a "cruciform" shape (*left* panel) in which the long axis of the cone assembly is elongated (Hayashi and Carthew, 2004). Compare to the more rounded shape seen in panel A. Separation of cone cells also occurred (*right* panel), a predicted outcome of a cone cell's decreased affinity for its companions. (C) Ommatidium from a 30-h-old pupa (Bao and Cagan, 2005). IPCs are pseudocolored green. Contacts between IPCs and 1°s are 'puckered' (e.g., arrows) to maximize 1°/IPC cell–cell contacts relative to IPC/IPC contacts. (D) Reduced *rst* activity led to poor assembly of the IPC lattice (Bao and Cagan, 2005). IPCs are pseudocolored green. (E) Reducing Cindr, an adaptor protein known to mediate events between the surface and the cytoskeleton, led to abnormal IPC patterning (from Johnson *et al.*, 2008). (F) Mutations in the ubiquitin conjugase *morgue* partially blocked programmed cell death but did not alter the overall hexagonal pattern. Examples of ectopic 2°s pseudocolored in red (from Hays *et al.*, 2002). (G) Overexpressing Rst in individual IPCs led to a single cell (green) taking over two niches. Compare with the normal 2°–3° pair (purple in schematic inset) (from Bao and Cagan, 2005). (See Color Insert.)

*Cells maximize their contacts with preferred neighbors.* Many of the mechanisms by which cells are assembled into precise patterns can be inferred simply by looking carefully at their contacts during the patterning process. Apical IPC cell profiles change from smooth to 'scalloped', effectively increasing their length of contact with neighboring 1°s (Fig. 5.5C). By contrast, IPCs do not exhibit 'scalloped' contacts with other IPCs (Fig. 5.5C) indicating low IPC–IPC adhesion.

*The pigment cell pattern does not count cells.* The cone cell quartet is a small, limited pattern element that depends heavily on the number of cells. By contrast, the "Preferential Adhesion" model makes the surprising prediction that heterophilic adhesion drives IPC patterning independent of cell number. We see this prediction verified in wild-type eyes. The normal adult hexagonal array of ommatidia is precise even though it contains many ectopic 2°s that are not removed during development. These ectopic 2°s do not affect the overall pattern, but rather two 2°s can share a hexagonal face without distorting the pattern by reducing their apical profile. Reducing cell death produces still more ectopic 2°s, again with minimal affect on the overall pattern (Fig. 5.5F).

How can an epithelium create a precisely patterned structure with little regard for cell number? The answer can be deduced when Rst—the adhesion factor expressed in IPCs—is strongly over-expressed well above its normal levels in single isolated cells. These cells expand their apical profiles and take over a 2° *plus* a 3° niche (Fig. 5.5G). Apparently the patterning system does not count cells but rather expands membranes to equilibrate and maximize Rst/Hbs-mediated contacts. When the concentration of the adhesion molecule is out of balance with the number of cells, adhesion wins.

*Nuclear movements.* The movement of a cell's nucleus also affects surface patterning. Cells in the eye are very elongate and thin: each larval cell is perhaps 30–40 $\mu$m long and just a few microns in width. The exception is the nucleus, which is often more than 10 $\mu$m in width and is less flexible than the cell body (schematized in Fig. 5.3B). As a result, when a cell's nucleus is basal the cell's apical profile tends to be thin; as the nucleus rises toward the surface during differentiation and patterning the apical surface can more than double in a few hours. This change dramatically alters the patterning landscape. For example, the nuclear movement-dependent expansion of ommatidia's apical profiles help 'push' neighboring IPCs into hexagonal patterning (D. Larson *et al.*, unpublished data).

## 4. CONCLUSION

The role of nuclear positioning is an interesting example of how intrinsic cell biological properties can influence a cell's positioning in surprising ways. It is also a reminder that we have some work to do to be able to answer the deceptively simple question: how is an epithelium patterned? Since publication of the seminal work by Ready, Hansen, and Benzer three decades ago (Ready et al., 1976), the field has shown remarkable success by selecting very specific questions to address: cell fate induction, morphogenetic furrow progression, cell movements and patterning, etc. As a result, we view events as occurring along a strict timeline in which each step directs the next as neighbors collide and signal locally.

As we expand our understanding of specific aspects of cell biology that direct cell morphogenesis and epithelial patterning, this view will change. Cells are capable of reaching 'down' from overlying peripodial membrane to signal the eye disk (Cho et al., 2000; Gibson and Schubiger, 2000) and across developmental time and space as cells within the morphogenetic furrow influence planar cell polarity by extending Scabrous-containing processes across the eye field (Chou and Chien, 2002). The outcome of eye morphogenesis and patterning is appealing in its simplicity. But the developmental processes required to build a fly eye are sufficiently complex to provide useful models for many of the basic aspects of epithelial maturation and, eventually, a more integrated view of the process.

## ACKNOWLEDGMENTS

Thanks to my laboratory members and my peers over the years for sharing their critical thinking on these topics. This chapter was supported by NIH-R01EY011495 from the National Eye Institute.

## REFERENCES

Baker, N. E. (2001). Cell proliferation, survival, and death in the *Drosophila* eye. *Semin. Cell Dev. Biol.* **12,** 499–507.
Bao, S., and Cagan, R. (2005). Preferential adhesion mediated by Hibris and Roughest regulates morphogenesis and patterning in the *Drosophila* eye. *Dev. Cell* **8,** 925–935.
Baonza, A., and Freeman, M. (2005). Control of cell proliferation in the *Drosophila* eye by Notch signaling. *Dev. Cell* **8,** 529–539.
Benlali, A., Draskovic, I., Hazelett, D. J., and Treisman, J. E. (2000). Act up controls actin polymerization to alter cell shape and restrict Hedgehog signaling in the *Drosophila* eye disc. *Cell* **101,** 271–281.
Brachmann, C. B., and Cagan, R. L. (2003). Patterning the fly eye: The role of apoptosis. *Trends Genet.* **19,** 91–96.
Cagan, R. L., and Ready, D. F. (1989a). Notch is required for successive cell decisions in the developing *Drosophila* retina. *Genes Dev.* **3,** 1099–1112.
Cagan, R. L., and Ready, D. F. (1989b). The emergence of order in the *Drosophila* pupal retina. *Dev. Biol.* **136,** 346–362.
Chang, T., Mazotta, J., Dumstrei, K., Dumitrescu, A., and Hartenstein, V. (2001). Dpp and Hh signaling in the *Drosophila* embryonic eye field. *Development* **128,** 4691–4704.

Chang, T., Shy, D., and Hartenstein, V. (2003). Antagonistic relationship between Dpp and EGFR signaling in *Drosophila* head patterning. *Dev. Biol.* **263**, 103–113.
Cho, K. O., Chern, J., Izaddoost, S., and Choi, K. W. (2000). Novel signaling from the peripodial membrane is essential for eye disc patterning in *Drosophila*. *Cell* **103**, 331–342.
Choi, K. W., and Benzer, S. (1994). Rotation of photoreceptor clusters in the developing *Drosophila* eye requires the nemo gene. *Cell* **78**, 125–136.
Chou, Y. H., and Chien, C. T. (2002). Scabrous controls ommatidial rotation in the *Drosophila* compound eye. *Dev. Cell* **3**, 839–850.
Cordero, J., Jassim, O., Bao, S., and Cagan, R. (2004). A role for wingless in an early pupal cell death event that contributes to patterning the *Drosophila* eye. *Mech. Dev.* **121**, 1523–1530.
Cordero, J. B., Larson, D. E., Craig, C. R., Hays, R., and Cagan, R. (2007). Dynamic decapentaplegic signaling regulates patterning and adhesion in the *Drosophila* pupal retina. *Development* **134**, 1861–1871.
Corrigall, D., Walther, R. F., Rodriguez, L., Fichelson, P., and Pichaud, F. (2007). Hedgehog signaling is a principal inducer of Myosin-II-driven cell ingression in *Drosophila* epithelia. *Dev. Cell* **13**, 730–742.
Escudero, L. M., Bischoff, M., and Freeman, M. (2007). Myosin II regulates complex cellular arrangement and epithelial architecture in *Drosophila*. *Dev. Cell* **13**, 717–729.
Fiehler, R. W., and Wolff, T. (2008). Nemo is required in a subset of photoreceptors to regulate the speed of ommatidial rotation. *Dev. Biol.* **313**, 533–544.
Firth, L. C., and Baker, N. E. (2005). Extracellular signals responsible for spatially regulated proliferation in the differentiating *Drosophila* eye. *Dev. Cell* **8**, 541–551.
Flores, G. V., Duan, H., Yan, H., Nagaraj, R., Fu, W., Zou, Y., Noll, M., and Banerjee, U. (2000). Combinatorial signaling in the specification of unique cell fates. *Cell* **103**, 75–85.
Freeman, M. (1996). Reiterative use of the EGF receptor triggers differentiation of all cell types in the *Drosophila* eye. *Cell* **87**, 651–660.
Freeman, M. (1997). Cell determination strategies in the *Drosophila* eye. *Development* **124**, 261–270.
Gibson, M. C., and Schubiger, G. (2000). Peripodial cells regulate proliferation and patterning of *Drosophila* imaginal discs. *Cell* **103**, 343–350.
Hayashi, T., and Carthew, R. W. (2004). Surface mechanics mediate pattern formation in the developing retina. *Nature* **431**, 647–652.
Hays, R., Wickline, L., and Cagan, R. (2002). Morgue mediates apoptosis in the *Drosophila melanogaster* retina by promoting degradation of DIAP1. *Nat. Cell Biol.* **4**, 425–431.
Heberlein, U., Wolff, T., and Rubin, G. M. (1993). The TGF beta homolog dpp and the segment polarity gene hedgehog are required for propagation of a morphogenetic wave in the *Drosophila* retina. *Cell* **75**, 913–926.
Hilgenfeldt, S., Erisken, S., and Carthew, R. W. (2008). Physical modeling of cell geometric order in an epithelial tissue. *Proc. Natl. Acad. Sci. USA* **105**, 907–911.
Johnson, R. I., Seppa, M. J., and Cagan, R. L. (2008). The *Drosophila* CD2AP/CIN85 orthologue Cindr regulates junctions and cytoskeleton dynamics during tissue patterning. *J. Cell. Biol.* **180**, 1191–1204.
Kafer, J., Hayashi, T., Maree, A. F., Carthew, R. W., and Graner, F. (2007). Cell adhesion and cortex contractility determine cell patterning in the *Drosophila* retina. *Proc. Natl. Acad. Sci. USA* **104**, 18549–18554.
Kenyon, K. L., Ranade, S. S., Curtiss, J., Mlodzik, M., and Pignoni, F. (2003). Coordinating proliferation and tissue specification to promote regional identity in the *Drosophila* head. *Dev. Cell* **5**, 403–414.
Kramer, H., Cagan, R. L., and Zipursky, S. L. (1991). Interaction of bride of sevenless membrane-bound ligand and the sevenless tyrosine-kinase receptor. *Nature* **352**, 207–212.
Kumar, J. P. (2008). The molecular circuitry governing retinal determination. *Biochim. Biophys. Acta* **1789**(4), 306–314.

Kumar, J. P., and Moses, K. (2001). EGF receptor and Notch signaling act upstream of Eyeless/Pax6 to control eye specification. *Cell* **104,** 687–697.

Larson, D. E., Liberman, Z., and Cagan, R. L. (2008). Cellular behavior in the developing *Drosophila* pupal retina. *Mech. Dev.* **125,** 223–232.

Ma, C., Zhou, Y., Beachy, P. A., and Moses, K. (1993). The segment polarity gene hedgehog is required for progression of the morphogenetic furrow in the developing *Drosophila* eye. *Cell* **75,** 927–938.

Nagaraj, R., and Banerjee, U. (2004). The little R cell that could. *Int. J. Dev. Biol.* **48,** 755–760.

Nowel, M. S. (1980). Ommatidium assembly and formation of the retina-lamina projection in interspecific chimeras of cockroach. *J. Embryol. Exp. Morphol.* **60,** 345–358.

Pappu, K. S., and Mardon, G. (2004). Genetic control of retinal specification and determination in *Drosophila*. *Int. J. Dev. Biol.* **48,** 913–924.

Pedraza, L. G., Stewart, R. A., Li, D. M., and Xu, T. (2004). *Drosophila* Src-family kinases function with Csk to regulate cell proliferation and apoptosis. *Oncogene* **23,** 4754–4762.

Pichaud, F., Treisman, J., and Desplan, C. (2001). Reinventing a common strategy for patterning the eye. *Cell* **105,** 9–12.

Ramos, R. G., Igloi, G. L., Lichte, B., Baumann, U., Maier, D., Schneider, T., Brandstatter, J. H., Frohlich, A., and Fischbach, K. F. (1993). The irregular chiasm C-roughest locus of *Drosophila*, which affects axonal projections and programmed cell death, encodes a novel immunoglobulin-like protein. *Genes Dev.* **7,** 2533–2547.

Read, R. D., Bach, E. A., and Cagan, R. L. (2004). *Drosophila* C-terminal Src kinase negatively regulates organ growth and cell proliferation through inhibition of the Src, Jun N-terminal kinase, and STAT pathways. *Mol. Cell. Biol.* **24,** 6676–6689.

Ready, D. F., Hanson, T. E., and Benzer, S. (1976). Development of the *Drosophila* retina, a neurocrystalline lattice. *Dev. Biol.* **53,** 217–240.

Reiter, C., Schimansky, T., Nie, Z., and Fischbach, K. F. (1996). Reorganization of membrane contacts prior to apoptosis in the *Drosophila* retina: The role of the IrreC-rst protein. *Development* **122,** 1931–1940.

Schlichting, K., and Dahmann, C. (2008). Hedgehog and Dpp signaling induce cadherin Cad86C expression in the morphogenetic furrow during *Drosophila* eye development. *Mech. Dev.* **125,** 712–728.

Seifert, J. R., and Mlodzik, M. (2007). Frizzled/PCP signalling: A conserved mechanism regulating cell polarity and directed motility. *Nat. Rev. Genet.* **8,** 126–138.

Seppa, M. J., Johnson, R. I., Bao, S., and Cagan, R. L. (2008). Polychaetoid controls patterning by modulating adhesion in the *Drosophila* pupal retina. *Dev. Biol.* **318,** 1–16.

Simons, M., and Mlodzik, M. (2008). Planar cell polarity signaling: From fly development to human disease. *Annu. Rev. Genet.* **42,** 517–540.

Steinberg, M. S. (1970). Does differential adhesion govern self-assembly processes in histogenesis? Equilibrium configurations and the emergence of a hierarchy among populations of embryonic cells. *J. Exp. Zool.* **173,** 395–433.

Steinberg, M. S., and Poole, T. J. (1981). Strategies for specifying form and pattern: Adhesion-guided multicellular assembly. *Philos. Trans. R. Soc. Lond. B Biol. Sci.* **295,** 451–460.

Tomlinson, A., and Ready, D. F. (1987). Neuronal differentiation in *Drosophila* ommatidium. *Dev. Biol.* **120,** 366–376.

Voas, M. G., and Rebay, I. (2004). Signal integration during development: Insights from the *Drosophila* eye. *Dev. Dyn.* **229,** 162–175.

Wernet, M. F., and Desplan, C. (2004). Building a retinal mosaic: Cell-fate decision in the fly eye. *Trends. Cell Biol.* **14,** 576–584.

Wieschaus, E., and Gehring, W. (1976). Clonal analysis of primordial disc cells in the early embryo of *Drosophila melanogaster*. *Dev. Biol.* **50,** 249–263.

Wolff, T., and Ready, D. F. (1991). Cell death in normal and rough eye mutants of *Drosophila*. *Development* **113,** 825–839.

Xu, C., Kauffmann, R. C., Zhang, J., Kladny, S., and Carthew, R. W. (2000). Overlapping activators and repressors delimit transcriptional response to receptor tyrosine kinase signals in the *Drosophila* eye. *Cell* **103,** 87–97.

Yang, X. J. (2004). Roles of cell-extrinsic growth factors in vertebrate eye pattern formation and retinogenesis. *Semin. Cell Dev. Biol.* **15,** 91–103.

Yang, L., and Baker, N. E. (2003). Cell cycle withdrawal, progression, and cell survival regulation by EGFR and its effectors in the differentiating *Drosophila* eye. *Dev. Cell* **4,** 359–369.

Yu, S. Y., Yoo, S. J., Yang, L., Zapata, C., Srinivasan, A., Hay, B. A., and Baker, N. E. (2002). A pathway of signals regulating effector and initiator caspases in the developing *Drosophila* eye. *Development* **129,** 3269–3278.

CHAPTER SIX

# Cellular and Molecular Mechanisms Underlying the Formation of Biological Tubes

Magdalena M. Baer,* Helene Chanut-Delalande,*,†
and Markus Affolter*

## Contents

1. Introduction 137
2. Architecture and Mechanisms of Formation of Tubular Structures 138
3. Molecular Basis of Cellular Aspects of Tube Formation 139
   3.1. Forming a tube from prepolarized epithelia 141
   3.2. Tube formation from nonpolarized tissue 144
   3.3. Novel modes of tube formation 154
4. Summary and Perspectives 156
5. Open Questions 158
Acknowledgments 159
References 159

## Abstract

Biological tubes are integral components of many organs. Based on their cellular organization, tubes can be divided into three types: multicellular, unicellular, and intracellular. The mechanisms by which these tubes form during development vary significantly, in many cases even for those sharing a similar final architecture. Here, we present recent advances in studying cellular and molecular aspects of tubulogenesis in different organisms.

## 1. Introduction

Tubular structures are found in all multicellular animals from worms to humans. Organs that are made up of structures of tubular architecture include the intestine, the lung, the kidney, and the cardiovascular system.

* Biozentrum der Universität Basel, Klingelbergstrasse, Basel, Switzerland
† Centre de Biologie du Développement, CNRS UMR5547, Toulouse, France

In all these cases, biological tubes confer important aspects of function by providing the environment for physiological activities, transport ways for liquids, gases, or cells, or by facilitating the exchange of the latter across the epithelia. Biological tubes come in ample varieties. In the simplest case they consist of pipes made up by a single cell layer, in other cases they can form complex three-dimensional networks with tubes of different architecture and function. The enormous diversity among tubular organs in form and function imposes specific requirements for the tubular architecture, the organization of the cells within theses tubes, as well as the morphogenetic processes that lead to the formation of a tubular epithelium. What are the morphogenetic mechanisms that shape biological tubes and how are they controlled and executed at the cellular and molecular level? And are different tubes made by similar mechanisms? These are not only intriguing biological questions, but also important medical issues; proper development and function of tubular structures are crucial for any organism throughout lifespan.

Here, we would like to describe what is known about how different cellular architectures of biological tubes arise during development and discuss emerging similarities and differences in the cellular and molecular control of tubulogenesis.

## 2. Architecture and Mechanisms of Formation of Tubular Structures

Tubular structures vary considerably in size and shape as well as in their histological complexity. Based on their cellular architecture, epithelial tubes can be classified into three distinct types, depicted in Fig. 6.1. (A) Multicellular tubes—several cells constitute the cross section of the epithelial cylinder, and the tube has a characteristic pattern of intercellular adherens junctions (AJs). This type of tube is the one most commonly found in fully developed tubular organs in vertebrates. (B) Unicellular tubes—single cells cover the circumference of the luminal space and are sealed with autocellular AJs along the tube axis; ring-like intercellular AJs connect neighboring cells. Such tubes have been described in the tracheal system in *Drosophila*, and have been proposed to exist in the vertebrate vasculature. Autocellular AJs exist in other cell types, such as in glial cells (Stork *et al.*, 2008); however, these cells do not wrap around an open luminal space but rather around neuronal extensions. (C) Intracellular tubes—the lumen is formed within a single cell, resulting in a seamless tube without AJs along the tube. Such tubes are found in the digestive tract in *Caenorhabditis elegans*, in terminal and fusion cells of the tracheal system in *Drosophila* and possibly in vascular capillaries in vertebrates. All of these three tube types show similar cell polarity, with the apical side toward the lumen and the basal side away from it.

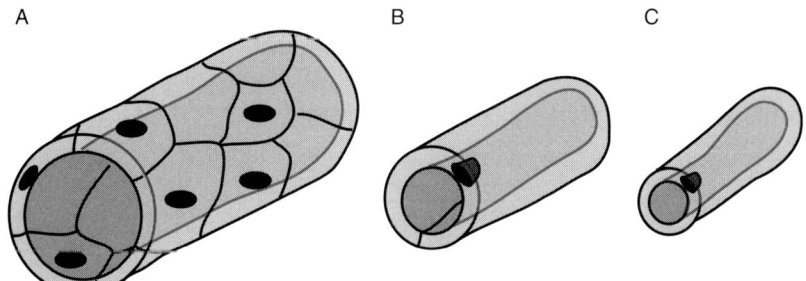

**Figure 6.1** Architecture of epithelial tubes. Based on cellular organization, tubular structures can be divided into three distinct types: (A) multicellular tube—several polarized cells constitute the cross section of the tube, which has characteristic pattern of intercellular adherence junctions (AJs); (B) unicellular tube—the luminal surface is formed by a single cell sealed along the tube axis with autocellular AJs; and (C) intracellular tube—the lumen is formed within a single cell, resulting in a seamless tube without AJs along its axis of extension. In all three tube types, cells show similar polarity, with the apical membrane facing the lumen (in green) and the basal membrane away from it. (See Color Insert.)

The morphogenetic mechanisms of tube formation, as outlined in a number of excellent reviews (Bryant and Mostov, 2008; Chung and Andrew, 2008; Hogan and Kolodziej, 2002; Lu and Werb, 2008; Lubarsky and Krasnow, 2003; Uv et al., 2003), are rather diverse and possibly controlled by different molecular pathways. There are several distinct ways by which different types of tubes develop (schematically displayed in Fig. 6.2): wrapping, budding, cavitation, cord hollowing, cell hollowing, as well as two recently described novel modes, cell wrapping/self-fusion and cell assembly. In the following sections, we will review recent data describing what is known about the molecular control of cell behavior involved in the generation of distinct tubes via these mechanisms. Due to the immense amount of work being done in the field, our review of the existing literature is not meant to be comprehensive, but covers selective examples outlining the substantial progress that has recently been made.

## 3. Molecular Basis of Cellular Aspects of Tube Formation

The different mechanism of tube formation can be divided into two general classes with respect to the state of tissue polarity at the onset of tubulogenesis: those remodeling prepolarized epithelia, and those in which polarity is established during lumen formation.

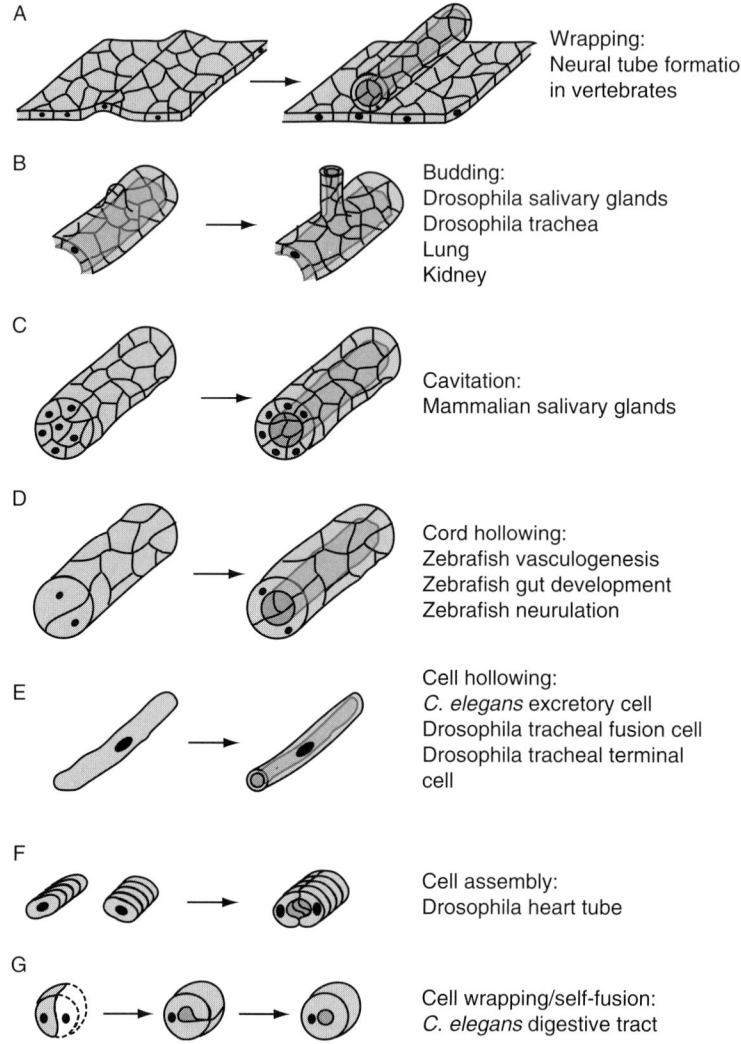

**Figure 6.2** Different ways to develop epithelial tubes. Different types of tubes can result from distinct modes of development. Scheme represents the "starting point" and the "final form" underlying each of the listed mechanisms: (A) wrapping—part of the epithelial sheet invaginates and rolls up until the edges of the invagination area meet and seal to form a tube; (B) budding—a group of cells from an existing epithelial tube (or sheet) migrates out or extends to form a new branch; (C) cavitation—the tube is formed by the elimination of cells in the center of a solid cylindrical cell mass, accompanied by polarization of the most outside layer of cells; (D) cord hollowing—in a solid cord-like structure, the lumen is formed by the generation of open space and remodeling of the apical surface of the cells facing the future lumen; (E) cell hollowing—the lumen forms within a single cell, spanning its length; (F) cell assembly—tube formation results from cell migration and cell shape changes, accompanied by specification of a nonadhesive luminal domain; and (G) cell wrapping/self-fusion—a cells wraps around the future lumen and fuses with itself. Apical (luminal) surface is marked in green. (See Color Insert.)

## 3.1. Forming a tube from prepolarized epithelia

### 3.1.1. Wrapping

Wrapping consists in rolling up of a flat epithelial sheet (Fig. 6.2A). This process has been described during neural tube formation in higher vertebrates (for review, see Colas and Schoenwolf, 2001). The neural plate, composed of specialized dorsal ectodermal cells, develops bilateral neural folds at its junction with non-neural ectoderm. The folds elevate, connect at the midline, and seal to form a tube, leading to the separation of the tubular structure from the rest of the epithelium that becomes the future epidermal ectoderm. What are the cellular activities that induce epithelial wrapping? Apical constriction has long been proposed to be a critical process driving the bending of epithelial sheets, and has been studied extensively in diverse systems, including neural tube wrapping. Recent studies demonstrate a crucial role for the PDZ protein Shroom3 in apical constriction during neural tube closure in *Xenopus* and chicken (Haigo et al., 2003; Nishimura and Takeichi, 2008). The Shroom3 protein is expressed apically in constricting regions. During neural tube closure in *Xenopus*, Shroom3 induces apical constriction of cells located at the hinge points by interacting with ROCK kinases and recruiting them to the apical junction complexes (AJCs) of neural epithelial cells (Nishimura and Takeichi, 2008). ROCKs are required to phosphorylate myosin regulatory light chains (MLCs), which subsequently induce Myosin2 activation (Riento and Ridley, 2003). Recent studies identified a Shroom3-binding domain, RII-C1, in the ROCK2 protein; when expressed ectopically, this domain antagonizes Shroom3–ROCK interaction during neural tube closure in chicken and leads to a constriction failure. Altogether, it is thought that Shroom3 recruits ROCKs proteins to the AJC of neuroepithelial cells to induce actin constriction via the MLC phosphorylation pathway. These studies point toward a mechanism leading to apical constriction, and more recent findings in flies have added even more twists to this process (Martin et al., 2009). It should be kept in mind, however, that apical constriction is just one of several mechanisms that drive the wrapping process eventually leading to the formation of the neural tube in *Xenopus* (and other vertebrates), and other processes such as convergent extension and epidermal pushing might also be involved and have to be analyzed in more detail at the molecular level.

### 3.1.2. Budding

Similar to the formation of tubes by wrapping, the formation of tubes or tubular networks by budding generally starts from a fully polarized epithelial sheet. Upon invagination of the epithelial sheet, branches are formed, often out of existing tubular structures, a process generally referred to as branching morphogenesis (Fig. 6.2B). Since budding followed by branching morphogenesis (reiterated budding in a temporally and spatially controlled manner) may be regarded as the principal mechanism to generate complex tubular organs, much work has been invested in trying to decipher mechanisms leading to the temporal and spatial control of budding. This process has been studied during lung, kidney and mammary gland development, as well as in *Drosophila* salivary gland and trachea development. Most of the studies concentrate on pathways controlling bud outgrowth and branching. Since our aim is to present more mechanistic aspects of tubulogenesis we refer the reader to some recent reviews on branching morphogenesis (Affolter and Caussinus, 2008; Horowitz and Simons, 2008; Lu and Werb, 2008). Here, we will only discuss cellular events involved in tube formation via budding, and not branching mechanisms.

The *Drosophila* salivary gland, which consists of two unbranched secretory tubes, develops from placodes of polarized epithelia that, upon invagination, form a tube that elongates due to cell shape changes and cell migration (reviewed in Kerman *et al.*, 2006). The elongation step also requires apical surface expansion, allowing lumen growth. This process is controlled by two transcription factors, Huckebein (Hkb) and Ribbon (Rib). Hkb acts through upregulation of two target genes: *crumbs* (*crb*), a determinant of the apical membrane, and *klarsicht* (*klar*), which is involved in microtubule-dependent vesicle transport (Myat and Andrew, 2002). Rib also upregulates *crb* expression and at the same time indirectly downregulates the activity of Moesin, which links the apical membrane to the actin cytoskeleton. Increased formation and delivery of apical membrane and decreased connection of the membrane to the cytoskeleton facilitate lumen expansion and tube elongation (Kerman *et al.*, 2008).

The cellular behavior underlying the budding process has been extensively studied during tracheal system development in *Drosophila*. The initial bud providing the cellular template for the future tracheal network is generated by apical constrictions, oriented cell divisions and cell rearrangements (Brodu and Casanova, 2006; Nishimura *et al.*, 2007). Upon invagination, tracheal cells migrate toward neighboring cells or toward tissues that express the Btl ligand, Branchless/Fgf (Bnl/Fgf); it is the spatial control of Bnl/Fgf secretion in the *Drosophila* embryo that ultimately controls the migratory behavior and the direction of tracheal cell movement (Sutherland *et al.*, 1996). Cells at the tip of tracheal branches appear to be highly dynamic when visualized in live embryos with confocal microscopy; they send out numerous filopodia in response to Breathless/Fgfr signaling. Stalk cells, which link the tip cells to the other tracheal branches, do not form such extensions (Ribeiro *et al.*, 2002). In the complete absence of Fgfr signaling, cells remain in the sac-like configuration and filopodia are not seen (Ribeiro *et al.*, 2002), demonstrating that Bnl/Fgf signaling regulates both the motility and directionality of tracheal cell movement in the embryo, and thus is the key regulator of tube formation via budding.

Epithelial tube formation via budding has also been studied during lung, kidney and mammary gland development (reviewed in Lu *et al.*, 2006). Strikingly, Fgf10 is expressed in a dynamic fashion in the mesenchyme surrounding the epithelial branch tips, and is intimately linked to the branching process in the lung (Cardoso and Lu, 2006; Hogan and Kolodziej, 2002; Warburton *et al.*, 2005). However, it remains unclear whether Fgf signaling regulates cell migration at the tip of the branches as does Fgf in the fly trachea. In the kidney, the secreted signaling molecule Glia cell line-derived neurotrophic factor (Gdnf) is required for budding. Since the tips of the ureteric buds are the growth centers in the developing kidney, it will be interesting to find out whether cell division is the driving force for budding, and whether cell rearrangements and cell migration contribute to bud extension (Shakya *et al.*, 2005).

Recent studies of the developing mammary gland have made use of high-resolution live imaging of cultured mammary glands (Ewald *et al.*, 2008). These studies showed that cells in elongating mammary ducts reorganize into a multilayered epithelium, migrate collectively, and rearrange dynamically, all without forming leading cell extensions. However, active movement of both luminal and myoepithelial cells were seen during budding. Fgfr2 is required for cells to be quantitatively retained in the tips of the buds, suggesting that Fgf signaling contributes to some cellular activities in tip cells (Lu and Werb, 2008); again, whether cell division/cell migration and/or other processes are the key players in bud growth remains to be discovered.

The development of certain blood vessels via angiogenesis also relies on budding or branching morphogenesis, that is, the sprouting of new vessels from pre-existing vessels. In this particular case, cell migration is the prime cellular event, with cells at the tips of outgrowing branches displaying numerous filopodial extensions induced under the control of vascular endothelial growth factor (Vegf) (Adams and Alitalo, 2007; Carmeliet, 2005). Vegf signaling also appears to control cell division, so that in this particular case the same signaling molecule regulates both directional outgrowth and cell proliferation (i.e., bud formation and elongation). The budding process in all known branched organs is thus controlled by cell signaling, and future studies will aim at the identification of the direct targets of these essential signaling cascades (Fgf and Gdnf signaling) and their direct and indirect effects on cell behavior.

Studies of the *Drosophila* tracheal system have also provided insight into how one particular type of the three biological tubes, namely those formed by single cells wrapped around a lumen and sealed by autocellular AJ (Fig. 6.1B), are generated *in vivo* during the budding process (Caussinus *et al.*, 2008; Ribeiro *et al.*, 2004; Samakovlis *et al.*, 1996a). During branch outgrowth tracheal stalk cells align in a pairwise fashion along the axis of extension and the apical sides of two cells contribute to the tube lumen circumference. Under the force generated by the migrating tip cells the stalk cells intercalate so that at the end of the branching process they are aligned in an end-to-end organization. During this intercalation process intercellular AJs of adjacent cells are progressively converted into autocellular AJs until, at the end of the process, the lumen circumference of the tube is made up from single tube-shaped cells and sealed with autocellular AJs. To our knowledge, tubes with autocellular AJs have not been documented unambiguously in any other tubular organ. If such tubes indeed exist elsewhere than in the trachea, it will be interesting to find out whether they develop via cell intercalation mechanisms, or whether cell wrapping is involved.

In the two mechanisms we described so far, wrapping and budding, tubes arise from a pre-existing polarized epithelium. By contrast, in the mechanisms we describe in the following sections, tube formation involves polarization or the formation of new junctions as the tubes are established.

## 3.2. Tube formation from nonpolarized tissue

### 3.2.1. Cavitation

Cavitation consists of hollowing out a tube through the elimination of cells located in the center of a cylinder of cells (Fig. 6.2C). This mechanism has been reported during mammary gland development, mammalian salivary gland formation, and proamniotic cavity formation (Lubarsky and Krasnow, 2003; Tucker, 2007). The clearance process has been attributed to apoptosis but is still poorly understood.

New studies on mammary gland development have allowed to better understand different aspects of the cavitation process (Mailleux *et al.*, 2007). The precise role of apoptosis during mammary gland development has been investigated using knockout mice for *bim*. Bim is a member of the proapoptotic BCL2 family and is an essential factor for apoptosis (Huang and Strasser, 2000). Mice deficient in *bim* display a delay in lumen formation during mammary gland development, confirming the important function of apoptosis during tube formation. Interestingly, the lumen forms correctly at later stages in *bim* mutant individuals, suggesting that the removal of cells is not only linked to apoptosis. Other mechanisms appear to be required during cavitation and Mailleux and collaborators suggest the involvement of autophagy in this process. However, the involvement of autophagy remains to be tested.

### 3.2.2. Cord hollowing

Another mechanism of forming a tube from a solid cylinder is cord hollowing. In this process, epithelial cells migrate and aggregate to form a solid, cord-like structure in which the lumen is subsequently made by the formation of an open space between the cells facing the future luminal side (Fig. 6.2D). Cord hollowing is observed in vasculogenesis, during gut development, and in neurulation. Recent studies on these processes provide insights into the cellular behavior involved and link molecules and biochemical functions to cellular activities. In contrast to cavitation, cell removal is not the driving force to generate the lumen in cord hollowing. In all tubes formed by cord hollowing, the generation of apical–basal polarity within the tissue is the key step. However, how this polarity is achieved can vary considerably between different tissues (reviewed by Belting and Affolter, 2007; see Fig. 6.3). Furthermore, as we will describe below using examples from zebrafish gut development and neurulation, the cellular behaviors that lead to lumen formation are also quite variable.

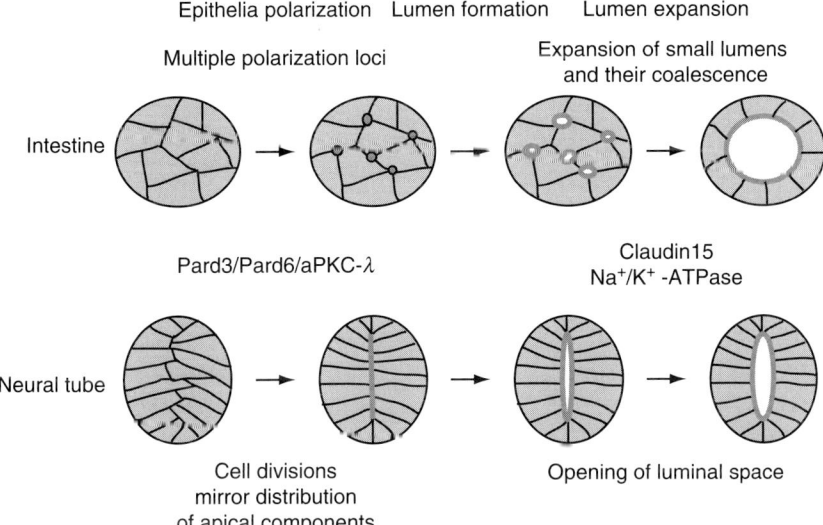

**Figure 6.3** Different ways of polarity induction in tube formation by cord hollowing. Epithelia polarization is an important step in tube formation by cord hollowing. Apical–basal polarity can be differently achieved in different tissues. During zebrafish intestine morphogenesis polarity is established at multiple places within the cord (upper panel, green dots) whereas in the neural tube apical–basal polarity is generated by mirror-symmetric cell divisions leading to distribution of apical proteins along the midline (lower panel, green line). In both tissues, the Pard3/Pard6/aPKC complex is essential for polarity establishment, but mutants show different phenotypes in different tissues; in the intestine, absence of aPKC leads to multiple lumens and in the neural tube no lumen forms. Interestingly, mutation in *pard6* results in multiple lumens within the neural tube, but this phenotype results from disruption of apical surface along the midline. Lumen expansion in both tissues is driven by water and ions influx, but formation of single lumen in the gut additionally requires coalescence of multiple lumens. (See Color Insert.)

**3.2.2.1. Lumen formation in zebrafish intestine** Analysis of intestine morphogenesis in zebrafish revealed three distinct steps of conversion from a solid cord to a tube with a single lumen (Fig. 6.3). First, proper apical–basal polarity of the epithelial cells within the cord has to be established. The localization of apical proteins, such as Zonula Occludens protein-1 (ZO-1) marking the tight junctions, or cadherin, marking the AJs, indicates that cell polarization occurs at multiple places within the cord. This requires the atypical protein kinase C λ (aPKC-λ); in its absence, in *heart and soul (has)* mutants, multiple lumens develop in the gut (Horne-Badovinac et al., 2001). During the next step, small local lumens form at the sites of polarization. Finally, these multiple lumens expand upon influx of ions and water and coalesce into a single lumen. Disruption of this later event results in multiple lumen phenotype. It has been shown that lumen coalescence is achieved by the action of Claudin15 and $Na^+/K^+$-ATPase, suggesting that an electrochemical gradient generated by $Na^+/K^+$-ATPases drives ion movement through ion-permeable pores in epithelial junctions formed by Claudin15. As a result, asymmetric ion distribution facilitates fluid accumulation that leads to expansion of small single lumens and possibly provides the force for their coalescence (Bagnat et al., 2007).

**3.2.2.2. Making a tube during neurulation** An important role for the induction of cell polarity as a necessary step in tube formation was also described during neurulation in zebrafish. However, polarity is achieved by different means, although a partial overlap in the involved molecules is evident (Fig. 6.3). Live imaging of the neurulation process (Ciruna et al., 2006; Geldmacher-Voss et al., 2003; Tawk et al., 2007) showed that cells on both sides of the midline of the neural rod divide to produce daughter cells with a mirror image polarity. After division, one of the two cells remains in place while the other one crosses the midline. The mirror polarity of the two sister cells is generated during cytokinesis by a novel mechanism that localizes apical polarity components to the cleavage furrow. One of the apical polarity proteins which localizes to the furrow is the zebrafish orthologue of Par3, Pard3. The localization of Pard3 to the cleavage plane and its mirror-symmetric inheritance to the prospective apical poles of the daughter cells is required for cells to cross the midline and establish tissue polarization prior to lumen formation (Tawk et al., 2007). The mirror distribution of apical components, such as the Pard3/Pard6/aPKC complex, on either side of the midline is necessary for the lumen to form, since in aPKC mutant *has*, where polarity is lost, no lumen forms (Geldmacher-Voss et al., 2003).

To form only a single apical surface leading to a single lumen opening, mirror-symmetric divisions have to be properly positioned within the tissue. This step might be regulated by the noncanonical Wnt/planar cell polarity (PCP) pathway. In the mutant *trilobite (tri)*, which affects the Van

Gogh-like 2 (Vangl2) protein, cells divide but the progenitors accumulate at the division site instead of intercalating on the other side of the midline (Ciruna et al., 2006). As a result the re-establishment of polarity after cell division occurs ectopically and two lumens start to form, on either side of the midline (Tawk et al., 2007).

Although the mode of primary lumen formation during zebrafish neurulation differs from the one in the gut (mirror image polarity versus *de novo* polarity induction, respectively), the process of lumen expansion is similar in both systems and similar to brain ventricle expansion (Lowery and Sive, 2005). In *snakehead* mutants, which carry a mutation in the $Na^+/K^+$ ATPase-encoding gene *atp1a1a.1*, no visible extracellular space is found within brain ventricles and the lumen does not open up; this suggests that for the lumen to form, an osmotic gradient is needed to drive water molecules into the closed ventricles after their morphogenesis, and the Atp1a1a.1 ion pump is required for that process.

*3.2.2.3. In vitro studies of tube formation* Obviously, it will be very difficult to fully dissect the process of apical–basal polarization during the course of tube formation in any *in vivo* system. However, the three-dimensional Madin–Darby canine kidney (3D MDCK) epithelial cell system provides an excellent *in vitro* model to study this process, since MDCK cells embedded in a gel of extracellular matrix (ECM) form cysts, spherical epithelial monolayers with fluid-filled central lumen (Montesano et al., 1991). Extensive studies in this model system over the last years revealed a number of molecular aspects of apical–basal polarity establishment, lumen formation, and the role of the ECM in epithelial tube morphogenesis. Experiments carried out in this system dissected how Cdc42 regulates constitution of apical–basal polarity. The findings suggest a scenario in which Par3 localizes to tight junctions and presumably mediates their integrity. In addition, Par3 (together with other factors) appears to localize PTEN to the apical region. PTEN activity is required for enrichment of phosphatidylinositol 4,5-bisphosphate (PtdIns(4,5)p2) at the apical domain and restricts phosphatidylinositol 3,4,5-trisphosphate (PtdIns(3,4,5)p3) to the basolateral surface, which is essential for the organization of this domain. In turn, PtdIns(4,5)p2 targets and activates Cdc42 at the apical domain, leading to the activation of Par6/aPKC and other effectors (Gassama-Diagne et al., 2006; Martin-Belmonte and Mostov, 2007; Martin-Belmonte et al., 2007). Cdc42 has also been found to play a role in lumen formation. Its depletion from MDCK cells disturbs the exocytosis of the vacuolar apical compartments (VAC), proposed to be required for delivery of the luminal membrane and apically targeted proteins to the luminal surface (Martin-Belmonte et al., 2007). Analysis in the MDCK system also shed light on the role of the ECM in the orientation of cell polarity. It has been shown that in the absence of Rac1 activity cysts acquire inverted polarity with the apical surfaces facing the exterior of the cyst (O'Brien et al., 2001). Rac1 activity depends on the interaction of cells with collagen I and on $\beta$-integrin function (Yu et al., 2005). Also PI3K and protein kinase B were shown to be involved in regulating cell polarity orientation (Liu et al., 2007). The recent study of Martin-Belmonte et al. (2008) revealed an interesting dependence of the lumen formation mechanism on the degree of cell polarization within a solid cord. In the case of fast polarizing cells, the lumen is formed by cord hollowing, whereas a low degree of cell polarization promotes cell death/cavitation as a mode of lumen formation (Martin-Belmonte et al., 2008). Whether all these observations made with MDCK cells apply to *in vivo* situation of tubulogenesis discussed here, and if yes, to which ones, remains to be investigated.

### 3.2.3. Cell hollowing

The third type of tubular architecture found in the animal kingdom is formed by cell hollowing. The lumen of such tubes is created within single cells, resulting in seamless structures in which the lumen is not lined by AJ complexes (see Figs. 6.1C and 6.2E). This type of tubes has been described in the excretory cell in *C. elegans*, in two types of cells in the tracheal system in *Drosophila melanogaster*, the fusion cells and the terminal cells, as well as in vascular capillaries. The exact mechanisms of intracellular lumen formation are still poorly understood in all of these systems; however, studies in cell culture and in model systems shed light on some cellular and molecular aspects of cell hollowing.

*3.2.3.1. Cell hollowing in* C. elegans   The *C. elegans* excretory cell forms the major tubular component of the four-celled nematode excretory system. During embryogenesis, the excretory cell extends branched processes, so called canals and ultimately forms an H shape along the body on the basolateral surface of the epidermis. The extensions are tunneled by a single lumen with blind endings (Nelson and Riddle, 1984). At the onset of lumen formation a large vacuole is generated by pinocytosis. Subsequently this vacuole extends tubular arms that eventually coalesce and remodel to form a mature lumen (Berry et al., 2003; Buechner, 2002).

A distinct series of mutants affecting tube formation within the excretory cell have been isolated and referred to as *exc* mutants. In many of these *exc* mutants, defects in the canal's apical membrane have been observed; the apical surface loses the ability to maintain a narrow tubular structure and swells into a series of fluid-filled cysts (Buechner et al., 1999). Detailed analysis of the phenotypes and subsequent cloning of the mutated genes revealed that achievement of proper lumen morphology is regulated by ion channels and pumps (Berry et al., 2003; Liegeois et al., 2007), cytoskeletal proteins (Gao et al., 2001; Gobel et al., 2004; McKeown et al., 1998; Praitis et al., 2005; Suzuki et al., 2001), proteins affecting trafficking toward the apical side (Fujita et al., 2003), as well as an apically secreted protein (Jones and Baillie, 1995). Studies over the last years on genetic interactions between different *exc* mutants and the identification of their molecular nature (Berry *et al.*, 2003; Buechner, 2002; Tong and Buechner, 2008) allowed the partial dissection of a network controlling lumen maintenance (see Fig. 6.4). The chloride channel Exc-4, an orthologue of the human CLIC family of proteins, together with its putative partner Exc-2 (so far not cloned) promote the function of Exc-9, a small LIM-domain protein homologous to the mammalian cysteine-rich intestinal protein (CRIP). Exc-9 signals through RhoGEF Exc-5, and possibly Exc-1, to activate Cdc42 (and most likely other small GTPases) in order to maintain the flexibility of the apical epithelial surface. The Exc-1–Exc-5–Cdc42 pathway is also involved in the transport of $\beta_H$-spectrin *SMA-1* mRNA, regulated via the mRNA-binding protein Exc-7 (a homolog of ELAV), to the apical surface. Sma-1/spectrin itself is required to anchor a cytoskeletal terminal web to the luminal surface. The placement of Sma-1 at the apical surface may send a signal to Exc-9 to modulate the activity of the pathway in a feedback loop, while stretching or lack of spectrin at the luminal surface could increase the Exc-9 activity (Berry *et al.*, 2003; Buechner, 2002; Tong and Buechner, 2008). Although these studies lead to a molecular picture of how the intracellular lumen is kept in its exquisite shape in the excretory cell, how the lumen is initially established remains poorly characterized. It is thought that the lumen is generated by the pinocytic formation of a large vacuole, which then extends tubular arms that eventually coalesce and remodel to form a mature lumen; however, genetic and molecular evidence for this scenario is lacking. In addition, while luminal expansion is likely achieved via the directed fusion of vesicles, the involvement of this process also needs to be confirmed.

### 3.2.3.2. Cell hollowing in D. melanogaster

More information about how lumen form within single cells comes from studies on the fusion of neighboring tracheal metamers in *Drosophila* trachea development (Samakovlis *et al.*, 1996b; Tanaka-Matakatsu *et al.*, 1996). This process is an essential step leading from separate tubular structures to a functional interconnected network. The fusion cells (two cells per fusion event) undergo extensive remodeling of the cytoskeleton accompanied by the formation of new apical surface. Although the cells involved are called fusion cells, it is important to note that these cells do not *fuse*, but rather facilitate the fusion of two independent tubular networks by providing intracellular tunnels to connect the latter.

By means of analyzing the fusion defects in mutants for plakin (*short stop*) (Lee and Kolodziej, 2002), E-cadherin (*shotgun*) and $\beta$-catenin (*armadillo*), combined with detailed *in vivo* imaging of cell behavior (Lee *et al.*, 2003), a comprehensive description of the events occurring during tracheal fusion has emerged. The process of tracheal fusion can be divided into four phases:

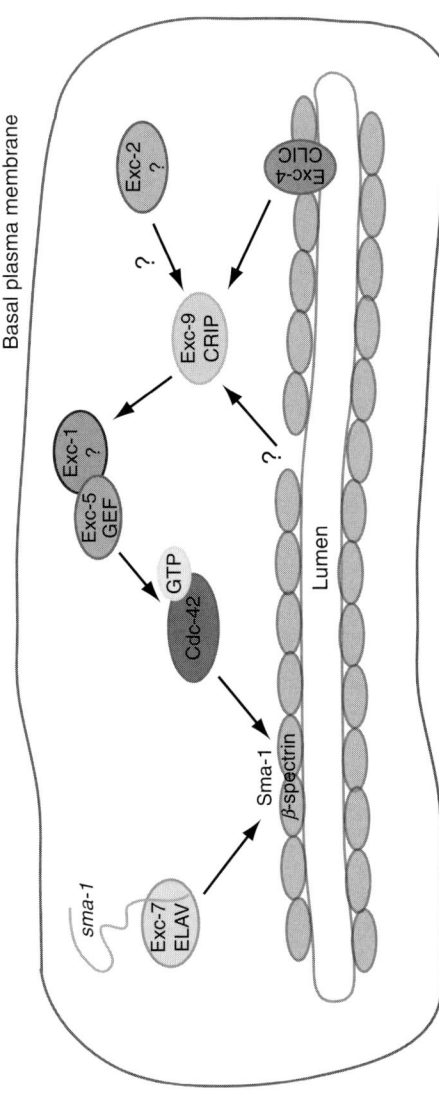

**Figure 6.4** Model of Exc proteins interaction in lumen maintenance in the excretory canals. The scheme represents an excretory cell of *C. elegans*. In the presence of the chloride channel Exc-4 and its putative partner Exc-2, Exc-9 signals through Exc-5 and Exc-1, which in turn is required to activate the Cdc42 pathway to maintenance apical surface flexibility. This is in part achieved by influence on *sma-1* mRNA localization to the luminal surface, regulated by Exc-7. Exc-9 activity is presumably increased by stretching of or by gaps in the apical cytoskeleton as well as by the absence of Sma-1. At the same time, apically localized Sma-1 is proposed to send a signal to downregulate Exc-9 activity.

contact, initiation, maturation/invagination, and fusion (Fig. 6.5). In a first step, tracheal cells at the tip of the neighboring branches touch each other through filopodial extensions. Next, a faint cytoskeletal track consisting of F-actin, microtubules, and the plakin Short Stop (Shot) appears in the fusion cells and becomes centered at the cell's contact points. This event is accompanied by *de novo* assembly of E-cadherin (DE-cad) at the contact site between the two fusion cells. In the maturation phase, the existing apical surfaces invaginate and the cytoskeletal track shrinks, leading to the formation of a bottleneck at the narrowest point. A fast transition follows and apical surfaces appear connected on one side and open on the other. In the final step, the bottleneck expands to form a tube. The initial opening (i.e., the fusion of the two apical membranes) occurs fast and is followed by slower expansion of the lumen required to obtain a diameter equal to the rest of the tube. This event is accompanied by the concentration of Shot into a ring at the AJ between the fusion cells.

How do the cytoskeleton and the cell junctions coordinate the cell remodeling taking place during the fusion process? Lee *et al.* (2003) proposed that DE-cad signals via $\beta$-catenin to recruit Shot to the new contact between the fusion cells, which in turn stabilizes it. Association of Shot with AJs initiates track assembly by promoting the accumulation of F-actin/microtubule and the track finally expands to span a fusion cell. The cytoskeletal track is required not only for stabilizing the remodeling of cellular junctions, but also to bridge the two apical membranes; the one facing the lumen tip and the other forming *de novo* at the contact site of the fusion cells. The cytoskeletal track is also believed to be necessary for directing/assembling vesicles needed for lumen formation.

Yet more detailed insight on the formation of new apical surface comes from studies on the role of Arf-like 3 protein (Arl3, also called Dead end—Dnd) in vesicle trafficking (Jiang *et al.*, 2007; Kakihara *et al.*, 2008). Arl3 can associate with vesicles and microtubules, which makes it a good candidate to play a role in bringing these two together, especially since Arl3 positive vesicles locate near microtubule tracks. Another protein that localizes near the microtubule tracks and accumulates at the position of apical membrane assembly is Sec5. Kakihara *et al.* (2008) showed that Arl3 is needed for Sec5 localization at the site of the internal membrane fusion and that both of them are genetically required for fusion to take place. Additionally, Arl3 and Sec5 distribution partially overlaps in vesicles at the fusion site. These results suggest that Arl3 is required for the recruitment of the exocyst complex to the fusion site. In their model, the authors propose that Arl3 regulates vesicle trafficking on the microtubules along the path of the future lumen and directs Sec5 positive exocystic vesicles to the closely apposed plasma membranes.

The cell culture experiments performed by Jiang *et al.* (2007) show that Arl3 positive vesicles are also positive for Rab11 and appear to traffic along the actin/microtubule-containing tracks. Activated Arl3 also causes

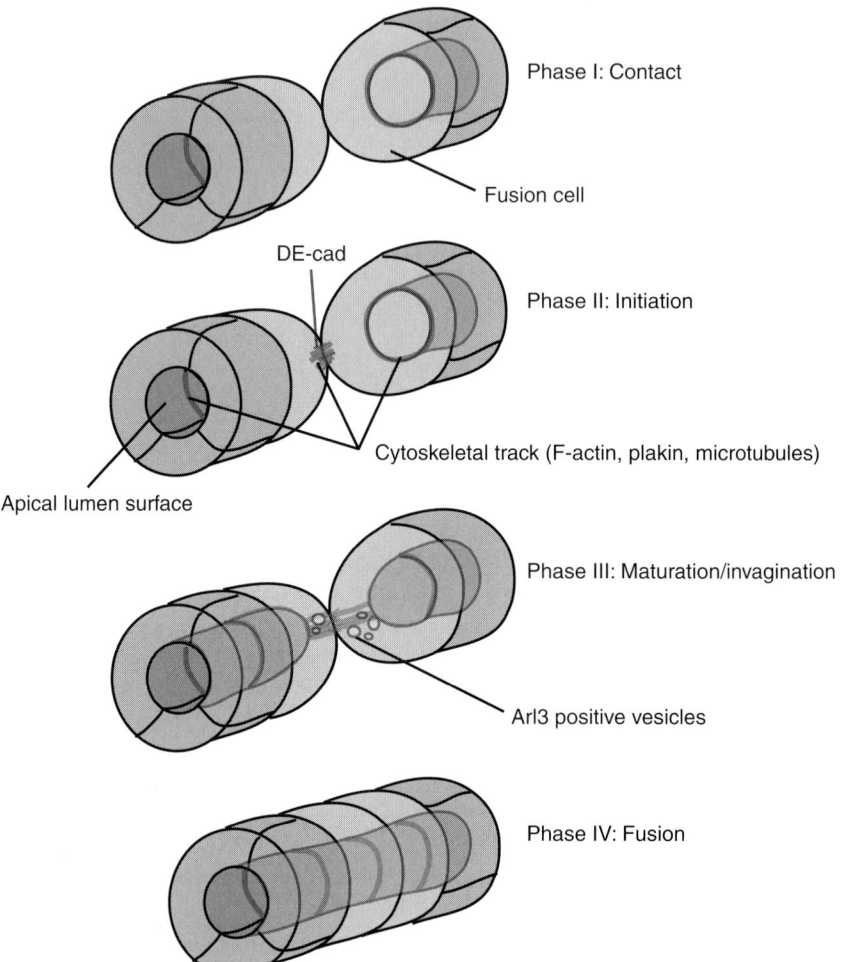

**Figure 6.5** Branch fusion in the trachea of *D. melanogaster*. The process of fusion between neighboring metamers during *Drosophila* tracheal system development can be divided into four steps. Phase I: contact—tip cells of branches touch each other through filopodia extensions. Phase II: initiation—cytoskeletal track consisting of F-actin, plakin, and microtubules forms at the cell's contact points, where E-cadherin is newly deposited. Phase III: maturation/invagination—the existing apical surface invaginates and the cytoskeletal track expands to span the cells and bridge the apical membrane facing the lumen and the one forming *de novo* at the contact site; the assembly of new apical membrane is facilitated by vesicle trafficking along the track requiring Arl3 and Sec5. Finally, the track shrinks resulting in formation of a bottleneck with apical surface connected on one side and open on the other. Phase IV: fusion—the bottleneck expands to form a tub.

dramatic changes in cellular morphology and cytoskeletal dynamics and Arl3 positive vesicles require an intact actin and microtubule cytoskeleton for trafficking. Additionally, due to the ability of Arl3 to affect F-acting depolymerization, Arl3 may be required for the cytoskeleton disassembly step during lumen formation; such a disassembly might be needed after formation of the membrane along the fusion track and final luminal fusion.

Taken together, the fusion of tracheal tubes requires tight cooperation and coordination between cytoskeletal proteins, cellular junctions, and vesicle trafficking. At the end of the process two doughnut-shaped cells are generated; these cells are the hallmark of fusion events in the tracheal system and allow fusion points to be unambiguously identified in the network using AJs markers.

Terminal cells of the *Drosophila* tracheal system can also serve as a model to study cell hollowing. They form numerous long extensions allowing gas diffusion between the trachea and virtually all cells in the larvae or the adult fly. Terminal cell development starts late in embryogenesis and continues throughout larval life. The initially compact cells grow thin cytoplasmic extensions in which a seamless lumen is subsequently formed (Guillemin et al., 1996). The mechanism leading to the formation of the lumen as a junction less, intracellular, membrane-bound channel is poorly understood. However, genetic mosaic analyses have identified a number of mutations that result in the absence of a lumen in terminal cells (Baer et al., 2007; Levi et al., 2006). The identification and characterization of the corresponding genes will provide important information as to the precise molecular mechanisms behind intracellular lumen formation in terminal tracheal cells, and will also reveal similarities and differences between intracellular lumen formation in terminal and fusion cells.

Some indications as to how intracellular lumen is formed in the developing vasculature come from live imaging of human endothelial cells (ECs), cultured *in vitro* in three-dimensional collagen gel matrices to promotes lumen formation, and from analyzing the *in vivo* development of intersegmental vessels in zebrafish (Kamei et al., 2006). Both *in vitro* and *in vivo* analyses revealed the existence of highly dynamic pinocytic vacuoles that appear individually and fuse together to form larger compartments, filling most of the cell. Based on these observations, a model for the establishment of vascular lumen has been proposed. In this model, the formation of pinocytic vesicles is followed first by their fusion within the cell and later by fusion between neighboring cells.

Although there are still many open question concerning molecular basis of cell hollowing, the data obtained from model systems described above indicate that, similarly to cord hollowing, apical membrane assembly and vesicle transport play a crucial role in lumen formation.

## 3.3. Novel modes of tube formation

### 3.3.1. Cell wrapping/self-fusion

Cell hollowing is not the only mechanism leading to formation of seamless tube. A novel way to form doughnut-shaped cells with a lumen inside and lacking an AJ seal (such as fusion cells in the *Drosophila* tracheal system) has recently been described in the *C. elegans* digestive tract (Rasmussen *et al.*, 2008) (Figs. 6.2G and 6.6). Two parts of the later, the pharynx and the intestine are multicellular tubes. However, at the junction between the two, the gut tube is formed by seamless single cells, known as pm8 and vpi1. During pharynx development, each of these cells has to move across the midline of the forming tube, wrap itself around the axis of the future lumen and finally fuse with itself. Rasmussen *et al.* (2008) have studied this process in detail. At the beginning, when the pharynx primordium forms a polarized cylindrical cyst, pm8, and vpi1 are wedge-shaped epithelial cells placed in the dorsal part of the cyst. Their apical side faces the midline where the lumen will form, and the basal surface is associated with the basal lamina at the cyst periphery. To obtain a C-shaped structure reaching around the future lumen, both pm8 and vpi1 enter in-between opposing ventral marginal cells. pm8 detaches from the dorsal basal lamina and remodels its apical surface into lamella that migrate ventrally. This movement is accompanied by the translocation of the nucleus and most of the cytoplasm toward the ventral side, leaving only a thin connection to the dorsal side. Subsequently, the cell gradually spreads across the cyst cross section. vpi1 extends processes into the ventral side, closely following the migrating cell body of pm8, but most of the cytoplasm remains on the dorsal side to be later redistributed symmetrically around the cyst midline. Essential for the ventral migration of the cells is Laminin, since in its absence migration is defective. The Laminin forms a tract between ventral cells prior to pm8 and vpi1 rearrangements, which functions as a transient path for pm8 migration and disappears as cell movement proceeds. To form a true doughnut or toroid, the two cells eventually have to self-fuse (Rasmussen *et al.*, 2008). An important aspect of this event is prevention of fusion with neighboring cells. This is achieved by using a different fusion mechanism in the two cells: vpi1 expresses the fusogen Eff-1 and pm8 the fusogen Aff-1. Each of the two proteins allows fusion only between cells expressing the same fusogen and thus vpi1 can only fuse with itself but not with pm8, and the same for pm8. Notch signaling is involved in this differential gene expression; the Notch pathway, activated only in pm8 cell, is required to induce Aff-1 expression and repress Eff-1, preventing a cross-fusion event. Whether a similar mechanism exists in vpi1 is not known. Also, the signals controlling pm8 and vpi1 migrations remain to be identified (Hardin, 2008; Rasmussen *et al.*, 2008).

These studies show that cells (doughnuts/toroids) of similar or identical structure can be generated in two very different ways, either through cell

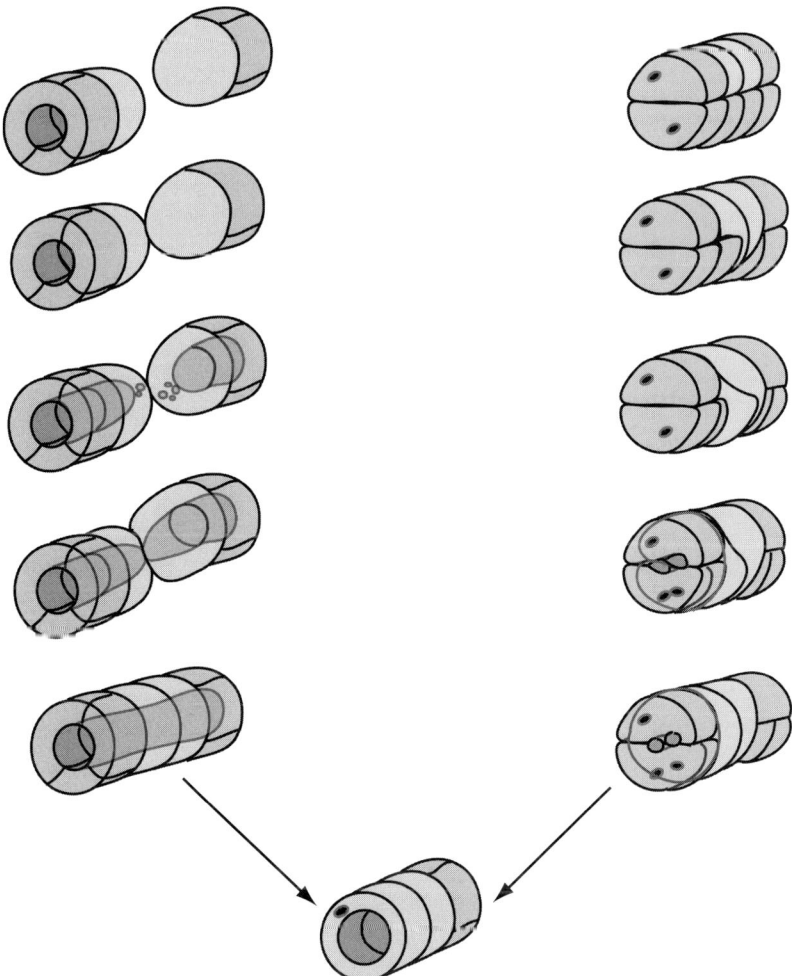

**Figure 6.6** Two ways to form toroid cells. Similar or identical architecture of tube cells can be achieved in different ways. Toroid cells can result from cell hollowing (i.e., fusion cells in *Drosophila* trachea, left panel) or cell wrapping/self-fusion (particular cells in the digestive track of *C. elegans*, right panel). In the first mode lumen within the cell is formed by invagination, *de novo* assembly of apical membrane and membrane fusion. The second mechanism involves cell movement or wrapping around the axis of the prospective lumen, followed by cell self-fusion. Cells to become toroids are marked in blue, apical surface in green. (See Color Insert.)

hollowing (fusion cell in the *Drosophila* trachea) or through cell wrapping/self-fusion (gut cells in *C. elegans*) (Fig. 6.6). Cell fusion occurs widely in *C. elegans*, so as such, it is not surprising to find this event also involved in tubulogenesis in this worm.

### 3.3.2. Cell assembly

All the mechanisms discussed in previous sections can be found in various organisms and could be considered as "canonical" modes of tube formation. But are these the only ways by which tubes can form? Recent studies on heart tube development in the *Drosophila* embryo show that this is not the case and describe a novel molecular mechanism of tube formation—cell assembly (Medioni *et al.*, 2008; Santiago-Martinez *et al.*, 2008) (Figs. 6.2F and 6.7).

In the initial step of heart formation, cardioblasts (CBs) are organized as two rows of cells on either side of the dorsal midline. Upon specification CBs undergo a mesenchyme-to-epithelium transition and migrate dorsally to meet at the midline to form the cardiac tube. Using time-lapse imaging, Medioni and colleagues demonstrated that during tube formation, cardioblasts undergo cell shape changes as well as distinct membrane specification and remodeling. As soon as cardioblasts establish contacts dorsally they define specific membrane domains, junctional (J) and luminal (L), by the region-specific localization of junctional and luminal components, respectively. The cells subsequently undergo cell shape changes to form crescent-like structures and remodel the membrane domains, which is crucial to eventually join ventrally and enclose the luminal space (see Fig. 6.7).

Slit/Robo signaling is essential for lumen formation since the localization of the two proteins at the luminal domain of cardioblasts restricts the size of the contact domains (Medioni *et al.*, 2008; Santiago-Martinez *et al.*, 2008). Indeed, the junctional domains of CBs are abnormally enlarged in *slit* and *robo* mutant embryos, which impairs proper contact between CBs and the subsequent cell shape changes (Medioni *et al.*, 2008). Consequently, only a very small lumen or no lumen forms. To understand how Slit/Robo signaling influences the formation of adherent junctions, Santiago-Martinez *et al.* (2008) examined the function of E-cadherin in heart tube formation. *shg* (E-cadherin) mutant cardioblasts do not succeed in contacting each other and do not form a lumen (Haag *et al.*, 1999), and the newer studies showed that *shg* and *robo* function in the same process to control the shape of the lumen. Altogether, these findings indicate a possible role of Robo in controlling the adherens junctions through E-cadherin (Helenius and Beitel, 2008).

Whether this novel mechanism of lumen formation by preventing cell adhesion is specific for *Drosophila* heart development or is found in other systems and tissues remains to be seen.

## 4. Summary and Perspectives

As outlined above, tubes of distinct architecture exist in different organs, and the cellular activities required for their formation vary to some extent. Many tubular networks arise from pre-existing epithelial

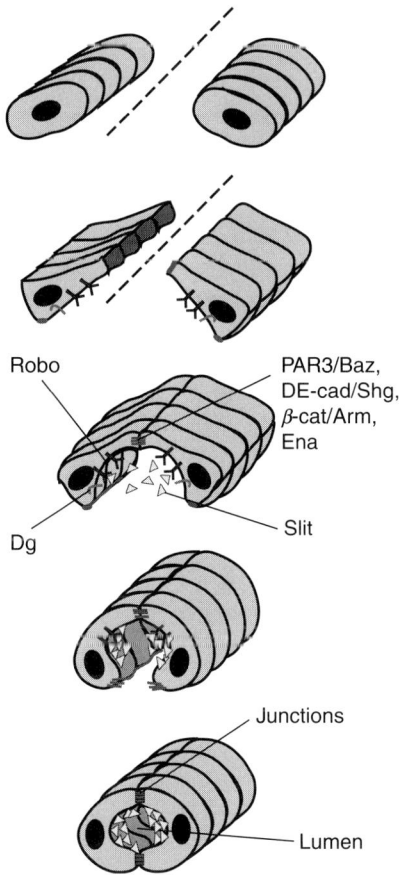

**Figure 6.7** Cell assembly in the fly heart. Cell assembly is a novel mechanism to form a tube. Cardioblasts (in blue) are initially organized in two rows on each side of the dorsal midline (dotted line). They undergo a mesenchyme-to-epithelium transition and migrate dorsally to meet at the midline. Upon the establishment of the contacts dorsally, the cells undergo shape changes and specific membrane domains are defined: junctional (in red), expressing junctional components and luminal (in green) expressing Robo receptor and its ligand Slit. Interaction of Robo with Slit, presumably regulated by dystroglycan (Dg), prevents adhesion between luminal domains and thus permits the formation of the cardiac tube. (See Color Insert.)

sheets. Their formation includes apical constrictions to form an initial luminal cavity; in other cases, cell migration leads to the formation of a new tube or branch, that is, to a new topology of an epithelial sheet. Branches might also form or at least extend by ordered cell rearrangements and/or cell division. All of these cellular activities are major driving forces for tubulogenesis by budding, and their spatial and temporal control regulate the architecture of tubular organs.

Quite obviously, when tubes are formed from nonpolarized cells the establishment of the apical (luminal) side becomes a major player in tube formation, and this is indeed seen in many cases. In several tissues, the aPKC/Par6/Par3 complex plays a central role in proper cell polarization; however, the mechanisms underlying the process differ (see Fig. 6.3) and mutations in polarity genes result in different phenotypes. For example, absence of aPKC prevents lumen formation in zebrafish neurulation (Geldmacher-Voss *et al.*, 2003), whereas in intestine morphogenesis multiple lumens form (Horne-Badovinac *et al.*, 2001). Interestingly, the multiple lumen phenotype was also observed during zebrafish neurulation, in *pard6* mutants; however, in this tissue it results from the interruption of apical surface and not its absence (Munson *et al.*, 2008).

Vesicle formation and vesicle fusion are thought to be driving forces in the formation of intracellular tubes, and they also contribute to tube expansion (Jayaram *et al.*, 2008). In addition, osmotic pressure is an important factor in the formation of continuous, open luminal spaces and evidence is accumulating that this holds true for tubes of different cellular architecture. Again, the same molecules can be involved in different aspects of lumen formation and maintenance. The $Na^+/K^+$ ATPase has been shown to be required for proper tubulogenesis in zebrafish gut development and neurulation and in *Drosophila* tracheal morphogenesis. Whereas in both zebrafish tissues the pump function is crucial for lumen to open up, $Na^+/K^+$ ATPase is required as a component of septate junctions and controls the size of the tube independently from the pump activity in the tracheal system (Paul *et al.*, 2007). Furthermore, the same ion channels can be involved in both lumen expansion and maintenance as it is in case of the chloride channel protein Clic (Exc-4) in *C. elegans* excretory cell (Berry *et al.*, 2003).

It thus looks like many different cellular activities are key to tubulogenesis and it is rather safe to say that no single tubular organ can rely on only one of the mechanisms generally involved in tube formation.

## 5. OPEN QUESTIONS

Apical constrictions initiate invagination processes that are crucial for tube formation by wrapping and budding. Recent studies analyzing mesoderm invagination have revealed new insight into the molecular and physical aspects of this process (Martin *et al.*, 2009), and similar analyses will have to be performed in tube formation processes. The cellular mechanisms driving the tube budding and elongation process in the lung, the kidney, and the mammary gland remain mysterious. Many processes are involved, including cell migration, cell rearrangement, and cell division but whether these processes contribute equally to the budding process needs to be determined experimentally. Another interesting question, not been addressed here, is how the budding process is controlled such that exquisite branching patterns are achieved.

The exact mechanisms generating *de novo* lumen are also still elusive. Vesicle generation and directional transport certainly plays a role but where these vesicles come from, how they are targeted to the prospective luminal side and whether they carry distinct proteins toward the luminal space or incorporate them into the apical membrane remains to be determined. Another interesting questions concerns unicellular tubes sealed with autocellular AJs. They do make up most of the tracheal system in *Drosophila*, but in which other tubular organ can they be found? And if such tubes are found elsewhere, do they form by cell intercalation or via cell wrapping? The continuous efforts to combine high-resolution live imaging with forward and reverse genetic approaches are expected to give answers to most of these questions in the near future.

## ACKNOWLEDGMENTS

We thank Heinz-Georg Belting, Emmanuel Caussinus, Amanda Ochoa Espinosa, and Alexandru Denes for comments on the manuscript. Work in our laboratory is supported by grants from the Swiss National Science Foundation, the FP6 NoE "Cells into Organs" of the European Community, and from the Kantons of Basel-Stadt and Basel-Land.

## REFERENCES

Adams, R. H., and Alitalo, K. (2007). Molecular regulation of angiogenesis and lymphangiogenesis. *Nat. Rev. Mol. Cell Biol.* **8,** 464–478.

Affolter, M., and Caussinus, E. (2008). Tracheal branching morphogenesis in *Drosophila*: New insights into cell behaviour and organ architecture. *Development* **135,** 2055–2064.

Baer, M. M., *et al.* (2007). A clonal genetic screen for mutants causing defects in larval tracheal morphogenesis in *Drosophila*. *Genetics* **176,** 2279–2291.

Bagnat, M., *et al.* (2007). Genetic control of single lumen formation in the zebrafish gut. *Nat. Cell Biol.* **9,** 954–960.

Belting, H. G., and Affolter, M. (2007). It takes guts to make a single lumen. *Nat. Cell Biol.* **9,** 880–881.

Berry, K. L., *et al.* (2003). A *C. elegans* CLIC-like protein required for intracellular tube formation and maintenance. *Science* **302,** 2134–2137.

Brodu, V., and Casanova, J. (2006). The RhoGAP crossveinless-c links tracheless and EGFR signaling to cell shape remodeling in *Drosophila* tracheal invagination. *Genes Dev.* **20,** 1817–1828.

Bryant, D. M., and Mostov, K. E. (2008). From cells to organs: Building polarized tissue. *Nat. Rev. Mol. Cell Biol.* **9,** 887–901.

Buechner, M., Hall, D. H., Bhatt, H., and Hedgecock, E. M. (1999). Cystic canal mutants in Caenorhabditis elegans are defective in the apical membrane domain of the renal (excretory) cell. *Dev. Biol.* **214,** 227–241.

Buechner, M. (2002). Tubes and the single *C. elegans* excretory cell. *Trends Cell Biol.* **12,** 479–484.
Cardoso, W. V., and Lu, J. (2006). Regulation of early lung morphogenesis: Questions, facts and controversies. *Development* **133,** 1611–1624.
Carmeliet, P. (2005). Angiogenesis in life, disease and medicine. *Nature* **438,** 932–936.
Caussinus, E., *et al.* (2008). Tip-cell migration controls stalk-cell intercalation during *Drosophila* tracheal tube elongation. *Curr. Biol.* **18,** 1727–1734.
Chung, S., and Andrew, D. J. (2008). The formation of epithelial tubes. *J. Cell Sci.* **121,** 3501–3504.
Ciruna, B., *et al.* (2006). Planar cell polarity signalling couples cell division and morphogenesis during neurulation. *Nature* **439,** 220–224.
Colas, J. F., and Schoenwolf, G. C. (2001). Towards a cellular and molecular understanding of neurulation. *Dev. Dyn.* **221,** 117–145.
Ewald, A. J., *et al.* (2008). Collective epithelial migration and cell rearrangements drive mammary branching morphogenesis. *Dev. Cell* **14,** 570–581.
Fujita, M., *et al.* (2003). The role of the ELAV homologue EXC-7 in the development of the *Caenorhabditis elegans* excretory canals. *Dev. Biol.* **256,** 290–301.
Gao, J., *et al.* (2001). The *Caenorhabditis elegans* homolog of FGD1, the human Cdc42 GEF gene responsible for faciogenital dysplasia, is critical for excretory cell morphogenesis. *Hum. Mol. Genet.* **10,** 3049–3062.
Gassama-Diagne, A., *et al.* (2006). Phosphatidylinositol-3,4,5-trisphosphate regulates the formation of the basolateral plasma membrane in epithelial cells. *Nat. Cell Biol.* **8,** 963–970.
Geldmacher-Voss, B., *et al.* (2003). A 90-degree rotation of the mitotic spindle changes the orientation of mitoses of zebrafish neuroepithelial cells. *Development* **130,** 3767–3780.
Gobel, V., *et al.* (2004). Lumen morphogenesis in *C. elegans* requires the membrane–cytoskeleton linker erm-1. *Dev. Cell* **6,** 865–873.
Guillemin, K., *et al.* (1996). The pruned gene encodes the *Drosophila* serum response factor and regulates cytoplasmic outgrowth during terminal branching of the tracheal system. *Development* **122,** 1353–1362.
Haag, T. A., *et al.* (1999). The role of cell adhesion molecules in *Drosophila* heart morphogenesis: Faint sausage, shotgun/DE-cadherin, and laminin A are required for discrete stages in heart development. *Dev. Biol.* **208,** 56–69.
Haigo, S. L., *et al.* (2003). Shroom induces apical constriction and is required for hingepoint formation during neural tube closure. *Curr. Biol.* **13,** 2125–2137.
Hardin, J. (2008). To thine own self be true: Self-fusion in single-celled tubes. *Dev. Cell* **14,** 465–466.
Helenius, I. T., and Beitel, G. J. (2008). The first "Slit" is the deepest: The secret to a hollow heart. *J. Cell Biol.* **182,** 221–223.
Hogan, B. L., and Kolodziej, P. A. (2002). Organogenesis: Molecular mechanisms of tubulogenesis. *Nat. Rev. Genet.* **3,** 513–523.
Horne-Badovinac, S., *et al.* (2001). Positional cloning of heart and soul reveals multiple roles for PKC lambda in zebrafish organogenesis. *Curr. Biol.* **11,** 1492–1502.
Horowitz, A., and Simons, M. (2008). Branching morphogenesis. *Circ. Res.* **103,** 784–795.
Huang, D. C., and Strasser, A. (2000). BH3-Only proteins-essential initiators of apoptotic cell death. *Cell* **103,** 839–842.
Jayaram, S. A., *et al.* (2008). COPI vesicle transport is a common requirement for tube expansion in *Drosophila*. *PLoS ONE* **3,** e1964.
Jiang, L., *et al.* (2007). The *Drosophila* Dead end Arf-like3 GTPase controls vesicle trafficking during tracheal fusion cell morphogenesis. *Dev. Biol.* **311,** 487–499.
Jones, S. J., and Baillie, D. L. (1995). Characterization of the let-653 gene in *Caenorhabditis elegans*. *Mol. Gen. Genet.* **248,** 719–726.

Kakihara, K., et al. (2008). Conversion of plasma membrane topology during epithelial tube connection requires Arf-like 3 small GTPase in *Drosophila*. *Mech. Dev.* **125,** 325–336.

Kamei, M., et al. (2006). Endothelial tubes assemble from intracellular vacuoles *in vivo*. *Nature* **442,** 453–456.

Kerman, B. E., et al. (2006). From fate to function: The *Drosophila* trachea and salivary gland as models for tubulogenesis. *Differentiation* **74,** 326–348.

Kerman, B. E., et al. (2008). Ribbon modulates apical membrane during tube elongation through Crumbs and Moesin. *Dev. Biol.* **320,** 278–288.

Lee, S., and Kolodziej, P. A. (2002). The plakin Short Stop and the RhoA GTPase are required for E-cadherin-dependent apical surface remodeling during tracheal tube fusion. *Development* **129,** 1509–1520.

Lee, M., et al. (2003). Distinct sites in E-cadherin regulate different steps in *Drosophila* tracheal tube fusion. *Development* **130,** 5989–5999.

Levi, B. P., et al. (2006). *Drosophila* talin and integrin genes are required for maintenance of tracheal terminal branches and luminal organization. *Development* **133,** 2383–2393.

Liegeois, S., et al. (2007). Genes required for osmoregulation and apical secretion in *Caenorhabditis elegans*. *Genetics* **175,** 709–724.

Liu, K. D., et al. (2007). Rac1 is required for reorientation of polarity and lumen formation through a PI 3-kinase-dependent pathway. *Am. J. Physiol. Renal Physiol.* **293,** F1633–F1640.

Lowery, L. A., and Sive, H. (2005). Initial formation of zebrafish brain ventricles occurs independently of circulation and requires the nagie oko and snakehead/atp1a1a.1 gene products. *Development* **132,** 2057–2067.

Lu, P., and Werb, Z. (2008). Patterning mechanisms of branched organs. *Science* **322,** 1506–1509.

Lu, P., et al. (2006). Comparative mechanisms of branching morphogenesis in diverse systems. *J. Mammary Gland Biol. Neoplasia* **11,** 213–228.

Lubarsky, B., and Krasnow, M. A. (2003). Tube morphogenesis: Making and shaping biological tubes. *Cell* **112,** 19–28.

Mailleux, A. A., et al. (2007). BIM regulates apoptosis during mammary ductal morphogenesis, and its absence reveals alternative cell death mechanisms. *Dev. Cell* **12,** 221–234.

Martin, A. C., et al. (2009). Pulsed contractions of an actin–myosin network drive apical constriction. *Nature* **457,** 495–499.

Martin-Belmonte, F., and Mostov, K. (2007). Phosphoinositides control epithelial development. *Cell Cycle* **6,** 1957–1961.

Martin-Belmonte, F., et al. (2007). PTEN-mediated apical segregation of phosphoinositides controls epithelial morphogenesis through Cdc42. *Cell* **128,** 383–397.

Martin-Belmonte, F., et al. (2008). Cell-polarity dynamics controls the mechanism of lumen formation in epithelial morphogenesis. *Curr. Biol.* **18,** 507–513.

McKeown, C., et al. (1998). sma-1 encodes a betaH-spectrin homolog required for *Caenorhabditis elegans* morphogenesis. *Development* **125,** 2087–2098.

Medioni, C., et al. (2008). Genetic control of cell morphogenesis during *Drosophila melanogaster* cardiac tube formation. *J. Cell Biol.* **182,** 249–261.

Montesano, R., et al. (1991). Induction of epithelial tubular morphogenesis *in vitro* by fibroblast-derived soluble factors. *Cell* **66,** 697–711.

Munson, C., et al. (2008). Regulation of neurocoel morphogenesis by Pard6 gamma b. *Dev. Biol.* **324,** 41–54.

Myat, M. M., and Andrew, D. J. (2002). Epithelial tube morphology is determined by the polarized growth and delivery of apical membrane. *Cell* **111,** 879–891.

Nelson, F. K., and Riddle, D. L. (1984). Functional study of the *Caenorhabditis elegans* secretory-excretory system using laser microsurgery. *J. Exp. Zool.* **231,** 45–56.

Nishimura, T., and Takeichi, M. (2008). Shroom3-mediated recruitment of Rho kinases to the apical cell junctions regulates epithelial and neuroepithelial planar remodeling. *Development* **135,** 1493–1502.

Nishimura, M., et al. (2007). A wave of EGFR signaling determines cell alignment and intercalation in the *Drosophila* tracheal placode. *Development* **134,** 4273–4282.

O'Brien, L. E., et al. (2001). Rac1 orientates epithelial apical polarity through effects on basolateral laminin assembly. *Nat. Cell Biol.* **3,** 831–838.

Paul, S. M., et al. (2007). A pump-independent function of the Na, K-ATPase is required for epithelial junction function and tracheal tube-size control. *Development* **134,** 147–155.

Praitis, V., et al. (2005). SMA-1 spectrin has essential roles in epithelial cell sheet morphogenesis in *C. elegans*. *Dev. Biol.* **283,** 157–170.

Rasmussen, J. P., et al. (2008). Notch signaling and morphogenesis of single-cell tubes in the *C. elegans* digestive tract. *Dev. Cell* **14,** 559–569.

Ribeiro, C., et al. (2002). In vivo imaging reveals different cellular functions for FGF and Dpp signaling in tracheal branching morphogenesis. *Dev. Cell* **2,** 677–683.

Ribeiro, C., et al. (2004). Genetic control of cell intercalation during tracheal morphogenesis in *Drosophila*. *Curr. Biol.* **14,** 2197–2207.

Riento, K., and Ridley, A. J. (2003). Rocks: Multifunctional kinases in cell behaviour. *Nat. Rev. Mol. Cell Biol.* **4,** 446–456.

Samakovlis, C., et al. (1996a). Development of the *Drosophila* tracheal system occurs by a series of morphologically distinct but genetically coupled branching events. *Development* **122,** 1395–1407.

Samakovlis, C., et al. (1996b). Genetic control of epithelial tube fusion during *Drosophila* tracheal development. *Development* **122,** 3531–3536.

Santiago-Martinez, E., et al. (2008). Repulsion by Slit and Roundabout prevents Shotgun/E-cadherin-mediated cell adhesion during *Drosophila* heart tube lumen formation. *J. Cell Biol.* **182,** 241–248.

Shakya, R., et al. (2005). The role of GDNF/Ret signaling in ureteric bud cell fate and branching morphogenesis. *Dev. Cell* **8,** 65–74.

Stork, T., et al. (2008). Organization and function of the blood–brain barrier in *Drosophila*. *J. Neurosci.* **28,** 587–597.

Sutherland, D., et al. (1996). branchless encodes a *Drosophila* FGF homolog that controls tracheal cell migration and the pattern of branching. *Cell* **87,** 1091–1101.

Suzuki, N., et al. (2001). A putative GDP–GTP exchange factor is required for development of the excretory cell in *Caenorhabditis elegans*. *EMBO Rep.* **2,** 530–535.

Tanaka-Matakatsu, M., et al. (1996). Cadherin-mediated cell adhesion and cell motility in *Drosophila* trachea regulated by the transcription factor Escargot. *Development* **122,** 3697–3705.

Tawk, M., et al. (2007). A mirror-symmetric cell division that orchestrates neuroepithelial morphogenesis. *Nature* **446,** 797–800.

Tong, X., and Buechner, M. (2008). CRIP homologues maintain apical cytoskeleton to regulate tubule size in *C. elegans*. *Dev. Biol.* **317,** 225–233.

Tucker, A. S. (2007). Salivary gland development. *Semin. Cell Dev. Biol.* **18,** 237–244.

Uv, A., et al. (2003). *Drosophila* tracheal morphogenesis: Intricate cellular solutions to basic plumbing problems. *Trends Cell Biol.* **13,** 301–309.

Warburton, D., et al. (2005). Molecular mechanisms of early lung specification and branching morphogenesis. *Pediatr. Res.* **57,** 26R–37R.

Yu, W., et al. (2005). Beta1-integrin orients epithelial polarity via Rac1 and laminin. *Mol. Biol. Cell* **16,** 433–445.

CHAPTER SEVEN

# Convergence and Extension Movements During Vertebrate Gastrulation

Chunyue Yin,* Brian Ciruna,[†] and Lilianna Solnica-Krezel[‡]

## Contents

| | |
|---|---|
| 1. Introduction | 164 |
| 2. The Regional and Temporal Pattern of C&E Movements in the Mesoderm | 165 |
| 3. Diverse Cellular Behaviors Underlie the Regional Differences of C&E Movements | 166 |
| 3.1. Directed migration | 166 |
| 3.2. Mediolateral intercalation | 170 |
| 3.3. Radial intercalation | 171 |
| 3.4. Oriented cell division | 171 |
| 4. Molecular Regulation of C&E Movements | 172 |
| 4.1. Stat3 signaling | 172 |
| 4.2. Noncanonical Wnt/PCP pathway | 174 |
| 4.3. Bmp signaling | 177 |
| 4.4. G protein-coupled receptors | 179 |
| 4.5. Molecular evidence suggests uncoupling of convergence and extension | 180 |
| 4.6. Interplay between different signaling pathways during C&E in the mesoderm | 181 |
| 5. Cell Polarization During C&E Movements | 183 |
| 6. Conclusion | 185 |
| Acknowledgments | 186 |
| References | 186 |

* Department of Biochemistry and Biophysics, Program in Developmental Biology, Genetics, and Human Genetics, University of California at San Francisco, California, USA
[†] Program in Developmental and Stem Cell Biology, The Hospital for Sick Children, Toronto, Ontario, Canada, and the Department of Molecular Genetics, University of Toronto, Ontario, Canada
[‡] Department of Biological Sciences, Vanderbilt University, Nashville, Tennessee, USA

## Abstract

During vertebrate gastrulation, coordinated cell movements shape the basic body plan. Key components of gastrulation are convergence and extension (C&E) movements, which narrow and lengthen the embryonic tissues, respectively. The rates of C&E movements differ significantly according to the position and the stage of gastrulation. Here, we review the distinct cellular behaviors that define the spatial and temporal patterns of C&E movements, with the special emphasis on zebrafish. We also summarize the molecular regulation of these cellular behaviors and the interplay between different signaling pathways that drive C&E. Finally, to ensure efficient C&E movements, cells must achieve mediolaterally-elongated cell morphology and polarize motile protrusions. We discuss the recent discoveries on the molecular and cellular mechanisms by which the mediolateral cell polarity is established.

## 1. INTRODUCTION

The animal body plan is established during gastrulation through coordinated morphogenetic movements of individual cells. Vertebrate gastrulation employs four types of evolutionarily conserved cell movements to generate and shape the three germ layers: endoderm, mesoderm, and ectoderm. Epiboly movements spread and thin the embryonic tissues. Internalization movements, emboly, bring the presumptive mesoderm and endoderm cells beneath the future ectoderm via the blastopore, an opening in the blastula. Convergence and extension (C&E) movements simultaneously narrow the germ layers mediolaterally and elongate the embryo from head to tail (Keller *et al.*, 2003; Solnica-Krezel, 2005). Upon internalization of the mesodermal and endodermal progenitors, all three germ layers are engaged in C&E movements in spatial and temporal specific manners. From studies in different vertebrate organisms, we now understand that C&E movements can be achieved by several distinct types of cellular behaviors depending on the stage of gastrulation and the region within the embryo. Here, we focus on different combinations of cellular behaviors that underlie the spatial and temporal pattern of C&E movements in vertebrates, with special emphasis on zebrafish. First, we describe patterns of C&E movements in different germ layers, subsequently we discuss the underlying cell behaviors and their molecular underpinnings.

## 2. THE REGIONAL AND TEMPORAL PATTERN OF C&E MOVEMENTS IN THE MESODERM

During the first 3 hours of development, the zebrafish embryo undergoes rapid synchronous cell divisions to form a mound of blastomeres atop a large syncytial yolk cell (Kimmel et al., 1995). Maternally contributed transcripts and proteins govern development during these cleavage stages. Zygotic transcription starts at the midblastula transition (MBT), which occurs at the 512/1024-cell stage (3-h postfertilization—hpf) (Kane and Kimmel, 1993). By MBT, the blastoderm consists of two cell types: the superficial enveloping layer (EVL) and deep cells. The EVL serves as a protective surface for the deep cells, which will give rise to all embryonic tissues (Kimmel et al., 1990; Warga and Kimmel, 1990).

When emboly initiates at 5.8 hpf in zebrafish, the mesodermal and endodermal progenitor cells located at the margin of deep cells internalize via the blastopore to lie underneath the prospective neural ectoderm on the dorsal side and epidermis on the ventral side (Keller et al., 2008; Kimmel et al., 1994, 1995). The internalized mesendodermal progenitors initially migrate away from the blastopore toward the animal pole without dorsal convergence (Sepich et al., 2005). Starting from midgastrulation (7.5 hpf), the progenitors of all three germ layers undergo C&E movements (Concha and Adams, 1998; Pezeron et al., 2008; Sepich et al., 2005). Whereas the pattern of endoderm and ectoderm C&E is less understood, four distinct C&E movement domains have been recognized in the mesoderm along the dorsoventral dimension of the zebrafish gastrulae (Figs. 7.1 and 7.2) (Myers et al., 2002a,b). First, at the most ventral region, known as the "no convergence no extension zone," the mesodermal cells do not participate in C&E movements, but rather migrate along the yolk into the tailbud region (Myers et al., 2002a). Second, in the lateral domain, C&E movements are initially slow, and then accelerate as cells move closer to the dorsal midline (Jessen et al., 2002; Myers et al., 2002b). Constituting the third C&E domain, the medial part of the presomitic mesoderm located within six-cell diameters to the axial mesoderm exhibits modest C&E rates (Glickman et al., 2003; Yin et al., 2008). Finally in the most dorsal domain, as originally described in the frog *Xenopus laevis* (Keller and Tibbetts, 1989), the axial mesoderm shows the same convergence rate as the medial presomitic mesoderm, but three-fold higher extension rate compared to the medial presomitic mesoderm three-fold (Glickman et al., 2003; Keller and Tibbetts, 1989)

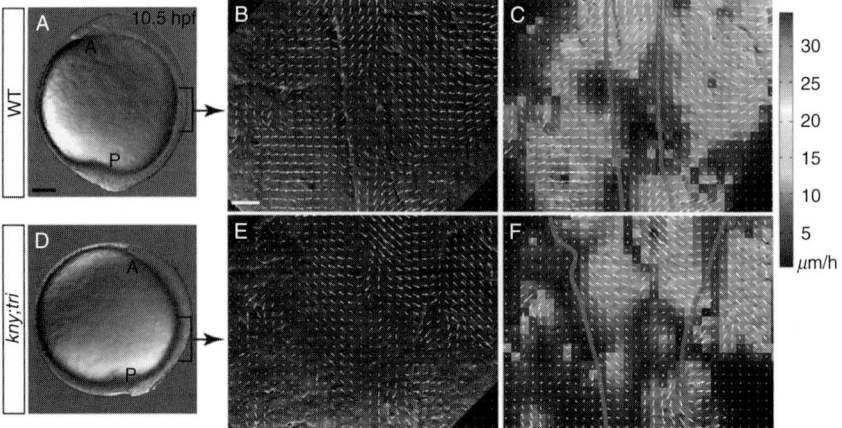

**Figure 7.1** Pattern of C&E movements in zebrafish gastrulae. (A, D) Lateral views of live zebrafish embryos at the one-somite stage (10.5 hpf). The regions shown in (B–F) are indicated by brackets. (B, E) Snapshots of particle image velocimetry analyses of C&E movements at the one-somite stage, showing the tissue displacements within a 9-min time window. Dorsal views, anterior to the top. The arrows indicate the direction of cell movements, whereas the length of the arrows represents the movement speed. (C, F) Same images as shown in (B) and (E) except that different movement speeds are presented in a color-coded manner. Purple lines delineate the axial mesoderm. (A–C) shows a wild-type (WT) embryo in which different regions exhibit distinct and organized C&E rates and patterns. (D–F) shows a *kny;tri* double mutant in which noncanonical Wnt/PCP signaling is severely perturbed. The stereotypic movement patterns seen in WT are largely lost in the mutant (Yin *et al.*, 2008). (See Color Insert.)

## 3. Diverse Cellular Behaviors Underlie the Regional Differences of C&E Movements

The differences in the magnitude of C&E movements in different regions and stages of gastrulation imply distinct underlying cellular mechanisms. Accordingly, current evidence indicates that different cellular behaviors or different combinations of cellular behaviors are employed in different territories of the gastrulae, leading to the spatial variations in the rate of C&E movements (Fig. 7.2).

### 3.1. Directed migration

One of the most basic cell behaviors that contribute to C&E movements is directed cell migration, whereby cells migrate either as individuals or in groups without significant neighbor exchanges (Fig. 7.2 and Table 7.1). Directed cell migration occurs in different germ layers of gastrulating embryos, as well as in various regions along the dorsoventral embryonic axis.

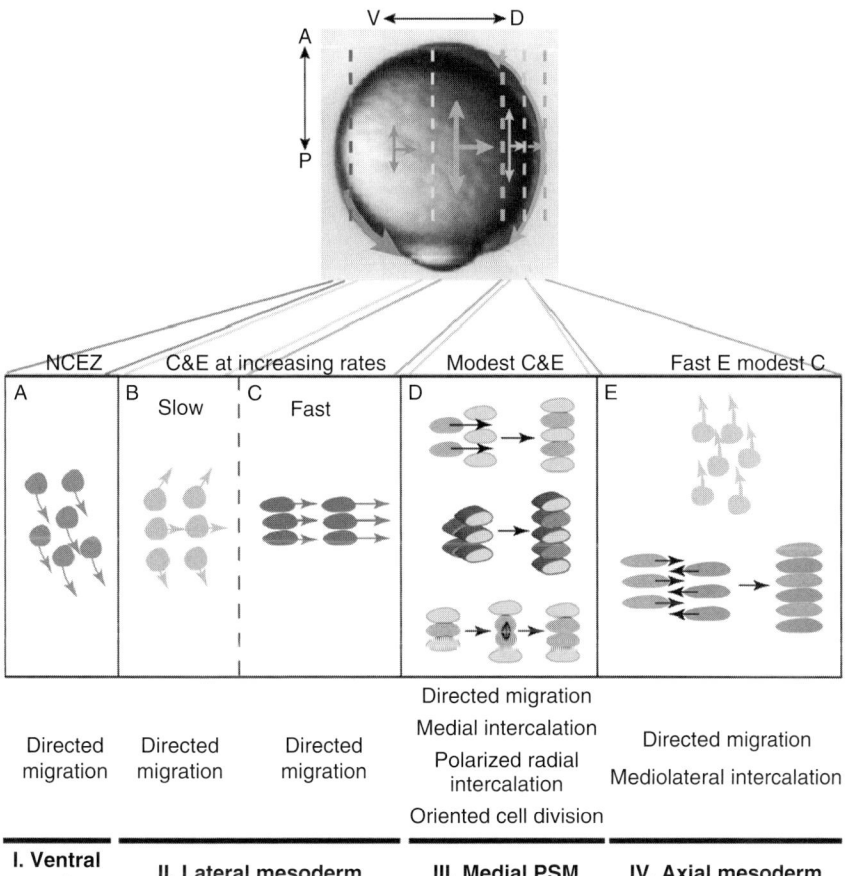

**Figure 7.2** Four distinct domains of C&E movements in the mesoderm of zebrafish gastrulae and the underlying cell movement behaviors. NCEZ, no convergence no extension zone; A, anterior; P, posterior; D, dorsal; V, ventral, PSM, presomitic mesoderm. (See Color Insert.)

In zebrafish, during the first half of gastrulation, the internalized endodermal cells engage in a nonoriented/noncoordinated random walk to rapidly disperse over the yolk surface. At midgastrulation, the endodermal cells change their behavior and migrate in a much more oriented fashion, as they move anteriorly toward the animal pole while converging dorsally toward the midline (Pezeron et al., 2008).

Mesodermal cells also undergo directed migration during C&E. In the axial region of the zebrafish gastrulae, the internalized mesodermal cells that give rise to the prechordal plate migrate as a cohesive group away from the blastopore and toward the animal pole, using the adjacent ectodermal layer

**Table 7.1** Molecular pathways regulating C&E and the cellular behaviors they control

| Cell behavior | Signaling pathway | References |
| --- | --- | --- |
| Slow dorsal-directed migration | Stat3 | Miyagi et al., (2004); Sepich et al., (2005) |
| Anterior-directed migration | Noncanonical Wnt/PCP; Edg5/Miles apart; Apelin; Stat3/Liv-1 | Heisenberg et al., (2000); Kai et al., (2008); Marlow et al., (1998); Ulrich et al., (2003); Yamashita et al., (2002, 2004); Zeng et al., (2007) |
| Fast dorsal-directed migration | Noncanonical Wnt/PCP; $G\alpha_{12/13}$; Stat3; Prostaglandin E2 (PGE2); Hyaluronan (HA) | Bakkers et al., (2004); Cha et al., (2006); Jessen et al., (2002); Lin et al., (2005); Miyagi et al., (2004); Myers et al., (2002b); Topczewski et al., (2001); Yamashita et al., (2002) |

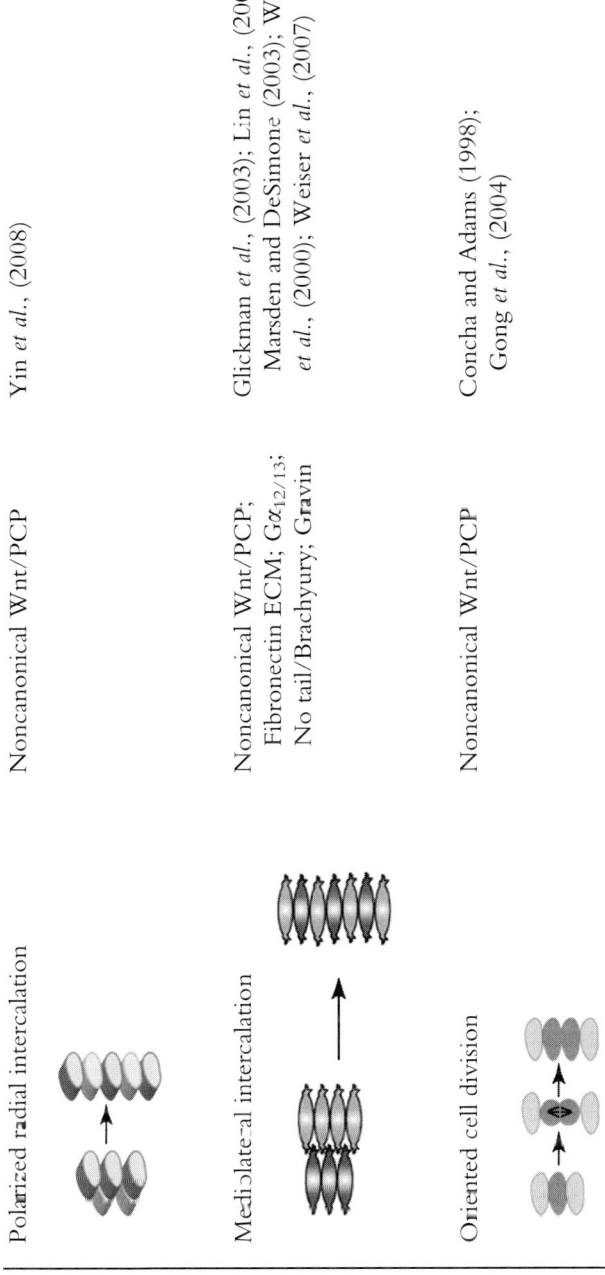

| | | |
|---|---|---|
| Polarized radial intercalation | Noncanonical Wnt/PCP | Yin et al., (2008) |
| Mediolateral intercalation | Noncanonical Wnt/PCP; Fibronectin ECM; $G\alpha_{12/13}$; No tail/Brachyury; Gravin | Glickman et al., (2003); Lin et al., (2005); Marsden and DeSimone (2003); Wallingford et al., (2000); Weiser et al., (2007) |
| Oriented cell division | Noncanonical Wnt/PCP | Concha and Adams (1998); Gong et al., (2004) |

as a substrate. This anterior migration contributes to the anterior extension of the axial mesendoderm and is evolutionarily conserved from fish, through frog and up to mouse (Heisenberg et al., 2000; Winklbauer and Nagel, 1991; Winklbauer and Selchow, 1992). The chordamesodermal precursors internalize after the prechordal mesendoderm. During early gastrulation, they move anteriorly away from the blastopore by directed migration, contributing to tissue extension (Gritsman et al., 2000). In the posterior midline of mouse and zebrafish embryos, the chordamesodermal precursor cells actively migrate posteriorly, thus contributing to the posterior extension of the axial mesoderm (Fig. 7.1) (Glickman et al., 2003; Yamanaka et al., 2007; Yin et al., 2008).

Dorsally directed cell migration is the main behavior contributing to the C&E of the lateral mesoderm during zebrafish gastrulation. However, the rates and directions of the lateral mesoderm cell migration differ significantly among different stages of gastrulation, as well as different gastrula regions (Fig. 7.2). Convergence of the lateral mesoderm initiates at midgastrulation, whereby cells migrate dorsally along complex trajectories with their direction biased either animally or vegetally according to their positions along the animal–vegetal axis. This fanning of complex cell trajectories leads to slow C&E movements (Jessen et al., 2002; Sepich et al., 2005). During late gastrulation, when cells reach more dorsal locations, they become densely packed, adopt a mediolaterally elongated cell morphology and converge as a cohort toward the midline along straighter trajectories and at higher speeds (Figs. 7.1 and 7.2) (Myers et al., 2002a,b).

## 3.2. Mediolateral intercalation

While migrating dorsally, cells may intercalate between one another in a polarized fashion, leading to tissue C&E. Mediolateral intercalation is the main driving force of convergent extension of the trunk axial mesoderm in many vertebrate species, including zebrafish (Glickman et al., 2003; Kimmel et al., 1994; Warga and Kimmel, 1990), amniotes (Schoenwolf and Alvarez, 1992), mouse (Yamanaka et al., 2007), and Xenopus as shown by pioneering studies of Keller and colleagues (Shih and Keller, 1992). During mediolateral intercalation, cells elongate and develop protrusive activity in the medial and lateral directions. Meanwhile, they wedge themselves between their immediate medial and lateral neighbors, resulting in rapid mediolateral narrowing and anteroposterior lengthening of the tissue (Fig. 7.2 and Table 7.1). In frog and fish, medial or lateral intercalation also contributes to C&E of neuroepithelial progenitor cells/dorsal ectoderm during gastrulation and ensures effective C&E of the neural plate prior to the onset of neurulation (Concha and Adams, 1998; Keller et al., 2000; Kimmel et al., 1994).

## 3.3. Radial intercalation

Whereas both directed cell migration and mediolateral intercalation occur within a plane of a single cell layer (planar) (Keller et al., 2000; Myers et al., 2002a), radial intercalations that occur between different cell layers can also contribute to C&E (Yin et al., 2008). In the process of radial intercalation, cells move between the cells in a deeper or more superficial layer. Unpolarized radial intercalation generates a thinner tissue with a greater area, and drives the epiboly movements that spread and thin the blastoderm in an isotropic fashion in frog and zebrafish gastrulae (Keller et al., 2003; Warga and Kimmel, 1990). However, studies in X. laevis reported that in the first half of gastrulation, deep mesenchymal cells in the dorsal mesoderm and the prospective posterior neural tissues undergo radial intercalations to extend the tissue (Keller et al., 2003; Wilson and Keller, 1991). Keller and colleagues thus hypothesized that to generate such an anisotropic expansion, radial intercalation has to occur in a polarized fashion, whereby cells intercalate primarily along one axis (Keller and Tibbetts, 1989; Wilson and Keller, 1991). Providing experimental support for this notion, in the medial presomitic mesoderm of the zebrafish gastrulae, cells undergo radial intercalations that preferentially separate anterior and posterior neighboring cells. Such polarized radial intercalation leads to anisotropic extension of the tissue and thus contributes to the anteroposterior extension of the nascent embryonic axis (Yin et al., 2008).

## 3.4. Oriented cell division

Besides directed cell migration and cell intercalation, oriented cell division also leads to tissue elongation in vertebrates (Fig. 7.2 and Table 7.1). In the notochord and neural plate of avian and mouse embryos (Schoenwolf and Alvarez, 1989, 1992; Schoenwolf and Yuan, 1995), and in the extending primitive streak of the avian gastrulae (Wei and Mikawa, 2000), the orientation of mitotic spindle is biased along the anteroposterior axis. In zebrafish, oriented cell division has also been observed during the C&E movements of the neuroepithelial progenitor cells (Concha and Adams, 1998; Gong et al., 2004). Cell lineage analyses performed during late gastrulation through early neurulation stages revealed that cell divisions in the dorsal ectoderm are oriented along the anteroposterior axis. Moreover, this non-random anteroposterior orientation of cell division occurs despite clear mediolateral polarized cell movement and elongation during interphase (Concha and Adams, 1998; Kimmel et al., 1994). These results suggest that neither the net direction of cell movements nor the direction of cell elongation is the direct determinant of cleavage orientation (Concha and Adams, 1998), and that an active orientation of the mitotic spindle contributes to elongation of the zebrafish axis. Indeed, randomization of the cell division plane impairs anteroposterior extension of the ectoderm (Gong *et al.*, 2004). Our preliminary studies indicate that at the end of gastrulation, cell divisions in the medial presomitic mesoderm are also oriented anteroposteriorly (C. Yin, D. Sepich, and L. Solnica-Krezel, unpublished data). Therefore, oriented cell division is employed in the C&E movements of various tissues.

In summary, migrating mesodermal cells in different territories of the zebrafish gastrulae are engaged in different cellular behaviors, accounting for the distinct rates of C&E movements in each domain (Fig. 7.2). First, at the most ventral region, so called "no convergence no extension zone," cells migrate toward the vegetal pole, without contributing to C&E. Second, in the lateral domain, cell populations converge dorsally and extend by directed migration at increasing speed as they move closer to the nascent axial mesoderm. Third, in the medial presomitic mesoderm adjacent to the axial mesoderm, cells participate in multiple motile behaviors, including directed migration, medially directed intercalation, polarized radial intercalation, and oriented cell division. This combination of cellular behaviors results in modest C&E of this tissue. Finally, at the most dorsal region, cells undergo mediolateral intercalation to achieve strong extension and modest convergence of the axial mesoderm.

## 4. Molecular Regulation of C&E Movements

The dynamic spatial and temporal pattern of C&E gastrulation movements and the involvement of distinct cell behaviors reflect the complexity of underlying molecular mechanisms. Indeed, numerous pathways and molecules have been shown either to control the directionality of cell movements or to regulate the distinct cell movement behaviors within the different C&E domains (Fig. 7.3 and Table 7.1), and this list is likely to grow.

### 4.1. Stat3 signaling

In the zebrafish gastrulae, mesodermal cells initially travel on meandering but dorsally oriented paths, which for cells in more animal locations are biased animally and for cells closer to the vegetal pole are biased vegetally. This fanning of cell trajectories culminates in slow C&E movements, and is reminiscent of active cell migration along a chemoattractant gradient (Jessen *et al.*, 2002; Myers *et al.*, 2002b; Sepich *et al.*, 2005). Mathematical modeling predicted that a minimum of two chemoattractant cues, one at the prechordal mesoderm and a second one located more vegetally in the dorsal midline,

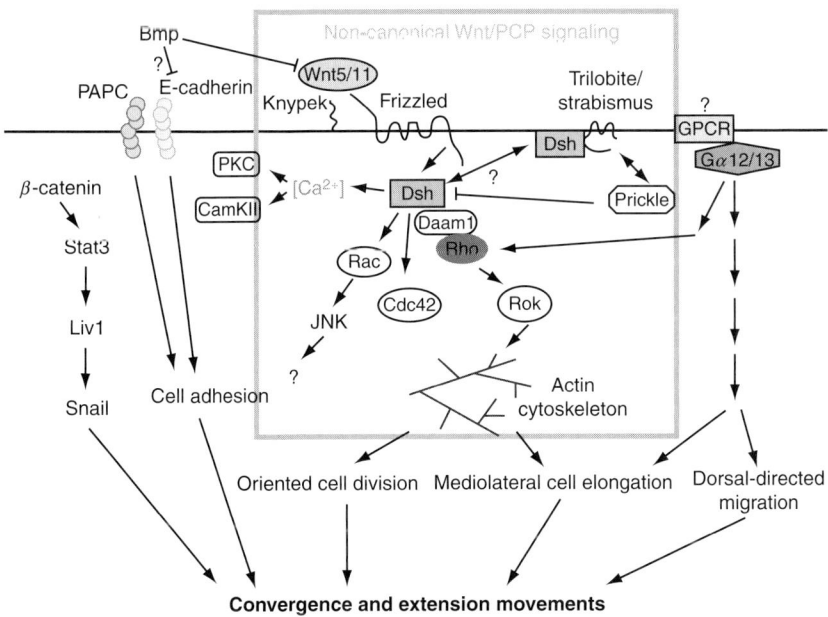

**Figure 7.3** Molecular regulation of C&E movements. (See Color Insert.)

are sufficient to orient the cell movement paths toward dorsal, as well as explain the fanning of the cell migration trajectories toward the animal and vegetal pole (Sepich et al., 2005). Work in chick embryos demonstrated that different FGFs produced in the primitive streak (blastopore) and the embryo midline serve as chemorepellents or chemoattractants to regulate the movements of the internalized mesoderm away from the blastopore or later its convergence toward the midline, respectively (Yang et al., 2002). Similar roles of FGF signaling during gastrulation have not been reported in zebrafish or mammals. Instead, the guiding cues in the zebrafish gastrulae are likely to be controlled by JAK/Stat3 (Janus kinase/signal transducer and activator of transcription) signaling (Yamashita et al., 2002). Although Stat3 is expressed ubiquitously in the embryo, it is only activated by phosphorylation downstream of maternal $\beta$-catenin in the dorsal blastula region shortly after MBT (Yamashita et al., 2002). Stat3 function is required in the prechordal and anterior chordamesoderm, where it acts in cell nonautonomous fashion to provide attractive cues for dorsal convergence of the lateral mesodermal cells (Miyagi et al., 2004; Yamashita et al., 2002). In Stat3-depleted embryos, the initiation of dorsal convergence in the lateral mesoderm is delayed (Sepich et al., 2005), and the directed cell migration is reduced (Yamashita et al., 2002). Such defects can be suppressed by restoring Stat3 function in the axial mesoderm only (Miyagi et al., 2004; Yamashita et al., 2002).

In addition to its role in dorsal convergence of the lateral mesodermal cells, Stat3 activity is also required cell-autonomously for the anterior migration of the prechordal mesendodermal cells (Yamashita et al., 2002). One intriguing downstream target of Stat3 signaling in the prechordal mesendoderm is LIV1, a breast cancer-associated zinc transporter protein (Yamashita et al., 2004). LIV1 is essential for the nuclear localization of the zinc-finger transcription factor Snail, a master regulator of epithelial to mesenchymal transition and cell motility, in part via its repression of E-cadherin expression (Blanco et al., 2007). Thus Stat3 signaling likely regulates the anterior migration of the prechordal mesendodermal cells by mediating their motile properties.

## 4.2. Noncanonical Wnt/PCP pathway

In vertebrates, noncanonical Wnt signaling, the vertebrate equivalent of the *Drosophila melanogaster* planar cell polarity (PCP) pathway that polarizes cells within the plane of epithelium (Klein and Mlodzik, 2005), is the key regulator of C&E gastrulation movements. Both gain- and loss-of-function manipulations of noncanonical Wnt/PCP components impair C&E without affecting cell fate (Carreira-Barbosa et al., 2003; Heisenberg et al., 2000; Jessen et al., 2002; Kilian et al., 2003; Sokol, 1996; Tada and Smith, 2000; Topczewski et al., 2001; Wallingford et al., 2000), indicating that a certain level of noncanonical Wnt signaling is essential for normal C&E movements. Noncanonical Wnt/PCP signaling mediates mediolateral elongation and alignment of the mesodermal and ectodermal cells, as well as the stability and polarized orientation of the cellular protrusions of these cells (Jessen et al., 2002; Keller, 2005; Myers et al., 2002b; Tada et al., 2002; Topczewski et al., 2001; Wallingford et al., 2000). Zebrafish embryos deficient in noncanonical Wnt/PCP signaling exhibit anteroposteriorly shortened and mediolaterally broadened bodies. The stereotyped C&E movement pattern seen in wild-type embryos is impaired or lost (Fig. 7.1D–F). In these embryos, multiple cellular behaviors, including the asymmetric division of ectodermal cells, the anterior migration of prechordal mesodermal cells, the dorsal-directed migration of lateral mesodermal cells, the polarized radial intercalation of presomitic cells, and the mediolateral intercalation of axial mesodermal cells, are all impaired due to defects in the mediolateral PCP (Table 7.1) (Gong et al., 2004; Heisenberg et al., 2000; Jessen et al., 2002; Topczewski et al., 2001; Ulrich et al., 2003; Wallingford et al., 2000; Yin et al., 2008).

In vertebrates, noncanonical Wnt/PCP signaling is activated upon binding of Wnt5 and Wnt11 ligands to the transmembrane receptor Frizzled7 (Fz7) (Djiane et al., 2000; Kilian et al., 2003), which requires the GPI-anchored extracellular heparan sulfate proteoglycan Knypek/Glypican4 (Kny) (Fig. 7.3) (Topczewski et al., 2001). Activated Fz7 recruits the docking protein Dishevelled (Dsh) to the cell membrane and stimulates downstream signaling events

(Axelrod et al., 1998; Heisenberg et al., 2000; Wallingford et al., 2000). The translocation of Dsh to the cell membrane requires the function of several regulators of the pathway. In Xenopus, the cell-surface transmembrane heparan sulfate proteoglycan xSyndecan-4 interacts functionally and biochemically with Fz7 and Dsh (Munoz and Larrain, 2006). It is necessary and sufficient for translocation of Dsh to the membrane. Also in Xenopus, protein kinase C δ (PKCδ) is recruited to the plasma membrane in response to Fz activation and forms a complex with Dsh (Kinoshita et al., 2003). Loss of PKCδ function inhibits translocation of Dsh and activation of downstream components.

Transduction of noncanonical Wnt/PCP signaling relies on the activities of cofactors that are highly conserved between fruit fly and vertebrates, including the transmembrane PDZ domain-binding protein Trilobite/Strabismus/Van Gogh-like 2 (Tri/Stbm/Vangl2) (Jessen et al., 2002; Park and Moon, 2002), the cytoplasmic protein Prickle (Pk) (Carreira-Barbosa et al., 2003; Veeman et al., 2003b), the ankyrin-repeat protein Diego (Dg)/Diversin (Moeller et al., 2006), and the serpentine cadherin domain-containing receptor Flamingo (Fmi) (Formstone and Mason, 2005a). In both fly and vertebrates, Stbm/Vangl2 forms a complex with Pk at the cell membrane (Jenny et al., 2003) (Fig. 7.4A). Pk in turn inhibits Fz-mediated Dsh localization to the membrane (Carreira-Barbosa et al., 2003; Tree et al., 2002). In Drosophila, the ankyrin-repeat protein Diego stimulates PCP signaling and prevents Pk from binding to Dsh (Das et al., 2004; Jenny et al., 2005). Consistently, Diversin, the vertebrate ortholog of Diego, also promotes noncanonical Wnt signaling in zebrafish (Moeller et al., 2006). The transmembrane protein Flamingo localizes to the proximal and distal sides of the fly wing epithelial cells and functions in both promoting and inhibiting Fz and Dsh activity (Usui et al., 1999). In vertebrates, impairing Flamingo function results in defective anterior migration of the prechordal mesendodermal cells in zebrafish and disrupts neural tube closure in mice and chick (Curtin et al., 2003; Formstone and Mason, 2005a,b).

Downstream of Dsh, the noncanonical Wnt/PCP pathway signals through small guanosine triphosphatases (GTPases), including Cdc42, Rac, and Rho, to regulate cytoskeleton rearrangements (Habas et al., 2001, 2003; Marlow et al., 2002; Winter et al., 2001; Zhu et al., 2006). Studies in cell culture established that each small GTPase has a distinct role in regulating cytoskeleton, cell polarity, and protrusive activities. Cdc42 mediates cell polarity and formation of filopodia. Rac is essential for lamellipodia formation, whereas Rho promotes formation of stress fibers and focal adhesions (Hall and Nobes, 2000). Noncanonical Wnt signaling activates these small GTPases through independent and parallel pathways (Veeman et al., 2003a; Wallingford and Habas, 2005). Downstream of Dsh, the Formin homology protein Daam1 binds to the PDZ domain of Dsh and activates RhoA (Habas et al., 2001). RhoA in turn activates the

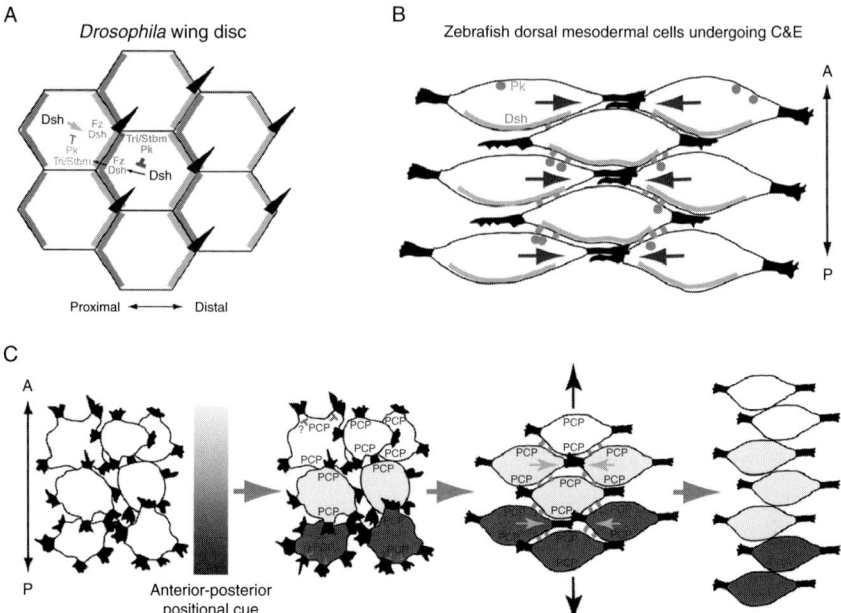

**Figure 7.4** Noncanonical Wnt signaling regulates the polarized orientation of cell intercalations during C&E. (A) Cellular interactions between noncanonical Wnt/PCP components that establish the cell polarity in the fly wing epithelium. (B) In zebrafish, the cells in the axial and medial presomitic mesoderm exhibit asymmetric localization of noncanonical Wnt components during C&E, with Pk localized to anterior edge and Dsh localized to the posterior edge. The subcellular localization of PK and Dsh in zebrafish is reminiscent of the Pk and Dsh localization in the fly wing disk (A). (C) A model of anteroposterior (AP) polarity cues activating noncanonical Wnt signaling to bias the orientation of cell intercalations. In zebrafish gastrulae, mesodermal cells initially exhibit round morphology and send out protrusions in random directions. Upon unknown anteroposterior polarity cues, noncanonical Wnt/PCP signaling is activated at the anterior and posterior sides of the cell. It may function to constrain the lamellipodia protrusions to the medial and lateral ends of the cell or establish the differential cell–cell adhesions in this region. Polarized cellular protrusions and differential cell adhesions subsequently define the biased orientation of cell intercalations. A, anterior; P, posterior.

Rho-associated kinase, Rok (Habas *et al.*, 2001; Katoh *et al.*, 2001; Marlow *et al.*, 2002; Winter *et al.*, 2001). Overexpression of either RhoA or Rok in zebrafish partially suppresses C&E defects in embryos lacking Wnt11 or Wnt5 function (Marlow *et al.*, 2002; Zhu *et al.*, 2006), arguing that they are downstream components of the pathway. Rac, on the other hand, acts in parallel to RhoA through interaction with the DEP domain of Dsh independent of Daam1 (Fanto *et al.*, 2000; Habas *et al.*, 2001, 2003). Finally, Cdc42 is activated by Wnt11/Fz through G protein and protein kinase C (Penzo-Mendez *et al.*, 2003). Other studies have placed Cdc42 downstream of the Wnt/$Ca^{2+}$ signaling that affects C&E movements and tissue separation during gastrulation (Choi and Han, 2002). Delineation of specific pathways downstream of noncanonical Wnt signaling that determine specific motile cell properties should be an important future goal.

How the noncanonical Wnt/PCP signaling promotes C&E remains elusive. The existing data argue for a permissive rather than an instructive role for this pathway in C&E. In zebrafish *kny* and *tri* mutants, dorsal convergence of the lateral mesoderm initiates normally, arguing that the dorsal chemoattractant cues guiding the direction of C&E is independent of the noncanonical Wnt signaling (Sepich *et al.*, 2005). Furthermore, ubiquitous ectopic expression of the noncanonical Wnt signaling components can rescue the C&E phenotype of zebrafish *wnt11* mutants (Heisenberg *et al.*, 2000; Kilian *et al.*, 2003; Marlow *et al.*, 2002; Topczewski *et al.*, 2001). This supports the notion that instead of providing a localized polarization cue, noncanonical Wnt signaling is required for the mesodermal cells to interpret an independent polarizing signal and acquire several properties necessary for effective C&E movements (Sepich *et al.*, 2005; Solnica-Krezel, 2006). However, the role for the Wnt signals in regulating C&E behaviors may be context dependent, since ubiquitous ectopic Wnt expression cannot rescue the C&E defects associated with zebrafish *wnt5* mutant embryos (Heisenberg *et al.*, 2000, Kilian *et al.*, 2003).

Consistent with the notion that noncanonical Wnt signaling is required for cells to interpret and respond to polarizing signals, it has been shown that this pathway modulates cell–cell adhesion and extracellular matrix assembly (Ulrich *et al.*, 2005; Ungar *et al.*, 1995; Witzel *et al.*, 2006). In zebrafish, Wnt11 controls cohesion of the prechordal plate progenitors required for directed and coherent cell migration. It mediates E-cadherin endocytosis and recycling through a Rab5c-dependent pathway (Ulrich *et al.*, 2005). In *Xenopus* gastrulae, noncanonical Wnt/PCP signaling is required for the assembly of polarized extracellular matrix on the inner and outer surfaces of the intercalating mesodermal tissues essential for C&E (Goto *et al.*, 2005). Perturbing the expression of Wnt/PCP signaling components Stbm/Vangl2, Fz, and Pk affects polarized Fibronectin fibril assembly and, in the case of Fz or Stbm/Vangl2, the ability of the mesodermal cells to move in a polarized way on Fibronectin substrates (Goto *et al.*, 2005). Similarly, in zebrafish, Kny and Tri/Vangl2 function are required for polarized Fibronectin deposition along the notochord surface. In *kny;tri* double mutants, Fibronectin is mislocalized inside the notochord and the polarity of the notochord cells is severely perturbed (Yin and Solnica-Krezel, 2007).

## 4.3. Bmp signaling

In vertebrate gastrulae, Bone morphogenetic proteins (Bmps) form a ventral-to-dorsal gradient to specify cell fates in all germ layers (De Robertis and Kurado, 2004). Interestingly, Bmp also serves as a negative regulator of

C&E movements in zebrafish (Myers *et al.*, 2002a). Inactivation of Bmp ligands, their receptors, as well as other transducers of the signaling cascade results in dorsalized embryos, which exhibit an elongated morphology at the end of gastrulation and later lose ventral tail structures or the entire tail (Hammerschmidt *et al.*, 1996a,b; Mullins *et al.*, 1996; Myers *et al.*, 2002a; Solnica-Krezel *et al.*, 1996). These defects result from reduced Bmp activity and consequent excess and ectopic mediolateral intercalation of cells in the "no convergence no extension zone," at the expense of vegetal migration to the tailbud (Myers *et al.*, 2002a). Conversely, ventralized mutants that are defective in the function of Bmp antagonists, such as Chordin, exhibit a rounder morphology at the end of gastrulation (Hammerschmidt and Mullins, 2002; Hammerschmidt *et al.*, 1996b; Solnica-Krezel *et al.*, 1996), due to decreased C&E movements (Myers *et al.*, 2002a). Reduction of C&E movements in ventralized mutants is correlated with dorsally expanded Bmp activity (Myers *et al.*, 2002a). These findings have led to a model in which a single ventral–dorsal gradient of Bmp activity specifies cell fates and C&E movements, thus coordinating cell fate specification and movement during zebrafish gastrulation. High Bmp activity levels in the ventral region inhibit C&E movements, specifying the "no convergence no extension zone." In the lateral domain, reduction of Bmp activity is correlated with an increase in the rates of C&E driven by directed migration. The low Bmp activity near the dorsal region promotes strong extension with modest convergence driven by mediolateral intercalation (Myers *et al.*, 2002a).

Given that Bmp signaling also regulates dorsoventral patterning, it is important to dissociate the effect of Bmp on cell movements from that on cell fate specification. Several lines of evidence from studies in zebrafish support a proposal that Bmp regulates C&E movements and cell fates via parallel rather than linear pathways. First, Bmp inhibits C&E by negatively regulating the expression of noncanonical Wnt/PCP pathway components, thus mediating cell movements independent of cell fate specification (Myers *et al.*, 2002a). Second, the threshold of Bmp activity that regulates the expression of Wnt5/*ppt* and Wnt11/*slb* in the paraxial mesoderm is different from that used to regulate the expression of paraxial mesodermal specification markers (Myers *et al.*, 2002a). Third, Bmp signaling impacts C&E cell behaviors during midgastrulation stages, whereas it specifies the fate of these cells only at late gastrulation (Tucker *et al.*, 2008). Finally, the ventral–dorsal Bmp gradient has recently been shown to determine the direction of lateral mesodermal cell migration by limiting $Ca^{2+}$/cadherin-dependent cell–cell adhesiveness, independent of its role in dorsoventral patterning (von der Hardt *et al.*, 2007). In this study, the authors proposed that BMP inhibits stability of lamellipodial contacts that form between migrating mesodermal cells and are required for cell displacement during lamellipodial retraction. In the lateral region, decreasing Bmp activity leads to increasing cell–cell adhesiveness, making cells move into the dorsal regions, where adhesiveness is highest (von der Hardt et al., 2007). It still remains unclear whether Bmp signaling has a similar function in regulating the C&E movements in other vertebrate species.

## 4.4. G protein-coupled receptors

G protein-coupled receptors (GPCRs) regulate chemotaxis in *Dictyostelium discoideum* and in the mammalian immune system (Devreotes and Janetopoulos, 2003). Accumulating evidence suggests that GPCRs also control cell movements during vertebrate gastrulation. In the zebrafish gastrulae, Prostaglandin E2 (PGE2) signaling through the G protein-coupled PGE2 receptor (EP4) promotes the speed of lateral mesodermal cells during dorsally directed migration (Cha et al., 2006). Heterotrimeric G proteins $G\alpha_{12/13}$, on the other hand, ensure efficient dorsal-directed migration independent of the noncanonical Wnt/PCP signaling (Lin et al., 2005). These vertebrate orthologs of Concertina, which regulates Drosophila gastrulation, also function cell-autonomously to mediate mediolateral cell elongation essential for cell intercalation during notochord C&E (Table 7.1 and Fig. 7.3).

Interestingly, some GPCRs appear to regulate the movements of specific cell populations. For instance, the zebrafish heart anlage arises during gastrulation when the prospective cardiac precursor cell populations converge toward the embryonic midline and extend rostrally to form bilateral heart fields at late gastrulation (Keegan et al., 2004). The chemokine Apelin and its receptor Agtrl1b mediate the C&E movements of the cardiac precursors (Scott et al., 2007; Zeng et al., 2007). During gastrulation, *agtrl1b* is expressed in the lateral plate mesoderm where the prospective cardiac precursors reside, whereas *apelin* RNA expression is confined to the midline (Scott et al., 2007; Zeng et al., 2007). Reduction or excess expression of Agtrl1b or Apelin impairs cardiac precursor migration toward the heart fields, leading to failure of heart formation. Notably, Apelin does not appear to simply act as a chemoattractant or chemorepellant. Whereas the cardiac precursors converge toward the *apelin*-expressing midline during gastrulation (Zeng et al., 2007), they stop short of the midline and form bilateral heart fields at early somitogenesis (Keegan et al., 2004). Therefore, it is tempting to speculate that Apelin has concentration-dependent effects on cardiac precursor migration (Zeng et al., 2007). Another GPCR that was implicated in C&E movements is Edg5, a sphingosine-1-phosphate (S1P) receptor belonging to a family of five GPCRs. Edg5 specifically mediates the anterior migration of prechordal mesodermal cells in zebrafish by regulating their motility and polarization (Kai et al., 2008). It is also essential for movements and fusion of the bilateral heart primordia during somitogenesis (Kupperman et al., 2000). The heterotrimeric G proteins that mediate signaling downstream of Apelin or S1P during gastrulation have not yet been identified.

Recent studies have also implicated GPCR signaling in the regulation of endoderm movements in zebrafish. Chemokine receptor and GPCR Cxcr4a is expressed in the endoderm during gastrulation, whereas its ligand Sdf1b/Cxcl12b is expressed in the mesoderm (Mizoguchi et al., 2008; Nair and Schilling, 2008). Knocking-down either the receptor or the ligand function specifically interrupts the migration of endodermal cells without affecting the cell movements in the mesoderm or ectoderm. In these *cxcr4a*- or *cxcl12b*-deficient embryos, endodermal cells migrate faster and more anteriorly ahead of the overlaying mesoderm. Based on the expression patterns of *cxcr4a* and *cxcl12b*, two plausible mechanisms were proposed to explain their roles in coupling the C&E movements of the mesoderm and endoderm. First, Cxcl12b may act as a chemoattractant for the *cxcr4a*-expressing endodermal cells. In support of this notion, *cxcl12b*-overexpressing donor cells were able to attract endodermal cells when positioned in ectopic positions in the host embryos (Mizoguchi et al., 2008). However, such chemoattractant function of Cxcl12 is incongruent with the proposed random walk of the endoderm during early gastrulation (Pezeron et al., 2008). Alternatively, Cxcr4a/Cxcl12b signaling might promote endodermal cell migration by negatively regulating adhesion of endodermal cells to the extracellular matrix. Consistently, disruption of Fibronectin–Integrin signaling in the zebrafish embryo causes anterior displacement of endoderm and delayed endoderm convergence toward the midline, similar to the phenotypes observed in *cxcl12b*- and *cxcr4a*-deficient embryos (Nair and Schilling, 2008). Furthermore, *integrinb1b* mRNA expression is downregulated in *cxcl12b/cxcr4a*-deficient embryos and the endodermal migration defects in these embryos can be suppressed by overexpressing *integrinb1b*.

## 4.5. Molecular evidence suggests uncoupling of convergence and extension

Although many C&E mutants, including those deficient in noncanonical Wnt/PCP signaling, exhibit defects in both convergence and extension movements, several lines of evidence suggest that convergence and extension can be mechanistically distinct processes. In dorsalized zebrafish mutants, the lateral mesoderm exhibits reduced convergence but normal or increased extension (Myers et al., 2002a). Similarly, zebrafish embryos harboring a mutation in the T-box transcription factor *no tail/brachyury* lack convergence of the chordamesoderm due to impaired mediolateral intercalation, whereas the notochord extension is unaffected in these mutants (Glickman et al., 2003). Hyaluronic acid synthesizing enzyme 2 (Has2) acts through the small GTPase Rac1 to regulate the formation of lamellipodia. It is required cell-autonomously for dorsal convergence, but not extension (Bakkers et al., 2004). Gravin tumor suppressor, a member of the AKAP (a kinase anchoring protein) family of scaffolding proteins, is required specifically for extension of the presomitic mesoderm (Weiser et al., 2007). In embryos with disrupted Gravin function, dorsal-directed convergence of the lateral presomitic cells, as well as extension of the chordamesoderm appear to be unaffected. However, in the lateral presomitic mesoderm, cells lacking Gravin function fail to shut down highly protrusive activity as they reach the dorsal side of the embryo. Consequently, they are unable to undergo normal mediolateral intercalation required for extension. Finally, cell groups lacking the function of noncanonical Wnt component Flamingo converge normally but fail to extend when transplanted into WT hosts, indicating that Flamingo is required cell-autonomously for extension but not convergence (Formstone and Mason, 2005a). How convergence and extension are differently affected in the above mutant embryos is still not clear. It is plausible that impairment of different pathways have differential effects on the distinct cell behaviors that contribute to convergence and/or extension.

### 4.6. Interplay between different signaling pathways during C&E in the mesoderm

It is intriguing that such a variety of signaling pathways are involved in C&E movements. Although they might be regulating different cellular properties of the migrating cells, these pathways crosstalk with one another, thus adding more complexity to the molecular mechanisms underlying C&E.

#### 4.6.1. Signaling pathways interacting with the noncanonical Wnt/PCP pathway

Epistatic analyses have placed several signaling pathways upstream or in parallel to the noncanonical Wnt/PCP pathway, the key regulator of C&E in vertebrates. In *X. laevis* chordamesoderm, the graded activin-like signaling that establishes anteroposterior tissue patterning has been shown to be necessary for the orientation of mediolateral cell intercalation (Ninomiya et al., 2004). Whereas blocking noncanonical Wnt/PCP signaling also interferes with mediolateral intercalation, it does not affect anteroposterior patterning, suggesting that the activin gradient acts in parallel or upstream of the noncanonical Wnt pathway.

In zebrafish, Stat3 signaling functions cell nonautonomously to regulate the directionality of lateral mesoderm C&E (Yamashita et al., 2002). Interestingly, in Stat3-depleted embryos, the lateral cells also exhibit severe polarity defects similar to the noncanonical Wnt/PCP mutants. Activation of the noncanonical Wnt signaling component Dsh in Stat3-depleted embryos is sufficient to suppress the defects in cell elongation, suggesting that noncanonical Wnt signaling functions at least partially downstream of Stat3 during lateral mesoderm C&E (Miyagi et al., 2004).

Bmp signaling, which guides the lateral mesoderm C&E by establishing graded cell–cell adhesion, also appears to act in parallel and upstream of noncanonical Wnts. This notion is supported by the fact that Bmp-mediated guidance of lateral mesodermal cells is not impaired in embryos deficient in Wnt5a and Wnt11 ligand function (von der Hardt et al., 2007). Moreover, in *somitabun/smad5* mutants with reduced Bmp signaling, the expression of *wnt11* mRNA in the dorsolateral mesoderm is expanded, whereas in *dino/chordin* mutants harboring a mutation in Bmp antagonist *chordin*, the expression of both *wnt5* and *wnt11* transcripts is greatly reduced (Myers et al., 2002a). These results suggest that Bmp signaling mediates C&E movements in part by negatively regulating noncanonical Wnt/PCP signaling.

Several signaling pathways that regulate the cytoskeleton have been shown to functionally interact with the noncanonical Wnt/PCP pathway. $G\alpha_{12/13}$ heterotrimeric proteins regulate Rho-mediated cytoskeletal rearrangements and E-cadherin-mediated cell adhesion, thus affecting cell shape and migration (Lin et al., 2009). In zebrafish, $G\alpha_{12/13}$ signaling controls mediolateral cell elongation underlying cell intercalation during notochord extension (Lin et al., 2005). Given that overexpression of $G\alpha_{12/13}$ fails to suppress C&E defects in the noncanonical Wnt mutants, and that loss of their function exacerbates C&E defects in *kny* mutants, $G\alpha_{12/13}$ likely influence gastrulation by acting in parallel to the noncanonical Wnt signaling pathway (Lin et al., 2005). Finally, two Src family kinases, Fyn and Yes, known regulators of the cytoskeleton, are required for epiboly and C&E movements during vertebrate gastrulation. Blocking the function of Fyn and Yes halts gastrulation movements in frog, resulting in the inability to close the blastopore (Denoyelle et al., 2001). In zebrafish, Fyn/Yes acts in parallel to the noncanonical Wnt/PCP pathway to regulate C&E. Interestingly, active RhoA rescued the phenotypes of the Fyn/Yes-deficient embryos. Therefore, similar to $G\alpha_{12/13}$ signaling, Fyn/Yes and noncanonical Wnt signaling may converge on RhoA (Jopling and den Hertog, 2005).

### 4.6.2. Canonical Wnt/β-catenin pathway in C&E movements

Wnt signaling is categorized as either canonical or noncanonical, depending on the intracellular effectors. Whereas the noncanonical Wnt/PCP signaling regulates cell polarity and morphogenesis, the canonical Wnt or so-called Wnt/β-catenin pathway functions mainly in patterning and proliferation of embryonic and adult tissues. In the canonical Wnt signaling, Wnt ligands activate Dsh, which in turn prevents the degradation of β-catenin by the APC/Axin/GSK3 complex (Moon et al., 2004). β-catenin then translocates into the nucleus to regulate the transcription of Wnt target genes. Although the canonical and noncanonical Wnt pathways have distinct roles in embryonic development, accumulating evidence indicates crosstalk between these two pathways. For instance, most mutant studies in zebrafish and gain-of-function studies in *Xenopus* have identified Wnt11 as a ligand involved in noncanonical Wnt signaling. However, maternal *wnt11* mRNA has been shown to be both necessary and sufficient for activation of the canonical Wnt/$\beta$-catenin pathway during axis formation in Xenopus (Tao *et al.*, 2005).

Conversely, components of the canonical Wnt/$\beta$-catenin pathway can also play roles in C&E movements during vertebrate gastrulation. Dickkopf-1 (Dkk1) is a secreted protein that negatively modulates the canonical Wnt/$\beta$-catenin pathway. Lack of Dkk1 function affects head formation in frog and mice (Glinka *et al.*, 1998; Kazanskaya *et al.*, 2000; Mukhopadhyay *et al.*, 2001). However, according to a recent report in zebrafish, blocking Dkk1 function accelerates internalization and anterior migration of the mesendoderm, whereas overexpressing Dkk1 slows down both internalization and C&E movements (Caneparo *et al.*, 2007). Intriguingly, upregulation of the canonical Wnt/$\beta$ catenin pathway does not induce similar gastrulation movement defects observed in the Dkk1-deficient embryos, suggesting that the function of Dkk1 in gastrulation movements is independent of $\beta$-catenin. Instead, Dkk1 modulates the noncanonical Wnt/PCP pathways by directly binding to the Wnt/PCP pathway component Kny, thus concomitantly repressing the Wnt/$\beta$-catenin and activating the Wnt/PCP signaling (Caneparo *et al.*, 2007).

Indeed, the presence of molecular switches between the canonical Wnt/$\beta$-catenin and noncanonical Wnt/PCP pathways have been reported in several other studies. Casein kinase 1ε (Strutt *et al.*, 2006) and metastasis-associated kinase (MAK) (Kibardin *et al.*, 2006) both promote the PCP pathway and inhibit the $\beta$-catenin pathway at the level of Dsh. The protein collagen triple helix repeat containing 1 (Cthrc1) positively influences PCP and suppresses $\beta$-catenin signaling by promoting the formation of Wnt/Fz complexes specific for the PCP pathway (Yamamoto *et al.*, 2008). These studies also establish novel connections between pathways that control embryonic patterning and those that regulate cell movements during gastrulation.

## 5. Cell Polarization During C&E Movements

Effective C&E movements rely on the proper establishment of mediolaterally elongated cell shape and protrusive activities within the migrating cells (Keller *et al.*, 2000; Myers *et al.*, 2002b; Wallingford *et al.*, 2000). In the early zebrafish gastrulae, before C&E movements initiate, the lateral mesodermal cells have a relatively round and nonpolarized morphology and extend both short bleb-like and longer lamelliform protrusions

(Solnica-Krezel, 2006). These cells meander in all directions but exhibit a slight dorsal bias in net motion (Sepich et al., 2005). By late gastrulation, the lateral cells become mediolaterally elongated and their protrusive activities are also biased mediolaterally (Myers et al., 2002b). They migrate with a very strong dorsal bias, achieving a fast net dorsal speed. As the mesodermal cells approach the dorsal midline, they become more adherent and protrusive activity is reduced (Weiser et al., 2007). The most dorsally located cells then switch from a monopolar mode of migration to exhibit primarily mediolaterally intercalative behavior (Keller et al., 2000; Myers et al., 2002b).

The noncanonical Wnt/PCP signaling pathway is the key regulator of mediolateral cell polarity. Reduction of noncanonical Wnt signaling results in round cell morphology, randomized protrusions, and consequently impaired C&E movements (Jessen et al., 2002; Myers et al., 2002b; Topczewski et al., 2001; Wallingford et al., 2000). The mechanism by which noncanonical Wnt signaling regulates mediolateral cell polarity during C&E remains elusive. In Drosophila, components of the PCP pathway accumulate asymmetrically to establish cell polarity within an epithelium (Adler, 2002; Klein and Mlodzik, 2005). In the fly wing epithelium, each cell orients itself with respect to the proximodistal axis, and develops a distally pointing actin-based hair near the distal cell edge (Fig. 7.4A) (Adler, 2002; Mlodzik, 2002). Upon activation of the pathway, PCP components, which are initially distributed evenly in the cell, become first localized uniformly on the apical cell edge, and then asymmetrically localized at either the distal or proximal cell surface (Strutt, 2002). Fz recruits and forms a complex with Dsh at the distal edge of the wing cell. Vang/Stbm, on the other hand, brings Pk to the membrane and the Vang/Stbm/Pk complex becomes enriched at the proximal cell membrane (Fig. 7.4A) (Adler, 2002; Klein and Mlodzik, 2005). The Vang/Stbm/Pk complex inhibits the formation of the Fz/Dsh complex on the proximal cell edge, thereby functioning in a feedback loop that amplifies differences between Fz/Dsh levels on the adjacent cell surfaces (Fig. 7.4A) (Tree et al., 2002). The core PCP components rely on each other for correct asymmetric localization (reviewed in Klein and Mlodzik, 2005).

In vertebrates, asymmetric localization of the noncanonical Wnt components has been linked to the establishment of PCP in the organ of Corti and the embryonic epidermal basal layer during hair follicle development (Deans et al., 2007; Devenport and Fuchs, 2008; Montcouquiol et al., 2006; Wang et al., 2005, 2006), whereas the localization of PCP components during gastrulation is just beginning to be investigated. In *X. laevis*, Dsh was reported to be highly expressed at the medial and lateral ends of the intercalating cells in the dorsal marginal zone explants (Kinoshita et al., 2003). Pk, on the other hand, was shown to be localized at the anterior edge of the cells in the notochord and neural tube during zebrafish somitogenesis (Ciruna et al., 2006). We recently observed using fluorescent fusion proteins that in the medial presomitic mesoderm and axial mesoderm of the zebrafish gastrulae, Pk is localized to the anterior side of the cell, whereas Dsh is transiently enriched near the posterior cell membrane (Fig. 7.4B) (Yin *et al.*, 2008). The anterior localization of Pk and posterior localization of Dsh in the zebrafish dorsal mesoderm are reminiscent of the asymmetric distribution of the fly homologs of these proteins in the wing epithelium, where Pk exhibits proximal distribution and Dsh is localized distally (Strutt, 2002). The asymmetric localization of Pk and Dsh is dependent on noncanonical Wnt signaling and is correlated with the mediolateral cell elongation at late gastrulation (Yin *et al.*, 2008). These results suggest that the interactions between these molecules during tissue polarization are evolutionarily conserved. It will be interesting to see whether similar asymmetric localization of the noncanonical Wnt/PCP pathway components also occurs in other domains of the gastrulae that undergo active C&E.

It is intriguing that while the mesodermal cells engaged in C&E are polarized along the mediolateral axis, the noncanonical Wnt signaling components are localized at the anterior and posterior cell edges. Our current understanding of C&E and its molecular regulation leads to a plausible model presented in Fig. 7.4C. In the early zebrafish gastrulae, cells in the medial presomitic mesoderm and axial mesoderm, for instance, display a round morphology, produce lamellipodia in all directions, and migrate slowly with little dorsal bias. At late gastrulation, the noncanonical Wnt signaling becomes activated at the anterior and posterior cell edges by an unknown anteroposterior positional cue, similar to the graded activin-like signal reported in Xenopus (Ninomiya *et al.*, 2004). This would lead to the restriction of lamellipodia protrusions to the medial and lateral ends of the cell, probably through localized regulation of different small Rho GTPases. The elimination of extensive protrusions on the anterior and posterior sides of the cells also requires other signaling pathways, such as Gravin and Hyaluronic acid synthesizing enzyme 2 (Bakkers *et al.*, 2004; Weiser *et al.*, 2007). Meanwhile, activation of noncanonical Wnt/PCP signaling along the anteroposterior axis may also establish differential cell adhesion so that the contacts between anteroposterior neighboring cells are different from the mediolateral neighbor contacts. The mediolaterally biased protrusive activities and differential cell adhesion consequently underlie the mediolateral cell elongation and polarized cell intercalations that drive C&E.

## 6. CONCLUSION

Convergence and extension are highly conserved morphogenetic movements that shape all germ layers during vertebrate gastrulation. In distinct vertebrates C&E movements are driven by various combinations of highly polarized cell behaviors including directed migration, polarized radial and planar intercalations, and oriented cell divisions. The noncanonical Wnt/PCP pathway emerged as an evolutionarily conserved regulator of these polarized cell behaviors in all studied vertebrates. It will be important to determine whether pathways interacting with the Wnt/PCP pathway that are being discovered in various model systems will be also universally employed during vertebrate gastrulation.

## ACKNOWLEDGMENTS

C.Y. wishes to acknowledge support of Dr. D. Stainier and Juvenile Diabetes Research Foundation Postdoctoral Fellowship. B.C. is supported by the Canada Research Chairs Program, and by operating grants from the Terry Fox Foundation and the Natural Sciences and Engineering Research Council of Canada. Work on C&E in the Solnica-Krezel's laboratory is supported by grant GM55101 from the National Institutes of Health, Human Frontiers in Science Program, and Martha Rivers Ingram Endowment.

## REFERENCES

Adler, P. N. (2002). Planar signaling and morphogenesis in *Drosophila*. *Dev. Cell* **2,** 525–535.

Axelrod, J. D., Miller, J. R., Shulman, J. M., Moon, R. T., and Perrimon, N. (1998). Differential recruitment of Dishevelled provides signaling specificity in the planar cell polarity and Wingless signaling pathways. *Genes Dev.* **12,** 2610–2622.

Bakkers, J., Kramer, C., Pothof, J., Quaedvlieg, N. E., Spaink, H. P., and Hammerschmidt, M. (2004). Has2 is required upstream of Rac1 to govern dorsal migration of lateral cells during zebrafish gastrulation. *Development* **131,** 525–537.

Blanco, M. J., Barrallo-Gimeno, A., Acloque, H., Reyes, A. E., Tada, M., Allende, M. L., Mayor, R., and Nieto, M. A. (2007). Snail1a and Snail1b cooperate in the anterior migration of the axial mesendoderm in the zebrafish embryo. *Development* **134,** 4073–4081.

Caneparo, L., Huang, Y. L., Staudt, N., Tada, M., Ahrendt, R., Kazanskaya, O., Niehrs, C., and Houart, C. (2007). A sequence-specific, single-strand binding protein activates the far upstream element of c-myc and defines a new DNA-binding motif. *Genes Dev.* **21,** 465–480.

Carreira-Barbosa, F., Concha, M. L., Takeuchi, M., Ueno, N., Wilson, S. W., and Tada, M. (2003). Prickle 1 regulates cell movements during gastrulation and neuronal migration in zebrafish. *Development* **130,** 4037–4046.

Cha, Y. I., Kim, S. H., Sepich, D., Buchanan, F. G., Solnica-Krezel, L., and DuBois, R. N. (2006). Cyclooxygenase-1-derived PGE2 promotes cell motility via the G-protein-coupled EP4 receptor during vertebrate gastrulation. *Genes Dev.* **20,** 77–86.

Choi, S. C., and Han, J. K. (2002). *Xenopus* Cdc42 regulates convergent extension movements during gastrulation through Wnt/Ca2+ signaling pathway. *Dev. Biol.* **244,** 342–357.

Ciruna, B., Jenny, A., Lee, D., Mlodzik, M., and Schier, A. F. (2006). Planar cell polarity signalling couples cell division and morphogenesis during neurulation. *Nature* **439,** 220–224.

Concha, M. L., and Adams, R. J. (1998). Oriented cell divisions and cellular morphogenesis in the zebrafish gastrula and neurula: A time-lapse analysis. *Development* **125,** 983–994.

Curtin, J. A., Quint, E., Tsipouri, V., Arkell, R. M., Cattanach, B., Copp, A. J., Henderson, D. J., Spurr, N., Stanier, P., Fisher, E. M., Nolan, P. M., Steel, K. P., et al. (2003). Mutation of Celsr1 disrupts planar polarity of inner ear hair cells and causes severe neural tube defects in the mouse. *Curr. Biol.* **13,** 1129–1133.

Das, G., Jenny, A., Klein, T. J., Eaton, S., and Mlodzik, M. (2004). Diego interacts with Prickle and Strabismus/Van Gogh to localize planar cell polarity complexes. *Development* **131,** 4467–4476.

Deans, M. R., Antic, D., Suyama, K., Scott, M. P., Axelrod, J. D., and Goodrich, L. V. (2007). Asymmetric distribution of prickle-like 2 reveals an early underlying polarization of vestibular sensory epithelia in the inner ear. *J. Neurosci.* **27,** 3139–3147.

Denoyelle, M., Valles, A. M., Lentz, D., Thiery, J. P., and Boyer, B. (2001). Mesoderm-independent regulation of gastrulation movements by the src tyrosine kinase in *Xenopus* embryo. *Differentiation* **69,** 38–48.

De Robertis, E. M., and Kurado, H. (2004). Dorsal-ventral patterning and neural induction in *Xenopus* embryos. *Annu. Rev. Cell Dev. Biol.* **20,** 285–308.

Devenport, D., and Fuchs, E. (2008). Planar polarization in embryonic epidermis orchestrates global asymmetric morphogenesis of hair follicles. *Nat. Cell Biol.* **10,** 1257–1268.

Devreotes, P., and Janetopoulos, C. (2003). Eukaryotic chemotaxis: distinctions between directional sensing and polarization. *J. Biol. Chem.* **278,** 20445–20448.

Djiane, A., Riou, J., Umbhauer, M., Boucaut, J., and Shi, D. (2000). Role of frizzled 7 in the regulation of convergent extension movements during gastrulation in *Xenopus laevis*. *Development* **127,** 3091–3100.

Fanto, M., Weber, U., Strutt, D. I., and Mlodzik, M. (2000). Nuclear signaling by Rac and Rho GTPases is required in the establishment of epithelial planar polarity in the *Drosophila* eye. *Curr. Biol.* **10,** 979–988.

Formstone, C. J., and Mason, I. (2005a). Combinatorial activity of Flamingo proteins directs convergence and extension within the early zebrafish embryo via the planar cell polarity pathway. *Dev. Biol.* **282,** 320–335.

Formstone, C. J., and Mason, I. (2005b). Expression of the Celsr/flamingo homologue, c-fmi1, in the early avian embryo indicates a conserved role in neural tube closure and additional roles in asymmetry and somitogenesis. *Dev. Dyn.* **232,** 408–413.

Glickman, N. S., Kimmel, C. B., Jones, M. A., and Adams, R. J. (2003). Shaping the zebrafish notochord. *Development* **130,** 873–887.

Glinka, A., Wu, W., Delius, H., Monaghan, A. P., Blumenstock, C., and Niehrs, C. (1998). Dickkopf-1 is a member of a new family of secreted proteins and functions in head induction. *Nature* **391,** 357–362.

Gritsman, K., Talbot, W. S., and Schier, A. F. (2000). Nodal signaling patterns the organizer. *Development* **127,** 921–932.

Gong, Y., Mo, C., and Fraser, S. E. (2004). Planar cell polarity signalling controls cell division orientation during zebrafish gastrulation. *Nature* **430,** 689–693.

Goto, T., Davidson, L., Asashima, M., and Keller, R. (2005). Planar cell polarity genes regulate polarized extracellular matrix deposition during frog gastrulation. *Curr. Biol.* **15,** 787–793.

Habas, R., Kato, Y., and He, X. (2001). Wnt/Frizzled activation of Rho regulates vertebrate gastrulation and requires a novel Formin homology protein Daam1. *Cell* **107,** 843–854.

Habas, R., Dawid, I. B., and He, X. (2003). Coactivation of Rac and Rho by Wnt/Frizzled signaling is required for vertebrate gastrulation. *Genes Dev.* **17,** 295–309.

Hall, A., and Nobes, C. D. (2000). Rho GTPases: molecular switches that control the organization and dynamics of the actin cytoskeleton. *Philos. Trans. R. Soc. Lond. B Biol. Sci.* **355,** 965–970.

Hammerschmidt, M., and Mullins, M. C. (2002). Dorsoventral patterning in the zebrafish: bone morphogenetic proteins and beyond. *Results Probl. Cell Differ.* **40,** 72–95.

Hammerschmidt, M., Pelegri, F., Mullins, M. C., Kane, D. A., Brand, M., van Eeden, F. J., Furutani-Seiki, M., Granato, M., Haffter, P., Heisenberg, C. P., Jiang, Y. J., Kelsh, R. N., et al. (1996a). Mutations affecting morphogenesis during gastrulation and tail formation in the zebrafish, *Danio rerio*. *Development* **123,** 143–151.

Hammerschmidt, M., Serbedzija, G. N., and McMahon, A. P. (1996b). Genetic analysis of dorsoventral pattern formation in the zebrafish: Requirement of a BMP-like ventralizing activity and its dorsal repressor. *Genes Dev.* **10,** 2452–2461.

Heisenberg, C. P., Tada, M., Rauch, G. J., Saude, L., Concha, M. L., Geisler, R., Stemple, D. L., Smith, J. C., and Wilson, S. W. (2000). Silberblick/Wnt11 mediates convergent extension movements during zebrafish gastrulation. *Nature* **405,** 76–81.

Jenny, A., Darken, R. S., Wilson, P. A., and Mlodzik, M. (2003). Prickle and Strabismus form a functional complex to generate a correct axis during planar cell polarity signaling. *EMBO J.* **22,** 4409–4420.

Jenny, A., Reynolds-Kenneally, J., Das, G., Burnett, M., and Mlodzik, M. (2005). Diego and Prickle regulate Frizzled planar cell polarity signalling by competing for Dishevelled binding. *Nat. Cell Biol.* **7,** 691–697.

Jessen, J. R., Topczewski, J., Bingham, S., Sepich, D. S., Marlow, F., Chandrasekhar, A., and Solnica-Krezel, L. (2002). Zebrafish trilobite identifies new roles for Strabismus in gastrulation and neuronal movements. *Nat. Cell Biol.* **4,** 610–615.

Jopling, C., and den Hertog, J. (2005). Fyn/Yes and non-canonical Wnt signalling converge on RhoA in vertebrate gastrulation cell movements. *EMBO Rep.* **6,** 426–431.

Kai, M., Heisenberg, C. P., and Tada, M. (2008). Sphingosine-1-phosphate receptors regulate individual cell behaviours underlying the directed migration of prechordal plate progenitor cells during zebrafish gastrulation. *Development* **135,** 3043–3051.

Kane, D. A., and Kimmel, C. B. (1993). The zebrafish midblastula transition. *Development* **119,** 447–456.

Katoh, K., Kano, Y., Amano, M., Onishi, H., Kaibuchi, K., and Fujiwara, K. (2001). Rho-kinase–mediated contraction of isolated stress fibers. *J. Cell Biol.* **153,** 569–584.

Kazanskaya, O., Glinka, A., and Niehrs, C. (2000). The role of *Xenopus* dickkopf1 in prechordal plate specification and neural patterning. *Development* **127,** 4981–4992.

Keegan, B. R., Meyer, D., and Yelon, D. (2004). Organization of cardiac chamber progenitors in the zebrafish blastula. *Development* **131,** 3081–3091.

Keller, R. (2005). Cell migration during gastrulation. *Curr. Opin. Cell Biol.* **17,** 533–541.

Keller, R., and Tibbetts, P. (1989). Mediolateral cell intercalation in the dorsal, axial mesoderm of *Xenopus laevis*. *Dev. Biol.* **131,** 539–549.

Keller, R., Davidson, L., Edlund, A., Elul, T., Ezin, M., Shook, D., and Skoglund, P. (2000). Mechanisms of convergence and extension by cell intercalation. *Philos. Trans. R. Soc. Lond. B Biol. Sci.* **355,** 897–922.

Keller, R., Davidson, L. A., and Shook, D. R. (2003). How we are shaped: The biomechanics of gastrulation. *Differentiation* **71,** 171–205.

Keller, P. J., Schmidt, A. D., Wittbrodt, J., and Stelzer, E. H. (2008). Reconstruction of zebrafish early embryonic development by scanned light sheet microscopy. *Science* **322,** 1065–1069.

Kibardin, A., Ossipova, O., and Sokol, S. Y. (2006). Metastasis-associated kinase modulates Wnt signaling to regulate brain patterning and morphogenesis. *Development* **133,** 2845–2854.

Kilian, B., Mansukoski, H., Barbosa, F. C., Ulrich, F., Tada, M., and Heisenberg, C. P. (2003). The role of Ppt/Wnt5 in regulating cell shape and movement during zebrafish gastrulation. *Mech. Dev.* **120,** 467–476.

Kimmel, C. B., Warga, R. M., and Schilling, T. F. (1990). Origin and organization of the zebrafish fate map. *Development* **108,** 581–594.

Kimmel, C. B., Warga, R. M., and Kane, D. A. (1994). Cell cycles and clonal strings during formation of the zebrafish central nervous system. *Development* **120,** 265–276.

Kimmel, C. B., Ballard, W. W., Kimmel, S. R., Ullmann, B., and Schilling, T. F. (1995). Stages of embryonic development of the zebrafish. *Dev. Dyn.* **203,** 253–310.

Kinoshita, N., Iioka, H., Miyakoshi, A., and Ueno, N. (2003). PKC delta is essential for Dishevelled function in a noncanonical Wnt pathway that regulates *Xenopus* convergent extension movements. *Genes Dev.* **17,** 1663–1676.

Klein, T. J., and Mlodzik, M. (2005). Planar cell polarization: An emerging model points in the right direction. *Annu. Rev. Cell Dev. Biol.* **21,** 155–176.

Kupperman, E., An, S., Osborne, N., Waldron, S., and Stainier, D. Y. (2000). A sphingosine-1-phosphate receptor regulates cell migration during vertebrate heart development. *Nature* **406,** 192–195.

Lin, F., Sepich, D. S., Chen, S., Topczewski, J., Yin, C., Solnica-Krezel, L., and Hamm, H. (2005). Essential roles of $G\alpha_{12/13}$ signaling in distinct cell behaviors driving zebrafish convergence and extension gastrulation movements. *J. Cell Biol.* **169,** 777–787.

Lin, F., Chen, S., Sepich, D. S., Panizzi, J. R., Clendenon, S. G., Marrs, J. A., Hamm, H., and Solnica-krezel, L. (2009). G(alpha)$_{12/13}$ regulate epiboly by inhibiting E-cadherin activity and modulating the actin cytoskeleton. *J. Cell Biol.* **184,** 909–921.

Marlow, F., Topczewski, J., Sepich, D., and Solnica-Krezel, L. (2002). Zebrafish Rho kinase 2 acts downstream of Wnt11 to mediate cell polarity and effective convergence and extension movements. *Curr. Biol.* **12,** 876–884.

Marsden, M., and DeSimone, D. W. (2003). Integrin-ECM interactions regulate cadherin-dependent cell adhesion and are required for convergent extension in *Xenopus*. *Curr. Biol.* **13,** 1182–1191.

Miyagi, C., Yamashita, S., Ohba, Y., Yoshizaki, H., Matsuda, M., and Hirano, T. (2004). STAT3 noncell-autonomously controls planar cell polarity during zebrafish convergence and extension. *J. Cell Biol.* **166,** 975–981.

Mizoguchi, T., Verkade, H., Heath, J. K., Kuroiwa, A., and Kikuchi, Y. (2008). Sdf1/Cxcr4 signaling controls the dorsal migration of endodermal cells during zebrafish gastrulation. *Development* **135,** 2521–2529.

Mlodzik, M. (2002). Planar cell polarization: Do the same mechanisms regulate *Drosophila* tissue polarity and vertebrate gastrulation? *Trends Genet.* **18,** 564–571.

Moeller, H., Jenny, A., Schaeffer, H. J., Schwarz-Romond, T., Mlodzik, M., Hammerschmidt, M., and Birchmeier, W. (2006). Diversin regulates heart formation and gastrulation movements in development. *Proc. Natl. Acad. Sci. USA* **103,** 15900–15905.

Montcouquiol, M., Sans, N., Huss, D., Kach, J., Dickman, J. D., Forge, A., Rachel, R. A., Copeland, N. G., Jenkins, N. A., Bogani, D., Murdoch, J., Warchol, M. E., Wenthold, R. J., and Kelley, M. W. (2006). Asymmetric localization of Vangl2 and Fz3 indicate novel mechanisms for planar cell polarity in mammals. *J. Neurosci.* **26,** 5265–5275.

Moon, R. T., Kohn, A. D., De Ferrari, G. V., and Kaykas, A. (2004). WNT and {beta}-catenin signalling: diseases and therapies. *Nat. Rev. Genet.* **5,** 691–701.

Mukhopadhyay, M., Shtrom, S., Rodriguez Esteban, C., Chen, L., Tsukui, T., Gomer, L., Dorward, D. W., Glinka, A., Grinberg, A., Huang, S. P., Niehrs, C., Belmonte, J. C., and Westphal, H. (2001). Dickkopf1 is required for embryonic head induction and limb morphogenesis in the mouse. *Dev. Cell* **1,** 423–434.

Mullins, M. C., Hammerschmidt, M., Kane, D. A., Odenthal, J., Brand, M., van Eeden, F. J., Furutani-Seiki, M., Granato, M., Haffter, P., Heisenberg, C. P., Jiang, Y. J., Kelsh, R. N., and Nusslein-Volhard, C. (1996). Genes establishing dorso-ventral pattern formation in the zebrafish embryo: The ventral specifying genes. *Development* **123,** 81–93.

Munoz, R., and Larrain, J. (2006). xSyndecan-4 regulates gastrulation and neural tube closure in *Xenopus* embryos. *ScientificWorldJournal* **6,** 1298–1301.

Myers, D. C., Sepich, D. S., and Solnica-Krezel, L. (2002a). Bmp activity gradient regulates convergent extension during zebrafish gastrulation. *Dev. Biol.* **243,** 81–98.

Myers, D. C., Sepich, D. S., and Solnica-Krezel, L. (2002b). Convergence and extension in vertebrate gastrulae: cell movements according to or in search of identity? *Trends Genet.* **18,** 447–455.

Nair, S., and Schilling, T. F. (2008). Chemokine signaling controls endodermal migration during zebrafish gastrulation. *Science* **322,** 89–92.

Ninomiya, H., Elinson, R. P., and Winklbauer, R. (2004). Antero-posterior tissue polarity links mesoderm convergent extension to axial patterning. *Nature* **430,** 364–367.

Park, M., and Moon, R. T. (2002). The planar cell-polarity gene stbm regulates cell behaviour and cell fate in vertebrate embryos. *Nat. Cell Biol.* **4,** 20–25.

Penzo-Mendez, A., Umbhauer, M., Djiane, A., Boucaut, J. C., and Riou, J. F. (2003). Activation of Gbetagamma signaling downstream of Wnt-11/Xfz7 regulates Cdc42 activity during *Xenopus* gastrulation. *Dev. Biol.* **257,** 302–314.

Pezeron, G., Mourrain, P., Courty, S., Ghislain, J., Becker, T. S., Rosa, F. M., and David, N. B. (2008). Live analysis of endodermal layer formation identifies random walk as a novel gastrulation movement. *Curr. Biol.* **18,** 276–281.

Schoenwolf, G. C., and Alvarez, I. S. (1989). Roles of neuroepithelial cell rearrangement and division in shaping of the avian neural plate. *Development* **106,** 427–439.

Schoenwolf, G. C., and Alvarez, I. S. (1992). Role of cell rearrangement in axial morphogenesis. *Curr. Top. Dev. Biol.* **27,** 129–173.

Schoenwolf, G. C., and Yuan, S. (1995). Experimental analyses of the rearrangement of ectodermal cells during gastrulation and neurulation in avian embryos. *Cell Tissue Res.* **280,** 243–251.

Scott, I. C., Masri, B., D'Amico, L. A., Jin, S. W., Jungblut, B., Wehman, A. M., Baier, H., Audigier, Y., and Stainier, D. Y. (2007). The g protein-coupled receptor agtrl1b regulates early development of myocardial progenitors. *Dev. Cell* **12,** 403–413.

Sepich, D. S., Calmelet, C., Kiskowski, M., and Solnica-Krezel, L. (2005). Initiation of convergence and extension movements of lateral mesoderm during zebrafish gastrulation. *Dev. Dyn.* **234,** 279–292.

Shih, J., and Keller, R. (1992). Cell motility driving mediolateral intercalation in explants of Xenopus *laevis*. *Development* **116,** 901–914.

Sokol, S. Y. (1996). Analysis of Dishevelled signalling pathways during *Xenopus* development. *Curr. Biol.* **6,** 1456–1467.

Solnica-Krezel, L. (2005). Conserved patterns of cell movements during vertebrate gastrulation. *Curr. Biol.* **15,** R213–R228.

Solnica-Krezel, L. (2006). Gastrulation in zebrafish – All just about adhesion? *Curr. Opin. Genet. Dev.* **16,** 433–441.

Solnica-Krezel, L., Stemple, D. L., Mountcastle-Shah, E., Rangini, Z., Neuhauss, S. C., Malicki, J., Schier, A. F., Stainier, D. Y., Zwartkruis, F., Abdelilah, S., and Driever, W. (1996). Mutations affecting cell fates and cellular rearrangements during gastrulation in zebrafish. *Development* **123,** 67–80.

Strutt, D. I. (2002). The asymmetric subcellular localisation of components of the planar polarity pathway. *Semin. Cell Dev. Biol.* **13,** 225–231.

Strutt, H., Price, M. A., and Strutt, D. (2006). Planar polarity is positively regulated by casein kinase Iepsilon in *Drosophila*. *Curr. Biol.* **16,** 1329–1336.

Tada, M., and Smith, J. C. (2000). Xwnt11 is a target of *Xenopus* Brachyury: Regulation of gastrulation movements via Dishevelled, but not through the canonical Wnt pathway. *Development* **127,** 2227–2238.

Tada, M., Concha, M. L., and Heisenberg, C. P. (2002). Non-canonical Wnt signalling and regulation of gastrulation movements. *Semin. Cell Dev. Biol.* **13,** 251–260.

Tao, Q., Yokota, C., Puck, H., Kofron, M., Birsoy, B., Yan, D., Asashima, M., Wylie, C. C., Lin, X., and Heasman, J. (2005). Maternal wnt11 activates the canonical wnt signaling pathway required for axis formation in *Xenopus* embryos. *Cell* **120,** 857–871.

Topczewski, J., Sepich, D. S., Myers, D. C., Walker, C., Amores, A., Lele, Z., Hammerschmidt, M., Postlethwait, J., and Solnica-Krezel, L. (2001). The zebrafish glypican knypek controls cell polarity during gastrulation movements of convergent extension. *Dev. Cell* **1,** 251–264.

Tree, D. R., Shulman, J. M., Rousset, R., Scott, M. P., Gubb, D., and Axelrod, J. D. (2002). Prickle mediates feedback amplification to generate asymmetric planar cell polarity signaling. *Cell* **109**, 371–381.

Tucker, J. A., Mintzer, K. A., and Mullins, M. C. (2008). The BMP signaling gradient patterns dorsoventral tissues in a temporally progressive manner along the anteroposterior axis. *Dev. Cell* **14**, 108–119.

Ulrich, F., Concha, M. L., Heid, P. J., Voss, E., Witzel, S., Roehl, H., Tada, M., Wilson, S. W., Adams, R. J., Soll, D. R., and Heisenberg, C. P. (2003). Slb/Wnt11 controls hypoblast cell migration and morphogenesis at the onset of zebrafish gastrulation. *Development* **130**, 5375–5384.

Ulrich, F., Krieg, M., Schotz, E. M., Link, V., Castanon, I., Schnabel, V., Taubenberger, A., Mueller, D., Puech, P. H., and Heisenberg, C. P. (2005). Wnt11 functions in gastrulation by controlling cell cohesion through Rab5c and E-cadherin. *Dev. Cell* **9**, 555–564.

Ungar, A. R., Kelly, G. M., and Moon, R. T. (1995). Wnt4 affects morphogenesis when misexpressed in the zebrafish embryo. *Mech. Dev.* **52**, 153–164.

Usui, T., Shima, Y., Shimada, Y., Hirano, S., Burgess, R. W., Schwarz, T. L., Takeichi, M., and Uemura, T. (1999). Flamingo, a seven-pass transmembrane cadherin, regulates planar cell polarity under the control of Frizzled. *Cell* **98**, 585–595.

Veeman, M. T., Axelrod, J. D., and Moon, R. T. (2003a). A second canon. Functions and mechanisms of beta-catenin-independent Wnt signaling. *Dev. Cell* **5**, 367–377.

Veeman, M. T., Slusarski, D. C., Kaykas, A., Louie, S. H., and Moon, R. T. (2003b). Zebrafish prickle, a modulator of noncanonical Wnt/Fz signaling, regulates gastrulation movements. *Curr. Biol.* **13**, 680–685.

von der Hardt, S., Bakkers, J., Inbal, A., Carvalho, L., Solnica-Krezel, L., Heisenberg, C. P., and Hammerschmidt, M. (2007). The Bmp gradient of the zebrafish gastrula guides migrating lateral cells by regulating cell-cell adhesion. *Curr. Biol.* **17**, 475–487.

Wallingford, J. B., and Habas, R. (2005). The developmental biology of Dishevelled: An enigmatic protein governing cell fate and cell polarity. *Development* **132**, 4421–4436.

Wallingford, J. B., Rowning, B. A., Vogeli, K. M., Rothbacher, U., Fraser, S. E., and Harland, R. M. (2000). Dishevelled controls cell polarity during *Xenopus* gastrulation. *Nature* **405**, 81–85.

Wang, J., Mark, S., Zhang, X., Qian, D., Yoo, S. J., Radde-Gallwitz, K., Zhang, Y., Lin, X., Collazo, A., Wynshaw-Boris, A., and Chen, P. (2005). Regulation of polarized extension and planar cell polarity in the cochlea by the vertebrate PCP pathway. *Nat. Genet.* **37**, 980–985.

Wang, J., Hamblet, N. S., Mark, S., Dickinson, M. E., Brinkman, B. C., Segil, N., Fraser, S. E., Chen, P., Wallingford, J. B., and Wynshaw-Boris, A. (2006). Dishevelled genes mediate a conserved mammalian PCP pathway to regulate convergent extension during neurulation. *Development* **133**, 1767–1778.

Warga, R. M., and Kimmel, C. B. (1990). Cell movements during epiboly and gastrulation in zebrafish. *Development* **108**, 569–580.

Wei, Y., and Mikawa, T. (2000). Formation of the avian primitive streak from spatially restricted blastoderm: Evidence for polarized cell division in the elongating streak. *Development* **127**, 87–96.

Weiser, D. C., Pyati, U. J., and Kimelman, D. (2007). Gravin regulates mesodermal cell behavior changes required for axis elongation during zebrafish gastrulation. *Genes Dev.* **21**, 1559–1571.

Wilson, P., and Keller, R. (1991). Cell rearrangement during gastrulation of *Xenopus*: Direct observation of cultured explants. *Development* **112**, 289–300.

Winklbauer, R., and Nagel, M. (1991). Directional mesoderm cell migration in the *Xenopus* gastrula. *Dev. Biol.* **148**, 573–589.

Winklbauer, R., and Selchow, A. (1992). Motile behavior and protrusive activity of migratory mesoderm cells from the *Xenopus* gastrula. *Dev. Biol.* **150**, 335–351.

Winter, C. G., Wang, B., Ballew, A., Royou, A., Karess, R., Axelrod, J. D., and Luo, L. (2001). *Drosophila* Rho-associated kinase (Drok) links Frizzled-mediated planar cell polarity signaling to the actin cytoskeleton. *Cell* **105**, 81–91.

Witzel, S., Zimyanin, V., Carreira-Barbosa, F., Tada, M., and Heisenberg, C. P. (2006). Wnt11 controls cell contact persistence by local accumulation of Frizzled 7 at the plasma membrane. *J. Cell Biol.* **175**, 791–802.

Yamamoto, S., Nishimura, O., Misaki, K., Nishita, M., Minami, Y., Yonemura, S., Tarui, H., and Sasaki, H. (2008). Cthrc1 selectively activates the planar cell polarity pathway of Wnt signaling by stabilizing the Wnt-receptor complex. *Dev. Cell* **15**, 23–36.

Yamanaka, Y., Tamplin, O. J., Beckers, A., Gossler, A., and Rossant, J. (2007). Live imaging and genetic analysis of mouse notochord formation reveals regional morphogenetic mechanisms. *Dev. Cell* **13**, 884–896.

Yamashita, S., Miyagi, C., Carmany-Rampey, A., Shimizu, T., Fujii, R., Schier, A. F., and Hirano, T. (2002). Stat3 Controls Cell Movements during Zebrafish Gastrulation. *Dev. Cell* **2**, 363–375.

Yamashita, S., Miyagi, C., Fukada, T., Kagara, N., Che, Y. S., and Hirano, T. (2004). Zinc transporter LIVI controls epithelial-mesenchymal transition in zebrafish gastrula organizer. *Nature* **429**, 298–302.

Yang, X., Dormann, D., Munsterberg, A. E., and Weijer, C. J. (2002). Cell movement patterns during gastrulation in the chick are controlled by positive and negative chemotaxis mediated by FGF4 and FGF8. *Dev. Cell* **3**, 425–437.

Yin, C., and Solnica-Krezel, L. (2007). Convergence and extension movements affect dynamic notochord-somite interactions essential for zebrafish slow muscle morphogenesis. *Dev. Dyn.* **236**, 2742–2756.

Yin, C., Kiskowski, M., Pouille, P. A., Farge, E., and Solnica-Krezel, L. (2008). Cooperation of polarized cell intercalations drives convergence and extension of presomitic mesoderm during zebrafish gastrulation. *J. Cell Biol.* **180**, 221–232.

Zeng, X. X., Wilm, T. P., Sepich, D. S., and Solnica-Krezel, L. (2007). Apelin and its receptor control heart field formation during zebrafish gastrulation. *Dev. Cell* **12**, 391–402.

Zhu, S., Liu, L., Korzh, V., Gong, Z., and Low, B. C. (2006). RhoA acts downstream of Wnt5 and Wnt11 to regulate convergence and extension movements by involving effectors Rho kinase and Diaphanous: Use of zebrafish as an *in vivo* model for GTPase signaling. *Cell Signal* **18**, 359–372.

# Subject Index

## A

Abelson (Abl) kinase, 60, 66
*abl* mutants, 66
Actin cytoskeleton architecture, 3
Actomyosin
   constriction, 59
   contractility, 3, 59
      driving, apical constriction, 63–67
      driving, cell intercalation, 67–69
      link between MTs and, 79
      for membrane invagination, 60
   regulator at LE, 76
Adherens junctions
   and actin systems, 37
   assembly pathway, during cellularization, 61
   associated actomyosin, 45
   cadherin-based, role of, 35
   critical role, actin filaments, 44
   depletion of Fat1, 43
   essential for, 37
   integrity, 43
   mechanisms to modulate (*see* Cadherin, recycling; MicroRNAs; p120-catenin)
   molecular elements organizing, 35
   MT cytoskeleton interaction, 8
   myosins roles, 8
   patterning, process of *Drosophila* eye, 128–129
   and PCP signals, 41–42, 46, 49
   Rap1 dependent remodeling, 44
   regulators, 38
   reshaping of epithelial sheets and regulation, 48
   role in dorsal closure, 75
Adhesion belt, 34
Adhesion energy. *See* Intercellular adhesion energies
Adhesion modulation, 14
Adhesion proteins, 3–4, 7, 127
Adhesion receptors, 34, 41
Afadin, 7, 37
AJ maturation, 62
AJ–MT interactions, 62
AJs. *See* Adherens junctions
Amnioserosa (AS), 70
*Anacharis*, 94
Animal–vegetal axis, 170
Ankyrin repeats domain protein 5 (ANR5), 41
Apical–basal polarity, 22
Apical IPC cell, 131

Apical junction complexes (AJCs), 141
Apical polarity protein Baz (PAR-3), 60–62, 65, 69
*apkc* mutants, 62
Apoptosis, 71, 76–77, 105, 127–128, 144
Armadillo, 75
Arp2/3 activity, 11
Atomic force microscopy (AFM), 4–5

## B

Baz (PAR-3)-associated AJ puncta, 62
Biological tubes, 138
Bmp signaling, 177–179, 182
Bnl/Fgf signaling, 142
BTB/POZ domain, 73

## C

Cadherin. *See also* E-cadherin
   and adhesion energy, 18
   adhesion proteins, link myosin VII, 7
   based AJs, 35
   cytoplasmic domain, 6
   differential expression, subtypes, 12–13
   distribution, 17
   endocytosis, 46
   expression levels
      in cell sorting, 3
      regulation, 10–11
      *vs.* cytoskeleton remodeling, 10, 18–19
   extracellular domain, 4, 13
   homotypic and heterotypic binding, 18
   mediated
      adhesion, 5
      endocytosis, 40
   recycling, 39
   regulation, 47
   role in adhesion energy, 3
   stabilization, 39–40
   subtype expression, 12
   surface density, 17
   *trans*-binding affinity, 5
   *trans*-dimer complexes, 7
Cadherin–$\beta$-catenin–$\alpha$-catenin complex, 36
Cadherin–catenin complex, 3, 34
Cadherin–p120 binding, 6
*Caenorhabditis elegans*, 138
Cancer growth, 34

Canoe, adapter protein, 76
Canonical Wnt/$\beta$-catenin pathway, 182–183
Cardioblasts (CBs), 156
$\alpha$-E-Catenin, 6–7
$\beta$—Catenin, 6, 11, 34, 74, 90, 149, 151, 173
$\alpha/\beta$-Catenin complex, 6
CD2AP (Cindr), 129
Cdc42, regulators of actin polymerization, 11–12
Cdc42-specific GEF, 38
Cell–cell adhesion, 2, 7, 10
    molecules, 34
Cell–cell contacts, 34
Cell/cell interface, 8
Cell–cell junctions, 34
    epithelial topology, 91
Cell–cell signaling, 89, 103–104
    activating signal transduction, 116–117
Cell delamination, 2
Cell hollowing
    in *C. elegans*, 148–149
    in *D. melanogaster*, 149, 151–153
Cell interfacial tension, 21
Cell/non-cell interfacial tensions, 8
Cell rearrangements, 2
Cell sorting
    adhesion energy differences, 25
    and aggregate mutual envelopment, 17
    cadherin expression levels, 3
    dynamics, 22
    equilibrium states, and contribution of adhesion, 24
    and myosin regulation, 18–19
    and tissue envelopment, 15–16
    and total envelopment, 16–17
    *in vitro* assays, 3
Cellular topology, 92
C&E movement. *See* Convergence and extension movements
Centrosomal-nucleated MT arrays, 70
*Chongmague (chm)* mutant, 107
*Ciona intestinalis*, 107
*cis*-dimers, 5
c-Jun N-terminal kinase (JNK), 41, 71, 73–75, 173
Cleavage plane orientation, single cell, 97–99
Compartmentalization, 2
Compound eye, 116
Concertina (Cta), 64
Cone cells, 117, 120, 122, 125, 128–130
Conservation
    of epithelial architecture, 89–91
    of topological structure, epithelia, 92–94
Convergence and extension movements, 164
    cell polarization during, 183–185
    directed cell migration, 166–167, 170
    mediolateral intercalation, 170
    molecular regulation
        Bmp signaling, 177–179
    and cellular behaviors, 169–170
        G protein-coupled receptors (GPCRs), 179–180
        noncanonical Wnt/PCP pathway, 174–177
        Stat3 signaling, 172–174
        uncoupling of C&E, 180–181
    oriented cell division, 171–172
    radial intercalations, cell layers, 171
    regional and temporal pattern, mesoderm, 165
    signaling pathways, interplay between
        canonical Wnt/$\beta$-catenin pathway, 182–183
        with noncanonical Wnt/PCP pathway, 181–182
    zebrafish gastrulae, pattern, 165–166
        distinct C&E movement domains, 165, 167
        midblastula transition (MBT), 165
        superficial enveloping layer (EVL), 165
Cord hollowing, 144–145
    during neurulation, 146–147
    in vitro studies, 147–148
    zebrafish intestine, lumen formation, 146
*crb* expression, 142

## D

DE-cadherin, 75, 78, 90
Decapentaplegic (Dpp), 71, 117
Desmosomes, 34
Dick-kopf-1 (Dkk1) protein, 183
Differential adhesion mode, 125, 127
*Dino/chordin* mutants, 182
Directed cell migration, 166–167, 170
*Disheveled* causes defects, 107
Division orientation, nongeometric mechanisms, 99–100
Dok phosphorylation, 75
*dPax2* factor, 124
Dpp expression, 75
Dpp receptors, 71
Dpp signaling, 119
*Drosophila*, 19, 21, 43. *See also* Epithelial structure, establishment, morphogenesis
    ankyrin-repeat protein Diego stimulating, 175
    AP elongation during convergence–extension in germband, 19–20
    components of PCP pathway accumulation, 184
    E-cadherin, 36
    eye development, 116–117
        adult eye, compound eye, 117–118
        assembly, fundamental principles, 123–124
        eye field, establishment, 117, 119–120
        larval and pupal eye, 122
        morphogenetic furrow and, 119–121
        peripodial membrane, signaling from, 121
        planar cell polarity, 121, 123
    germ-band elongation, 97

# Subject Index

heart tube development in embryo, 156
hexagonal cell repacking in wing epithelium, 11
mesoderm invagination, 12, 22
morphogenesis and cell movements, 124–125
 differential adhesion, 125
 interommatidial lattice, emergence of, 126–127
 long-range patterning, 127–131
myosin VI activity for dorsal closure, 12
ommatidia, 21
planar cell polarity (PCP) pathway, 174
Rap1, 38
rearrangement in wing disk, 103, 105
salivary gland and trachea development, 142–144
septate junctions (SJs), 90
terminal cells of tracheal system, 153
wing development, 96
wing disk epithelium, 91
*Drosophila rho-associated kinase (Drok)* gene, 106
Dynamin, 57–58
Dynein, 60

## E

E-cadherin, 125
 depletion, 47
 downregulation, 47
 endocytosis, 40
  with Rab11-positive endosomes and recycling, 46
 from endoplasmic reticulum, 10
 expression and inhibition of miR-200, 40
 and F-actin distributed, 38
 HGF-induced downregulation, 38–39
 internalization, 39–40
 mediated adhesion, 8
  and Rap1 activation, 38
 and mutation of PCP signaling molecules, 46
 protein during gastrulation, 47–48
 regulation, 47
 repressors, 40
 stabilization, 44
  p120-dependent, 40
 tail, role in cadherin turnover, 39
 trafficking, 39
  for ubiquitinylation, 11
EC domains, 4
EC1 repeat domain, 5–6
*ed* mutants, 75–76
EGF-receptor pathways, 119
Eiger signaling, 73
Elastic tension, 125
EMT. *See* Epithelial to mesenchyme transition
Enabled (Ena), actin regulator, 66, 77
Ena downregulation, 66
Ena/VASP proteins, 43

Endosome regulator Rab5, 59
Enveloping cell layer (EVL), 22, 165
Epiboly movements, 164
Epithelial cell layers, morphological changes, 34
Epithelial structure
 establishment, 57
  connecting cells, 60–63
  forming first epithelial cells, 57–60
 morphogenesis
  actin and MT network coordination, 78–79
  cell protrusions in dorsal closure, roles for, 77–78
  dorsal closure, 70–71
  force driving closure, 76–77
  germband extension, 67–69
  internalizing mesoderm, 63–67
  JNK-dependent Dpp signaling, 71, 73–74
  microtubule arrays, contribution, 69–70
  nonreceptor tyrosine kinases, role of, 74–75
  roles for regulators, 75–76
 remodeling, 34
Epithelial-to-mesenchyme transition, 3, 11, 40
Epithelial tubes
 architecture, 139
 morphogenetic mechanisms, 139
 types of, 138
 ways to develop, 140
EPLIN
 actin-binding and-stabilizing protein, 35–37, 48
 actin depolymerization inhibitor, 7
E3 ubiquitin ligase, 40

## F

F-actin, 6–7, 36, 38–39, 44, 66, 68–69, 99, 151, 153
Fasciclin III, 90
Fat cadherins, 43–44
Fat1-dependent actin polymerization, 44
Fer kinase, 75
Fgfr signaling, 142
Fgf signaling, 143
Filopodial tethers, 78
Flamingo cadherin, 46, 181
Fog–Cta signaling, 64
Folded gastrulation (Fog), 64
Formin-1, 7
Fyn/Yes-deficient embryos, 182
Fz/Dsh complex, 184

## G

GBE. *See* Germ band extension
G-coupled receptor, Fog, 64
Gdnf signaling, 143
Germ band extension, 15, 67–69
Glia cell line-derived neurotrophic factor (Gdnf), 143

G protein α12/13 subunit, 64
G protein-coupled receptors (GPCRs), 179–180
Gravin, tumor suppressor, 181, 185
Green fluorescent protein (GFP), 125
GTPase-activating proteins (GAPs), 37
GTPase Rac1, 76, 180. *See also* Rac1
Guanine nucleotide exchange factors (GEFs), 37

## H

β-Heavy-spectrin, 62
Hedgehog (Hh) pathways, 119
Heterotypic adhesion energy, 16–18
HGF-induced Rab5 activation, 40
HGF receptor c-Met, 40
Hibris (Hbs), 126
Homophilic interaction, arcadlin molecules, 42
Huckebein (Hkb), 142
Hyaluronic acid synthesizing enzyme 2, 185

## I

Intercellular adhesion
　modulation, in development, 10
　　cadherin–catenin membrane levels, 10–11
　　cadherin subtypes, differential expression, 12–13
　　cortical cytoskeleton organization, 11–12
　molecular basis, 4
　　adhesion energy, rise of, 8–10
　　cell interfacial tensions, 8–10
　　cytoskeleton, changes, 7–8
　　intracellular signaling, cadherins, 6–7
　　trans-binding interactions, cadherins, 4–6
Intercellular adhesion energies, 3–4
　and cadherin expression levels, 18–19
　and cell interfacial tension, 20–21
　cell sorting and tissue envelopment, 15–16
　differences in, 21, 23
　dynamics, 22–24
　gradation, between cells, 3
　heterotypic adhesion energy, 16–18
　and polarity, 19–20
　and tissue
　　liquid-like behavior, 21–22
　　surface tension, 13–15
Interkinetic mode, cell division, 91
Interommatidial precursor cells, 126–127, 129
Intracellular tubes, 138
IPC–IPC adhesion, 131
IPCs. *See* Interommatidial precursor cells

## J

Jaguar, 76
JNKK activators, 73
JNK target genes, 71
Jun N-terminal kinase (JNK) pathway, 71, 73–75

## K

Kirre, nephrin superfamily member, 126–127

## L

Lamellipodia, 175–176, 178, 181, 185
Laminin, 154
Leading edge (LE) cells, 71–72, 74–75, 77

## M

Madin–Darby canine kidney (MDCK) cells, 77
Mediolateral intercalation, 170
Membrane–cytoskeleton mechanical coupling, 7
Metastasis-associated kinase (MAK), 183
Metazoans, life cycle, 2
MicroRNAs (miRNAs), 40
Microtubule
　coordination of actin, 78–79
　cytoskeleton interaction, adherens junction, 8
　disruption, 79
　downregulation, 70
　and dynein, 57
　function during dorsal closure, 78
　important framework, 57
　influence on AJ assembly, 60–61
　inhibitors, 70, 78
　noncentrosomal, 78
　polarity positions Baz (PAR-3) to organize, 61
　protrusion, 70
Midblastula transition (MBT), 165
miR-200, ectopic expression in cancer cell lines, 40
Mitosis, 102
Moesin, 62
Morphogenesis
　adhesion energy-tissue patterning, 4
　of amnioserosa (AS), 70
　cadherin, role in, 3
　cortical actomyosin activity, modulation, 11
　intercellular adhesion dynamics, 23
　isoforms of myosin, involved in, 12
　junctional remodeling
　　cadherin endocytosis, 46
　　cadherin regulation, 47–48
　　through actin modulation, 44–46
　localization, adhesion proteins complex, 22

potential role of Rap1, 39
and tethering of cytoskeleton, 7
of ventral furrow and germband, 65
MT. *See* Microtubule
MT–actin crosstalk functions, embryo, 79
Multicellular tubes, 138
Myosin II, 12, 45, 59, 64, 66–67, 74
  role in zipping, 76
Myosin IIA, 79
Myosin regulatory light chains (MLCs), 45, 141
Myosin VII, 7
Myosin XV, 78

## N

Nap–WAVE complex, 42
N-cadherins, 12, 25, 125, 129–130
  internalization, 42
Nectins, 35, 44
Nephrin family members, 126–127
Neural tube closure, 45
Neurexin IV, 90
Neurulation, 170
NF-protocadherin, 42
Noncanonical Wnt/PCP pathway, 174–177, 181–182, 184. *See also* PCP signaling
Nonreceptor tyrosine kinases, 74
Notch pathways, 119
*Notch*-plus-*Ras* signal, 124
Notochord, 184
N-terminal repeat EC1, 4
Nuclear movements, 131
Nullo mutants, 60

## O

OL-protocadherin, 42. *See also* Cadherin in regulation of cell motility, 43
Ommatidial core, 121. *See also* Cone cells
Ommatidium, 117
Oriented cell division, 171–172

## P

PAR-6/aPKC complex, 61
Paraxial protocadherin (PAPC), 41
  expression, 42
  role in PCP signaling, 42, 46
P-cadherin, 12
p120-catenin, 39
PCP signaling
  activity of PAPC, 42
  controling tissue patterns, 49
  *Dvk* gene to link, 106
  mutation, affecting E-cadherin recycling, 46
PDZ protein, 141
p120 expression
  on cadherin turnover, 40
  in SW40 carcinoma cells, 39

Pigment cells, 117, 121–122, 129–130
  pattern, 131
Planar cell polarity (PCP) genes, 69
Posterior midgut invagination (PMGI), 67
Prechordal plate, 167
Preferential adhesion mode, 126
Proliferating epithelia
  conservation of topological structure, 92–94
    cellular polygons, distributions, 93–94
  geometrical models and cellular mechanics, 100
    cellular Potts models, 102–103
    Dirichlet models, 100–102
    finite-element models, 103–104
    subcellular element models, 103
  topological inference, maximum entropy calculations, 94–95
Protein kinase B, 148
Proteins p120, 6, 11
Protocadherins, 41–43

## R

Rab5 activation, 40
Rab5c-mediated E-cadherin endocytosis, 47
Rac1
  and Cdc42 dependent cytoskeleton remodeling, 10
  and Cdc42 target regulators, 12
  cysts acquire inverted polarity, 12
  and interaction of cells with collagen I, 148
  mediated lamellipodial activity, 8
  and RhoA activities regulation through, 8
Radial intercalation, 171
*Rana pipiens*, 21
R-cadherin, 12
RhoA activities, 6, 8, 41, 175
Rho family GTPases, 11, 37
RhoGEF2
  bottleneck double mutants, 60
  ectopic expression of Fog, 64
Rho kinase
  dependent contraction of actomyosin, 45
  inhibitors, 45
  Shroom3 interaction, 141
Ribbon, DNA-binding protein, 73
Ribbon (Rib), 142
ROCK. *See* Rho kinase
Rosette formation, 45
Roughest (Rst), 126

## S

Shroom3-associated actin filaments, 45
Shroom3 protein, 38, 141
Simple camera eye, 116

Sisyphus, 78
SJ proteins, 90
Slipper (Slpr), 73
Slit/Robo signaling, 156
*Snail* mutant embryos, 66
Sns, nephrin superfamily members, 126–127, 129
Solid-like behavior, epithelia, 22
*Somitabun/smad5* mutants, 182
Somitogenesis, 184
Stat3 signaling, 172–174, 181–182
*String* (Cdc25) mutants, 70
Surface tension, 3

## T

TAO2$\beta$–MEK3 MAPKK–p38 MAPK pathway, 42
Tissue. *See also* Morphogenesis
  anteroposterior patterning, 181
  C&E (*see* Convergence and extension movements)
  cohesion, 2, 11
  compression, 9, 18
  comprise of, 16
  culture, 8
  differentiation, 8
  displacements, 166
  elongation
    during C–E completion, 20
    in vertebrates, 171
    of embryo explants *in vitro*, 21
  envelopment, 15–16 (*see also* Cell sorting)
  extension, 70, 170
  folding and cell division, 98
  growth, 100, 104
  integrity, 13
  interfacial tension, 15, 23
  level biophysical simulations, 106
  level coordination, 106
  liquid-like behavior, 21
  mechanics and cellular rearrangement, 109
  morphogenesis, 4, 7, 16, 89, 109
  to move dorsally, 78
  nonpolarized, 144
  patterning, 121, 127
    and polarity, 20
    and remodeling, 121
  polarization, 19, 185
    prior to lumen formation, 146
  reconstruction from, 128
  regeneration, 34
  remodeling, 3
    at gastrulation, 60
  repositioning, 2
  separation, 41
  shape, 2
  specific expression patterns, 12

surface area, 14
surface tension, 13–15, 20
types, 12
viscosity, 23–24
Topological models, epithelia, 95–97
Toroid cells, 155
*Trans*-dimerization, 4, 6
T48, transmembrane protein, 64
Tuba, 38
Tuba–Cdc42–N-WASP pathway, 38
Tube formation, cellular aspects, 139
  from nonpolarized tissue
    cavitation, 144
    cell hollowing, 148–153
    cord hollowing, 144–148
  novel modes
    cell assembly, 156
    cell wrapping/self-fusion, 154–155
  from prepolarized epithelia
    budding, 141–144
    wrapping, 141
Tubular epithelium, 138
Tubular structures, 138
Tumor necrosis factor (TNF), 73
*Twist* mutant embryos, 66
Tyrosine kinase, 40

## U

Ubiquitinated E-cadherin, 40
Unconventional myosin VI, 76
Unicellular tubes, 138

## V

Vang/Stbm/Pk complex, 184
Vascular endothelial growth factor (VEGF), 143
VE-cadherins, 12
v-Src activation, 40

## W

Wnt pathways, 47, 116
Wnt signaling, 71

## X

*Xenopus*, 19, 22, 41–42, 47, 97–98, 108, 141, 165, 170

## Y

Yolk syncytial layer (YSL), 22

## Z

ZA. *See* Zonula adherens
ZA proteins, *Drosophila*, 90
ZEB1 ($\delta$EF1) and ZEB2 (SIP2), involve in EMT, 40

# Subject Index

Zebrafish gastrulae, mesodermal cells migration. *See also* Convergence and extension movements
- directed migration during C&E., 167
- distinct C&E movement domains, 165, 167
- epiboly movements, radial intercalation, 171
- guiding cues, 173
- noncanonicalWnt/PCP pathway components, 185
- noncanonical Wnt signaling, 176
- pattern of C&E movements, 166
- PGE2 signaling through, 179
- Pk localization, 185
- Stat3 signaling, 172

*Zipper* expression, 74

Zonula adherens, 34, 38, 90

ZO-1 (Pyd), 129

# Contents of Previous Volumes

## Volume 47

1. **Early Events of Somitogenesis in Higher Vertebrates: Allocation of Precursor Cells during Gastrulation and the Organization of a Moristic Pattern in the Paraxial Mesoderm**
   *Patrick P. L. Tam, Devorah Goldman, Anne Camus, and Gary C. Shoenwolf*

2. **Retrospective Tracing of the Developmental Lineage of the Mouse Myotome**
   *Sophie Eloy-Trinquet, Luc Mathis, and Jean-François Nicolas*

3. **Segmentation of the Paraxial Mesoderm and Vertebrate Somitogenesis**
   *Olivier Pourqulé*

4. **Segmentation: A View from the Border**
   *Claudio D. Stern and Daniel Vasiliauskas*

5. **Genetic Regulation of Somite Formation**
   *Alan Rawls, Jeanne Wilson-Rawls, and Eric N. Olsen*

6. **Hox Genes and the Global Patterning of the Somitic Mesoderm**
   *Ann Campbell Burke*

7. **The Origin and Morphogenesis of Amphibian Somites**
   *Ray Keller*

8. **Somitogenesis in Zebrafish**
   *Scott A. Halley and Christiana Nüsslain-Volhard*

9. **Rostrocaudal Differences within the Somites Confer Segmental Pattern to Trunk Neural Crest Migration**
   *Marianne Bronner-Fraser*

## Volume 48

1. **Evolution and Development of Distinct Cell Lineages Derived from Somites**
   *Beate Brand-Saberi and Bodo Christ*

2. **Duality of Molecular Signaling Involved in Vertebral Chondrogenesis**
   Anne-Hélène Monsoro-Burq and Nicole Le Douarin
3. **Sclerotome Induction and Differentiation**
   Jennifer L. Docker
4. **Genetics of Muscle Determination and Development**
   Hans-Henning Arnold and Thomas Braun
5. **Multiple Tissue Interactions and Signal Transduction Pathways Control Somite Myogenesis**
   Anne-Gaëlle Borycki and Charles P. Emerson, Jr.
6. **The Birth of Muscle Progenitor Cells in the Mouse: Spatiotemporal Considerations**
   Shahragim Tajbakhsh and Margaret Buckingham
7. **Mouse–Chick Chimera: An Experimental System for Study of Somite Development**
   Josiane Fontaine-Pérus
8. **Transcriptional Regulation during Somitogenesis**
   Dennis Summerbell and Peter W. J. Rigby
9. **Determination and Morphogenesis in Myogenic Progenitor Cells: An Experimental Embryological Approach**
   Charles P. Ordahl, Brian A. Williams, and Wilfred Denetclaw

## Volume 49

1. **The Centrosome and Parthenogenesis**
   Thomas Küntziger and Michel Bornens
2. **$\gamma$-Tubulin**
   Berl R. Oakley
3. **$\gamma$-Tubulin Complexes and Their Role in Microtubule Nucleation**
   Ruwanthi N. Gunawardane, Sofia B. Lizarraga, Christiane Wiese, Andrew Wilde, and Yixian Zheng
4. **$\gamma$-Tubulin of Budding Yeast**
   Jackie Vogel and Michael Snyder
5. **The Spindle Pole Body of *Saccharomyces cerevisiae*: Architecture and Assembly of the Core Components**
   Susan E. Francis and Trisha N. Davis

6. **The Microtubule Organizing Centers of *Schizosaccharomyces pombe***
   Iain M. Hagan and Janni Petersen
7. **Comparative Structural, Molecular, and Functional Aspects of the *Dictyostelium discoideum* Centrosome**
   Ralph Gräf, Nicole Brusis, Christine Daunderer, Ursula Euteneuer, Andrea Hestermann, Manfred Schliwa, and Masahiro Ueda
8. **Are There Nucleic Acids in the Centrosome?**
   Wallace F. Marshall and Joel L. Rosenbaum
9. **Basal Bodies and Centrioles: Their Function and Structure**
   Andrea M. Preble, Thomas M. Giddings, Jr., and Susan K. Dutcher
10. **Centriole Duplication and Maturation in Animal Cells**
    B. M. H. Lange, A. J. Faragher, P. March, and K. Gull
11. **Centrosome Replication in Somatic Cells: The Significance of the $G_1$ Phase**
    Ron Balczon
12. **The Coordination of Centrosome Reproduction with Nuclear Events during the Cell Cycle**
    Greenfield Sluder and Edward H. Hinchcliffe
13. **Regulating Centrosomes by Protein Phosphorylation**
    Andrew M. Fry, Thibault Mayor, and Erich A. Nigg
14. **The Role of the Centrosome in the Development of Malignant Tumors**
    Wilma L. Lingle and Jeffrey L. Salisbury
15. **The Centrosome-Associated Aurora/Ipl-like Kinase Family**
    T. M. Goepfert and B. R. Brinkley
16. **Centrosome Reduction during Mammalian Spermiogenesis**
    G. Manandhar, C. Simerly, and G. Schatten
17. **The Centrosome of the Early *C. elegans* Embryo: Inheritance, Assembly, Replication, and Developmental Roles**
    Kevin F. O'Connell
18. **The Centrosome in *Drosophila* Oocyte Development**
    Timothy L. Megraw and Thomas C. Kaufman
19. **The Centrosome in Early *Drosophila* Embryogenesis**
    W. F. Rothwell and W. Sullivan

20. Centrosome Maturation

  Robert E. Palazzo, Jacalyn M. Vogel, Bradley J. Schnackenberg, Dawn R. Hull, and Xingyong Wu

## Volume 50

1. Patterning the Early Sea Urchin Embryo

  Charles A. Ettensohn and Hyla C. Sweet

2. Turning Mesoderm into Blood: The Formation of Hematopoietic Stem Cells during Embryogenesis

  Alan J. Davidson and Leonard I. Zon

3. Mechanisms of Plant Embryo Development

  Shunong Bai, Lingjing Chen, Mary Alice Yund, and Zinmay Rence Sung

4. Sperm-Mediated Gene Transfer

  Anthony W. S. Chan, C. Marc Luetjens, and Gerald P. Schatten

5. Gonocyte–Sertoli Cell Interactions during Development of the Neonatal Rodent Testis

  Joanne M. Orth, William F. Jester, Ling-Hong Li, and Andrew L. Laslett

6. Attributes and Dynamics of the Endoplasmic Reticulum in Mammalian Eggs

  Douglas Kline

7. Germ Plasm and Molecular Determinants of Germ Cell Fate

  Douglas W. Houston and Mary Lou King

## Volume 51

1. Patterning and Lineage Specification in the Amphibian Embryo

  Agnes P. Chan and Laurence D. Etkin

2. Transcriptional Programs Regulating Vascular Smooth Muscle Cell Development and Differentiation

  Michael S. Parmacek

3. Myofibroblasts: Molecular Crossdressers

  Gennyne A. Walker, Ivan A. Guerrero, and Leslie A. Leinwand

4. Checkpoint and DNA-Repair Proteins Are Associated with the Cores of Mammalian Meiotic Chromosomes
   Madalena Tarsounas and Peter B. Moens

5. Cytoskeletal and $Ca^{2+}$ Regulation of Hyphal Tip Growth and Initiation
   Sara Torralba and I. Brent Heath

6. Pattern Formation during *C. elegans* Vulval Induction
   Minqin Wang and Paul W. Sternberg

7. A Molecular Clock Involved in Somite Segmentation
   Miguel Maroto and Olivier Pourquié

## Volume 52

1. Mechanism and Control of Meiotic Recombination Initiation
   Scott Keeney

2. Osmoregulation and Cell Volume Regulation in the Preimplantation Embryo
   Jay M. Baltz

3. Cell–Cell Interactions in Vascular Development
   Diane C. Darland and Patricia A. D'Amore

4. Genetic Regulation of Preimplantation Embryo Survival
   Carol M. Warner and Carol A. Brenner

## Volume 53

1. Developmental Roles and Clinical Significance of Hedgehog Signaling
   Andrew P. McMahon, Philip W. Ingham, and Clifford J. Tabin

2. Genomic Imprinting: Could the Chromatin Structure Be the Driving Force?
   Andras Paldi

3. Ontogeny of Hematopoiesis: Examining the Emergence of Hematopoietic Cells in the Vertebrate Embryo
   Jenna L. Galloway and Leonard I. Zon

4. Patterning the Sea Urchin Embryo: Gene Regulatory Networks, Signaling Pathways, and Cellular Interactions
   Lynne M. Angerer and Robert C. Angerer

## Volume 54

1. **Membrane Type-Matrix Metalloproteinases (MT-MMP)**
   Stanley Zucker, Duanqing Pei, Jian Cao, and Carlos Lopez-Otin

2. **Surface Association of Secreted Matrix Metalloproteinases**
   Rafael Fridman

3. **Biochemical Properties and Functions of Membrane-Anchored Metalloprotease-Disintegrin Proteins (ADAMs)**
   J. David Becherer and Carl P. Blobel

4. **Shedding of Plasma Membrane Proteins**
   Joaquín Arribas and Anna Merlos-Suárez

5. **Expression of Meprins in Health and Disease**
   Lourdes P. Norman, Gail L. Matters, Jacqueline M. Crisman, and Judith S. Bond

6. **Type II Transmembrane Serine Proteases**
   Qingyu Wu

7. **DPPIV, Seprase, and Related Serine Peptidases in Multiple Cellular Functions**
   Wen-Tien Chen, Thomas Kelly, and Giulio Ghersi

8. **The Secretases of Alzheimer's Disease**
   Michael S. Wolfe

9. **Plasminogen Activation at the Cell Surface**
   Vincent Ellis

10. **Cell-Surface Cathepsin B: Understanding Its Functional Significance**
    Dora Cavallo-Medved and Bonnie F. Sloane

11. **Protease-Activated Receptors**
    Wadie F. Bahou

12. **Emmprin (CD147), a Cell Surface Regulator of Matrix Metalloproteinase Production and Function**
    Bryan P. Toole

13. **The Evolving Roles of Cell Surface Proteases in Health and Disease: Implications for Developmental, Adaptive, Inflammatory, and Neoplastic Processes**
    Joseph A. Madri

14. Shed Membrane Vesicles and Clustering of Membrane-Bound Proteolytic Enzymes
    M. Letizia Vittorelli

## Volume 55

1. The Dynamics of Chromosome Replication in Yeast
   Isabelle A. Lucas and M. K. Raghuraman

2. Micromechanical Studies of Mitotic Chromosomes
   M. G. Poirier and John F. Marko

3. Patterning of the Zebrafish Embryo by Nodal Signals
   Jennifer O. Liang and Amy L. Rubinstein

4. Folding Chromosomes in Bacteria: Examining the Role of Csp Proteins and Other Small Nucleic Acid-Binding Proteins
   Nancy Trun and Danielle Johnston

## Volume 56

1. Selfishness in Moderation: Evolutionary Success of the Yeast Plasmid
   Soundarapandian Velmurugan, Shwetal Mehta, and Makkuni Jayaram

2. Nongenomic Actions of Androgen in Sertoli Cells
   William H. Walker

3. Regulation of Chromatin Structure and Gene Activity by Poly(ADP Ribose) Polymerases
   Alexei Tulin, Yurli Chinenov, and Allan Spradling

4. Centrosomes and Kinetochores, Who needs 'Em? The Role of Noncentromeric Chromatin in Spindle Assembly
   Priya Prakash Budde and Rebecca Heald

5. Modeling Cardiogenesis: The Challenges and Promises of 3D Reconstruction
   Jeffrey O. Penetcost, Claudio Silva, Maurice Pesticelli, Jr., and Kent L. Thornburg

6. Plasmid and Chromosome Traffic Control: How ParA and ParB Drive Partition
   Jennifer A. Surtees and Barbara E. Funnell

## Volume 57

1. **Molecular Conservation and Novelties in Vertebrate Ear Development**
   B. Fritzsch and K. W. Beisel

2. **Use of Mouse Genetics for Studying Inner Ear Development**
   Elizabeth Quint and Karen P. Steel

3. **Formation of the Outer and Middle Ear, Molecular Mechanisms**
   Moisés Mallo

4. **Molecular Basis of Inner Ear Induction**
   Stephen T. Brown, Kareen Martin, and Andrew K. Groves

5. **Molecular Basis of Otic Commitment and Morphogenesis: A Role for Homeodomain-Containing Transcription Factors and Signaling Molecules**
   Eva Bober, Silke Rinkwitz, and Heike Herbrand

6. **Growth Factors and Early Development of Otic Neurons: Interactions between Intrinsic and Extrinsic Signals**
   Berta Alsina, Fernando Giraldez, and Isabel Varela-Nieto

7. **Neurotrophic Factors during Inner Ear Development**
   Ulla Pirvola and Jukka Ylikoski

8. **FGF Signaling in Ear Development and Innervation**
   Tracy J. Wright and Suzanne L. Mansour

9. **The Roles of Retinoic Acid during Inner Ear Development**
   Raymond Romand

10. **Hair Cell Development in Higher Vertebrates**
    Wei-Qiang Gao

11. **Cell Adhesion Molecules during Inner Ear and Hair Cell Development, Including Notch and Its Ligands**
    Matthew W. Kelley

12. **Genes Controlling the Development of the Zebrafish Inner Ear and Hair Cells**
    Bruce B. Riley

13. **Functional Development of Hair Cells**
    Ruth Anne Eatock and Karen M. Hurley

14. **The Cell Cycle and the Development and Regeneration of Hair Cells**
    *Allen F. Ryan*

## Volume 58

1. **A Role for Endogenous Electric Fields in Wound Healing**
   *Richard Nuccitelli*
2. **The Role of Mitotic Checkpoint in Maintaining Genomic Stability**
   *Song-Tao Liu, Jan M. van Deursen, and Tim J. Yen*
3. **The Regulation of Oocyte Maturation**
   *Ekaterina Voronina and Gary M. Wessel*
4. **Stem Cells: A Promising Source of Pancreatic Islets for Transplantation in Type 1 Diabetes**
   *Cale N. Street, Ray V. Rajotte, and Gregory S. Korbutt*
5. **Differentiation Potential of Adipose Derived Adult Stem (ADAS) Cells**
   *Jeffrey M. Gimble and Farshid Guilak*

## Volume 59

1. **The Balbiani Body and Germ Cell Determinants: 150 Years Later**
   *Malgorzata Kloc, Szczepan Bilinski, and Laurence D. Etkin*
2. **Fetal–Maternal Interactions: Prenatal Psychobiological Precursors to Adaptive Infant Development**
   *Matthew F. S. X. Novak*
3. **Paradoxical Role of Methyl-CpG-Binding Protein 2 in Rett Syndrome**
   *Janine M. LaSalle*
4. **Genetic Approaches to Analyzing Mitochondrial Outer Membrane Permeability**
   *Brett H. Graham and William J. Craigen*
5. **Mitochondrial Dynamics in Mammals**
   *Hsiuchen Chen and David C. Chan*
6. **Histone Modification in Corepressor Functions**
   *Judith K. Davie and Sharon Y. R. Dent*
7. **Death by Abl: A Matter of Location**
   *Jiangyu Zhu and Jean Y. J. Wang*

## Volume 60

1. **Therapeutic Cloning and Tissue Engineering**
   Chester J. Koh and Anthony Atala

2. **α-Synuclein: Normal Function and Role in Neurodegenerative Diseases**
   Erin H. Norris, Benoit I. Giasson, and Virginia M.-Y. Lee

3. **Structure and Function of Eukaryotic DNA Methyltransferases**
   Taiping Chen and En Li

4. **Mechanical Signals as Regulators of Stem Cell Fate**
   Bradley T. Estes, Jeffrey M. Gimble, and Farshid Guilak

5. **Origins of Mammalian Hematopoiesis: *In Vivo* Paradigms and *In Vitro* Models**
   M. William Lensch and George Q. Daley

6. **Regulation of Gene Activity and Repression: A Consideration of Unifying Themes**
   Anne C. Ferguson-Smith, Shau-Ping Lin, and Neil Youngson

7. **Molecular Basis for the Chloride Channel Activity of Cystic Fibrosis Transmembrane Conductance Regulator and the Consequences of Disease-Causing Mutations**
   Jackie F. Kidd, Ilana Kogan, and Christine E. Bear

## Volume 61

1. **Hepatic Oval Cells: Helping Redefine a Paradigm in Stem Cell Biology**
   P. N. Newsome, M. A. Hussain, and N. D. Theise

2. **Meiotic DNA Replication**
   Randy Strich

3. **Pollen Tube Guidance: The Role of Adhesion and Chemotropic Molecules**
   Sunran Kim, Juan Dong, and Elizabeth M. Lord

4. **The Biology and Diagnostic Applications of Fetal DNA and RNA in Maternal Plasma**
   Rossa W. K. Chiu and Y. M. Dennis Lo

5. **Advances in Tissue Engineering**
   Shulamit Levenberg and Robert Langer

6. **Directions in Cell Migration Along the Rostral Migratory Stream: The Pathway for Migration in the Brain**
   *Shin-ichi Murase and Alan F. Horwitz*

7. **Retinoids in Lung Development and Regeneration**
   *Malcolm Maden*

8. **Structural Organization and Functions of the Nucleus in Development, Aging, and Disease**
   *Leslie Mounkes and Colin L. Stewart*

## Volume 62

1. **Blood Vessel Signals During Development and Beyond**
   *Ondine Cleaver*

2. **HIFs, Hypoxia, and Vascular Development**
   *Kelly L. Covello and M. Celeste Simon*

3. **Blood Vessel Patterning at the Embryonic Midline**
   *Kelly A. Hogan and Victoria L. Bautch*

4. **Wiring the Vascular Circuitry: From Growth Factors to Guidance Cues**
   *Lisa D. Urness and Dean Y. Li*

5. **Vascular Endothelial Growth Factor and Its Receptors in Embryonic Zebrafish Blood Vessel Development**
   *Katsutoshi Goishi and Michael Klagsbrun*

6. **Vascular Extracellular Matrix and Aortic Development**
   *Cassandra M. Kelleher, Sean E. McLean, and Robert P. Mecham*

7. **Genetics in Zebrafish, Mice, and Humans to Dissect Congenital Heart Disease: Insights in the Role of VEGF**
   *Diether Lambrechts and Peter Carmeliet*

8. **Development of Coronary Vessels**
   *Mark W. Majesky*

9. **Identifying Early Vascular Genes Through Gene Trapping in Mouse Embryonic Stem Cells**
   *Frank Kuhnert and Heidi Stuhlmann*

## Volume 63

1. **Early Events in the DNA Damage Response**
   *Irene Ward and Junjie Chen*

2. **Afrotherian Origins and Interrelationships: New Views and Future Prospects**
   *Terence J. Robinson and Erik R. Seiffert*

3. **The Role of Antisense Transcription in the Regulation of X-Inactivation**
   *Claire Rougeulle and Philip Avner*

4. **The Genetics of Hiding the Corpse: Engulfment and Degradation of Apoptotic Cells in *C. elegans* and *D. melanogaster***
   *Zheng Zhou, Paolo M. Mangahas, and Xiaomeng Yu*

5. **Beginning and Ending an Actin Filament: Control at the Barbed End**
   *Sally H. Zigmond*

6. **Life Extension in the Dwarf Mouse**
   *Andrzej Bartke and Holly Brown-Borg*

## Volume 64

1. **Stem/Progenitor Cells in Lung Morphogenesis, Repair, and Regeneration**
   *David Warburton, Mary Anne Berberich, and Barbara Driscoll*

2. **Lessons from a Canine Model of Compensatory Lung Growth**
   *Connie C. W. Hsia*

3. **Airway Glandular Development and Stem Cells**
   *Xiaoming Liu, Ryan R. Driskell, and John F. Engelhardt*

4. **Gene Expression Studies in Lung Development and Lung Stem Cell Biology**
   *Thomas J. Mariani and Naftali Kaminski*

5. **Mechanisms and Regulation of Lung Vascular Development**
   *Michelle Haynes Pauling and Thiennu H. Vu*

6. **The Engineering of Tissues Using Progenitor Cells**
   *Nancy L. Parenteau, Lawrence Rosenberg, and Janet Hardin-Young*

7. **Adult Bone Marrow-Derived Hemangioblasts, Endothelial Cell Progenitors, and EPCs**
   Gina C. Schatteman

8. **Synthetic Extracellular Matrices for Tissue Engineering and Regeneration**
   Eduardo A. Silva and David J. Mooney

9. **Integrins and Angiogenesis**
   D. G. Stupack and D. A. Cheresh

## Volume 65

1. **Tales of Cannibalism, Suicide, and Murder: Programmed Cell Death in *C. elegans***
   Jason M. Kinchen and Michael O. Hengartner

2. **From Guts to Brains: Using Zebrafish Genetics to Understand the Innards of Organogenesis**
   Carsten Stuckenholz, Paul E. Ulanch, and Nathan Bahary

3. **Synaptic Vesicle Docking: A Putative Role for the Munc18/Sec1 Protein Family**
   Robby M. Weimer and Janet E. Richmond

4. **ATP-Dependent Chromatin Remodeling**
   Corey L. Smith and Craig L. Peterson

5. **Self-Destruct Programs in the Processes of Developing Neurons**
   David Shepherd and V. Hugh Perry

6. **Multiple Roles of Vascular Endothelial Growth Factor (VEGF) in Skeletal Development, Growth, and Repair**
   Elazar Zelzer and Bjorn R. Olsen

7. **G-Protein Coupled Receptors and Calcium Signaling in Development**
   Geoffrey E. Woodard and Juan A. Rosado

8. **Differential Functions of 14-3-3 Isoforms in Vertebrate Development**
   Anthony J. Muslin and Jeffrey M. C. Lau

9. **Zebrafish Notochordal Basement Membrane: Signaling and Structure**
   Annabelle Scott and Derek L. Stemple

10. **Sonic Hedgehog Signaling and the Developing Tooth**
    Martyn T. Cobourne and Paul T. Sharpe

## Volume 66

1. **Stepwise Commitment from Embryonic Stem to Hematopoietic and Endothelial Cells**
   *Changwon Park, Jesse J. Lugus, and Kyunghee Choi*

2. **Fibroblast Growth Factor Signaling and the Function and Assembly of Basement Membranes**
   *Peter Lonai*

3. **TGF-$\beta$ Superfamily and Mouse Craniofacial Development: Interplay of Morphogenetic Proteins and Receptor Signaling Controls Normal Formation of the Face**
   *Marek Dudas and Vesa Kaartinen*

4. **The Colors of Autumn Leaves as Symptoms of Cellular Recycling and Defenses Against Environmental Stresses**
   *Helen J. Ougham, Phillip Morris, and Howard Thomas*

5. **Extracellular Proteases: Biological and Behavioral Roles in the Mammalian Central Nervous System**
   *Yan Zhang, Kostas Pothakos, and Styliana-Anna (Stella) Tsirka*

6. **The Genetic Architecture of House Fly Mating Behavior**
   *Lisa M. Meffert and Kara L. Hagenbuch*

7. **Phototropins, Other Photoreceptors, and Associated Signaling: The Lead and Supporting Cast in the Control of Plant Movement Responses**
   *Bethany B. Stone, C. Alex Esmon, and Emmanuel Liscum*

8. **Evolving Concepts in Bone Tissue Engineering**
   *Catherine M. Cowan, Chia Soo, Kang Ting, and Benjamin Wu*

9. **Cranial Suture Biology**
   *Kelly A Lenton, Randall P. Nacamuli, Derrick C. Wan, Jill A. Helms, and Michael T. Longaker*

## Volume 67

1. **Deer Antlers as a Model of Mammalian Regeneration**
   *Joanna Price, Corrine Faucheux, and Steve Allen*

2. **The Molecular and Genetic Control of Leaf Senescence and Longevity in *Arabidopsis***
   *Pyung Ok Lim and Hong Gil Nam*

3. **Cripto-1: An Oncofetal Gene with Many Faces**
   *Caterina Bianco, Luigi Strizzi, Nicola Normanno, Nadia Khan, and David S. Salomon*

4. **Programmed Cell Death in Plant Embryogenesis**
   *Peter V. Bozhkov, Lada H. Filonova, and Maria F. Suarez*

5. **Physiological Roles of Aquaporins in the Choroid Plexus**
   *Daniela Boassa and Andrea J. Yool*

6. **Control of Food Intake Through Regulation of cAMP**
   *Allan Z. Zhao*

7. **Factors Affecting Male Song Evolution in *Drosophila montana***
   *Anneli Hoikkala, Kirsten Klappert, and Dominique Mazzi*

8. **Prostanoids and Phosphodiesterase Inhibitors in Experimental Pulmonary Hypertension**
   *Ralph Theo Schermuly, Hossein Ardeschir Ghofrani, and Norbert Weissmann*

9. **14-3-3 Protein Signaling in Development and Growth Factor Responses**
   *Daniel Thomas, Mark Guthridge, Jo Woodcock, and Angel Lopez*

10. **Skeletal Stem Cells in Regenerative Medicine**
    *Wataru Sonoyama, Carolyn Coppe, Stan Gronthos, and Songtao Shi*

## Volume 68

1. **Prolactin and Growth Hormone Signaling**
   *Beverly Chilton and Aveline Hewetson*

2. **Alterations in cAMP-Mediated Signaling and Their Role in the Pathophysiology of Dilated Cardiomyopathy**
   *Matthew A. Movsesian and Michael R. Bristow*

3. **Corpus Luteum Development: Lessons from Genetic Models in Mice**
   *Anne Bachelot and Nadine Binart*

4. **Comparative Developmental Biology of the Mammalian Uterus**
   *Thomas E. Spencer, Kanako Hayashi, Jianbo Hu, and Karen D. Carpenter*

5. Sarcopenia of Aging and Its Metabolic Impact
   *Helen Karakelides and K. Sreekumaran Nair*
6. Chemokine Receptor CXCR3: An Unexpected Enigma
   *Liping Liu, Melissa K. Callahan, DeRen Huang, and Richard M. Ransohoff*
7. Assembly and Signaling of Adhesion Complexes
   *Jorge L. Sepulveda, Vasiliki Gkretsi, and Chuanyue Wu*
8. Signaling Mechanisms of Higher Plant Photoreceptors: A Structure-Function Perspective
   *Haiyang Wang*
9. Initial Failure in Myoblast Transplantation Therapy Has Led the Way Toward the Isolation of Muscle Stem Cells: Potential for Tissue Regeneration
   *Kenneth Urish, Yasunari Kanda, and Johnny Huard*
10. Role of 14-3-3 Proteins in Eukaryotic Signaling and Development
    *Dawn L. Darling, Jessica Yingling, and Anthony Wynshaw-Boris*

## Volume 69

1. Flipping Coins in the Fly Retina
   *Tamara Mikeladze-Dvali, Claude Desplan, and Daniela Pistillo*
2. Unraveling the Molecular Pathways That Regulate Early Telencephalon Development
   *Jean M. Hébert*
3. Glia–Neuron Interactions in Nervous System Function and Development
   *Shai Shaham*
4. The Novel Roles of Glial Cells Revisited: The Contribution of Radial Glia and Astrocytes to Neurogenesis
   *Tetsuji Mori, Annalisa Buffo, and Magdalena Götz*
5. Classical Embryological Studies and Modern Genetic Analysis of Midbrain and Cerebellum Development
   *Mark Zervas, Sandra Blaess, and Alexandra L. Joyner*
6. Brain Development and Susceptibility to Damage; Ion Levels and Movements
   *Maria Erecinska, Shobha Cherian, and Ian A. Silver*

7. **Thinking about Visual Behavior; Learning about Photoreceptor Function**

   Kwang-Min Choe and Thomas R. Clandinin

8. **Critical Period Mechanisms in Developing Visual Cortex**

   Takao K. Hensch

9. **Brawn for Brains: The Role of MEF2 Proteins in the Developing Nervous System**

   Aryaman K. Shalizi and Azad Bonni

10. **Mechanisms of Axon Guidance in the Developing Nervous System**

    Céline Plachez and Linda J. Richards

## Volume 70

1. **Magnetic Resonance Imaging: Utility as a Molecular Imaging Modality**

   James P. Basilion, Susan Yeon, and René Botnar

2. **Magnetic Resonance Imaging Contrast Agents in the Study of Development**

   Angelique Louie

3. **$^1$H/$^{19}$F Magnetic Resonance Molecular Imaging with Perfluorocarbon Nanoparticles**

   Gregory M. Lanza, Patrick M. Winter, Anne M. Neubauer, Shelton D. Caruthers, Franklin D. Hockett, and Samuel A. Wickline

4. **Loss of Cell Ion Homeostasis and Cell Viability in the Brain: What Sodium MRI Can Tell Us**

   Fernando E. Boada, George LaVerde, Charles Jungreis, Edwin Nemoto, Costin Tanase, and Ileana Hancu

5. **Quantum Dot Surfaces for Use *In Vivo* and *In Vitro***

   Byron Ballou

6. ***In Vivo* Cell Biology of Cancer Cells Visualized with Fluorescent Proteins**

   Robert M. Hoffman

7. **Modulation of Tracer Accumulation in Malignant Tumors: Gene Expression, Gene Transfer, and Phage Display**

   Uwe Haberkorn

8. Amyloid Imaging: From Benchtop to Bedside
   *Chungying Wu, Victor W. Pike, and Yanming Wang*
9. *In Vivo* Imaging of Autoimmune Disease in Model Systems
   *Eric T. Ahrens and Penelope A. Morel*

## Volume 71

1. The Choroid Plexus-Cerebrospinal Fluid System: From Development to Aging
   *Zoran B. Redzic, Jane E. Preston, John A. Duncan, Adam Chodobski, and Joanna Szmydynger-Chodobska*
2. Zebrafish Genetics and Formation of Embryonic Vasculature
   *Tao P. Zhong*
3. Leaf Senescence: Signals, Execution, and Regulation
   *Yongfeng Guo and Susheng Gan*
4. Muscle Stem Cells and Regenerative Myogenesis
   *Iain W. McKinnell, Gianni Parise, and Michael A. Rudnicki*
5. Gene Regulation in Spermatogenesis
   *James A. MacLean II and Miles F. Wilkinson*
6. Modeling Age-Related Diseases in *Drosophila:* Can this Fly?
   *Kinga Michno, Diana van de Hoef, Hong Wu, and Gabrielle L. Boulianne*
7. Cell Death and Organ Development in Plants
   *Hilary J. Rogers*
8. The Blood-Testis Barrier: Its Biology, Regulation, and Physiological Role in Spermatogenesis
   *Ching-Hang Wong and C. Yan Cheng*
9. Angiogenic Factors in the Pathogenesis of Preeclampsia
   *Hai-Tao Yuan, David Haig, and S. Ananth Karumanchi*

## Volume 72

1. Defending the Zygote: Search for the Ancestral Animal Block to Polyspermy
   *Julian L. Wong and Gary M. Wessel*

2. **Dishevelled: A Mobile Scaffold Catalyzing Development**
   Craig C. Malbon and Hsien-yu Wang

3. **Sensory Organs: Making and Breaking the Pre-Placodal Region**
   Andrew P. Bailey and Andrea Streit

4. **Regulation of Hepatocyte Cell Cycle Progression and Differentiation by Type I Collagen Structure**
   Linda K. Hansen, Joshua Wilhelm, and John T. Fassett

5. **Engineering Stem Cells into Organs: Topobiological Transformations Demonstrated by Beak, Feather, and Other Ectodermal Organ Morphogenesis**
   Cheng-Ming Chuong, Ping Wu, Maksim Plikus, Ting-Xin Jiang, and Randall Bruce Widelitz

6. **Fur Seal Adaptations to Lactation: Insights into Mammary Gland Function**
   Julie A. Sharp, Kylie N. Cane, Christophe Lefevre, John P. Y. Arnould, and Kevin R. Nicholas

## Volume 73

1. **The Molecular Origins of Species-Specific Facial Pattern**
   Samantha A. Brugmann, Minal D. Tapadia, and Jill A. Helms

2. **Molecular Bases of the Regulation of Bone Remodeling by the Canonical Wnt Signaling Pathway**
   Donald A. Glass II and Gerard Karsenty

3. **Calcium Sensing Receptors and Calcium Oscillations: Calcium as a First Messenger**
   Gerda E. Breitwieser

4. **Signal Relay During the Life Cycle of *Dictyostelium***
   Dana C. Mahadeo and Carole A. Parent

5. **Biological Principles for *Ex Vivo* Adult Stem Cell Expansion**
   Jean-François Paré and James L. Sherley

6. **Histone Deacetylation as a Target for Radiosensitization**
   David Cerna, Kevin Camphausen, and Philip J. Tofilon

7. **Chaperone-Mediated Autophagy in Aging and Disease**
   Ashish C. Massey, Cong Zhang, and Ana Maria Cuervo

8. Extracellular Matrix Macroassembly Dynamics in Early Vertebrate Embryos

   *Andras Czirok, Evan A. Zamir, Michael B. Filla, Charles D. Little, and Brenda J. Rongish*

## Volume 74

1. Membrane Origin for Autophagy

   *Fulvio Reggiori*

2. Chromatin Assembly with H3 Histones: Full Throttle Down Multiple Pathways

   *Brian E. Schwartz and Kami Ahmad*

3. Protein–Protein Interactions of the Developing Enamel Matrix

   *John D. Bartlett, Bernhard Ganss, Michel Goldberg, Janet Moradian-Oldak, Michael L. Paine, Malcolm L. Snead, Xin Wen, Shane N. White, and Yan L. Zhou*

4. Stem and Progenitor Cells in the Formation of the Pulmonary Vasculature

   *Kimberly A. Fisher and Ross S. Summer*

5. Mechanisms of Disordered Granulopoiesis in Congenital Neutropenia

   *David S. Grenda and Daniel C. Link*

6. Social Dominance and Serotonin Receptor Genes in Crayfish

   *Donald H. Edwards and Nadja Spitzer*

7. Transplantation of Undifferentiated, Bone Marrow-Derived Stem Cells

   *Karen Ann Pauwelyn and Catherine M. Verfaillie*

8. The Development and Evolution of Division of Labor and Foraging Specialization in a Social Insect (*Apis mellifera* L.)

   *Robert E. Page Jr., Ricarda Scheiner, Joachim Erber, and Gro V. Amdam*

## Volume 75

1. Dynamics of Assembly and Reorganization of Extracellular Matrix Proteins

   *Sarah L. Dallas, Qian Chen, and Pitchumani Sivakumar*

2. Selective Neuronal Degeneration in Huntington's Disease

   *Catherine M. Cowan and Lynn A. Raymond*

3. **RNAi Therapy for Neurodegenerative Diseases**
   *Ryan L. Boudreau and Beverly L. Davidson*

4. **Fibrillins: From Biogenesis of Microfibrils to Signaling Functions**
   *Dirk Hubmacher, Kerstin Tiedemann, and Dieter P. Reinhardt*

5. **Proteasomes from Structure to Function: Perspectives from Archaea**
   *Julie A. Maupin-Furlow, Matthew A. Humbard, P. Aaron Kirkland, Wei Li, Christopher J. Reuter, Amy J. Wright, and G. Zhou*

6. **The Cytomatrix as a Cooperative System of Macromolecular and Water Networks**
   *V. A. Shepherd*

7. **Intracellular Targeting of Phosphodiesterase-4 Underpins Compartmentalized cAMP Signaling**
   *Martin J. Lynch, Elaine V. Hill, and Miles D. Houslay*

## Volume 76

1. **BMP Signaling in the Cartilage Growth Plate**
   *Robert Pogue and Karen Lyons*

2. **The CLIP-170 Orthologue Bik1p and Positioning the Mitotic Spindle in Yeast**
   *Rita K. Miller, Sonia D'Silva, Jeffrey K. Moore, and Holly V. Goodson*

3. **Aggregate-Prone Proteins Are Cleared from the Cytosol by Autophagy: Therapeutic Implications**
   *Andrea Williams, Luca Jahreiss, Sovan Sarkar, Shinji Saiki, Fiona M. Menzies, Brinda Ravikumar, and David C. Rubinsztein*

4. **Wnt Signaling: A Key Regulator of Bone Mass**
   *Roland Baron, Georges Rawadi, and Sergio Roman-Roman*

5. **Eukaryotic DNA Replication in a Chromatin Context**
   *Angel P. Tabancay, Jr. and Susan L. Forsburg*

6. **The Regulatory Network Controlling the Proliferation–Meiotic Entry Decision in the *Caenorhabditis elegans* Germ Line**
   *Dave Hansen and Tim Schedl*

7. **Regulation of Angiogenesis by Hypoxia and Hypoxia-Inducible Factors**
   *Michele M. Hickey and M. Celeste Simon*

## Volume 77

1. **The Role of the Mitochondrion in Sperm Function: Is There a Place for Oxidative Phosphorylation or Is this a Purely Glycolytic Process?**
   Eduardo Ruiz-Pesini, Carmen Díez-Sánchez, Manuel José López-Pérez, and José Antonio Enríquez

2. **The Role of Mitochondrial Function in the Oocyte and Embryo**
   Rémi Dumollard, Michael Duchen, and John Carroll

3. **Mitochondrial DNA in the Oocyte and the Developing Embryo**
   Pascale May-Panloup, Marie-Françoise Chretien, Yves Malthiery, and Pascal Reynier

4. **Mitochondrial DNA and the Mammalian Oocyte**
   Eric A. Shoubridge and Timothy Wai

5. **Mitochondrial Disease—Its Impact, Etiology, and Pathology**
   R. McFarland, R. W. Taylor, and D. M. Turnbull

6. **Cybrid Models of mtDNA Disease and Transmission, from Cells to Mice**
   Ian A. Trounce and Carl A. Pinkert

7. **The Use of Micromanipulation Methods as a Tool to Prevention of Transmission of Mutated Mitochondrial DNA**
   Helena Fulka and Josef Fulka, Jr.

8. **Difficulties and Possible Solutions in the Genetic Management of mtDNA Disease in the Preimplantation Embryo**
   J. Poulton, P. Oakeshott, and S. Kennedy

9. **Impact of Assisted Reproductive Techniques: A Mitochondrial Perspective from the Cytoplasmic Transplantation**
   A. J. Harvey, T. C. Gibson, T. M. Quebedeaux, and C. A. Brenner

10. **Nuclear Transfer: Preservation of a Nuclear Genome at the Expense of Its Associated mtDNA Genome(s)**
    Emma J. Bowles, Keith H. S. Campbell, and Justin C. St. John

## Volume 78

1. **Contribution of Membrane Mucins to Tumor Progression Through Modulation of Cellular Growth Signaling Pathways**
   *Kermit L. Carraway III, Melanie Funes, Heather C. Workman, and Colleen Sweeney*

2. **Regulation of the Epithelial Na$^+$ Channel by Peptidases**
   *Carole Planès and George H. Caughey*

3. **Advances in Defining Regulators of Cementum Development and Periodontal Regeneration**
   *Brian L. Foster, Tracy E. Popowics, Hanson K. Fong, and Martha J. Somerman*

4. **Anabolic Agents and the Bone Morphogenetic Protein Pathway**
   *I. R. Garrett*

5. **The Role of Mammalian Circadian Proteins in Normal Physiology and Genotoxic Stress Responses**
   *Roman V. Kondratov, Victoria Y. Gorbacheva, and Marina P. Antoch*

6. **Autophagy and Cell Death**
   *Devrim Gozuacik and Adi Kimchi*

## Volume 79

1. **The Development of Synovial Joints**
   *I. M. Khan, S. N. Redman, R. Williams, G. P. Dowthwaite, S. F. Oldfield, and C. W. Archer*

2. **Development of a Sexually Differentiated Behavior and Its Underlying CNS Arousal Functions**
   *Lee-Ming Kow, Cristina Florea, Marlene Schwanzel-Fukuda, Nino Devidze, Hosein Kami Kia, Anna Lee, Jin Zhou, David MacLaughlin, Patricia Donahoe, and Donald Pfaff*

3. **Phosphodiesterases Regulate Airway Smooth Muscle Function in Health and Disease**
   *Vera P. Krymskaya and Reynold A. Panettieri, Jr.*

4. Role of Astrocytes in Matching Blood Flow to Neuronal Activity
   *Danica Jakovcevic and David R. Harder*

5. Elastin-Elastases and Inflamm-Aging
   *Frank Antonicelli, Georges Bellon, Laurent Debelle, and William Hornebeck*

6. A Phylogenetic Approach to Mapping Cell Fate
   *Stephen J. Salipante and Marshall S. Horwitz*

## Volume 80

1. Similarities Between Angiogenesis and Neural Development: What Small Animal Models Can Tell Us
   *Serena Zacchigna, Carmen Ruiz de Almodovar, and Peter Carmeliet*

2. Junction Restructuring and Spermatogenesis: The Biology, Regulation, and Implication in Male Contraceptive Development
   *Helen H. N. Yan, Dolores D. Mruk, and C. Yan Cheng*

3. Substrates of the Methionine Sulfoxide Reductase System and Their Physiological Relevance
   *Derek B. Oien and Jackob Moskovitz*

4. Organic Anion-Transporting Polypeptides at the Blood–Brain and Blood–Cerebrospinal Fluid Barriers
   *Daniel E. Westholm, Jon N. Rumbley, David R. Salo, Timothy P. Rich, and Grant W. Anderson*

5. Mechanisms and Evolution of Environmental Responses in *Caenorhabditis elegans*
   *Christian Braendle, Josselin Milloz, and Marie-Anne Félix*

6. Molluscan Shell Proteins: Primary Structure, Origin, and Evolution
   *Frédéric Marin, Gilles Luquet, Benjamin Marie, and Davorin Medakovic*

7. Pathophysiology of the Blood–Brain Barrier: Animal Models and Methods
   *Brian T. Hawkins and Richard D. Egleton*

8. Genetic Manipulation of Megakaryocytes to Study Platelet Function
   *Jun Liu, Jan DeNofrio, Weiping Yuan, Zhengyan Wang, Andrew W. McFadden, and Leslie V. Parise*

9. Genetics and Epigenetics of the Multifunctional Protein CTCF
   *Galina N. Filippova*

## Volume 81

1. **Models of Biological Pattern Formation: From Elementary Steps to the Organization of Embryonic Axes**
   Hans Meinhardt

2. **Robustness of Embryonic Spatial Patterning in *Drosophila Melanogaster***
   David Umulis, Michael B. O'Connor, and Hans G. Othmer

3. **Integrating Morphogenesis with Underlying Mechanics and Cell Biology**
   Lance A. Davidson

4. **The Mechanisms Underlying Primitive Streak Formation in the Chick Embryo**
   Manli Chuai and Cornelis J. Weijer

5. **Grid-Free Models of Multicellular Systems, with an Application to Large-Scale Vortices Accompanying Primitive Streak Formation**
   T. J. Newman

6. **Mathematical Models for Somite Formation**
   Ruth E. Baker, Santiago Schnell, and Philip K. Maini

7. **Coordinated Action of N-CAM, N-cadherin, EphA4, and ephrinB2 Translates Genetic Prepatterns into Structure during Somitogenesis in Chick**
   James A. Glazier, Ying Zhang, Maciej Swat, Benjamin Zaitlen, and Santiago Schnell

8. **Branched Organs: Mechanics of Morphogenesis by Multiple Mechanisms**
   Sharon R. Lubkin

9. **Multicellular Sprouting during Vasculogenesis**
   Andras Czirok, Evan A. Zamir, Andras Szabo, and Charles D. Little

10. **Modelling Lung Branching Morphogenesis**
    Takashi Miura

11. **Multiscale Models for Vertebrate Limb Development**
    Stuart A. Newman, Scott Christley, Tilmann Glimm, H. G. E. Hentschel, Bogdan Kazmierczak, Yong-Tao Zhang, Jianfeng Zhu, and Mark Alber

12. **Tooth Morphogenesis *in vivo*, *in vitro* and *in silico***
    Isaac Salazar-Ciudad
13. **Cell Mechanics with a 3D Kinetic and Dynamic Weighted Delaunay-Triangulation**
    Michael Meyer-Hermann
14. **Cellular Automata as Microscopic Models of Cell Migration in Heterogeneous Environments**
    H. Hatzikirou and A. Deutsch
15. **Multiscale Modeling of Biological Pattern Formation**
    Ramon Grima
16. **Relating Biophysical Properties Across Scales**
    Elijah Flenner, Francoise Marga, Adrian Neagu, Ioan Kosztin, and Gabor Forgacs
17. **Complex Multicellular Systems and Immune Competition: New Paradigms Looking for a Mathematical Theory**
    N. Bellomo and G. Forni

## Volume 82

1. **Ontogeny of Erythropoiesis in the Mammalian Embryo**
   Kathleen McGrath and James Palis
2. **The Erythroblastic Island**
   Deepa Manwani and James J. Bieker
3. **Epigenetic Control of Complex Loci During Erythropoiesis**
   Ryan J. Wozniak and Emery H. Bresnick
4. **The Role of the Epigenetic Signal, DNA Methylation, in Gene Regulation During Erythroid Development**
   Gordon D. Ginder, Merlin N. Gnanapragasam, and Omar Y. Mian
5. **Three-Dimensional Organization of Gene Expression in Erythroid Cells**
   Wouter de Laat, Petra Klous, Jurgen Kooren, Daan Noordermeer, Robert-Jan Palstra, Marieke Simonis, Erik Splinter, and Frank Grosveld
6. **Iron Homeostasis and Erythropoiesis**
   Diedra M. Wrighting and Nancy C. Andrews

7. **Effects of Nitric Oxide on Red Blood Cell Development and Phenotype**
   Vladan P. Čokić and Alan N. Schechter

8. **Diamond Blackfan Anemia: A Disorder of Red Blood Cell Development**
   Steven R. Ellis and Jeffrey M. Lipton

## Volume 83

1. **Somatic Sexual Differentiation in *Caenorhabditis elegans***
   Jennifer Ross Wolff and David Zarkower

2. **Sex Determination in the *Caenorhabditis elegans* Germ Line**
   Ronald E. Ellis

3. **The Creation of Sexual Dimorphism in the *Drosophila* Soma**
   Nicole Camara, Cale Whitworth, and Mark Van Doren

4. ***Drosophila* Germline Sex Determination: Integration of Germline Autonomous Cues and Somatic Signals**
   Leonie U. Hempel, Rasika Kalamegham, John E. Smith III, and Brian Oliver

5. **Sexual Development of the Soma in the Mouse**
   Danielle M. Maatouk and Blanche Capel

6. **Development of Germ Cells in the Mouse**
   Gabriela Durcova-Hills and Blanche Capel

7. **The Neuroendocrine Control of Sex-Specific Behavior in Vertebrates: Lessons from Mammals and Birds**
   Margaret M. McCarthy and Gregory F. Ball

## Volume 84

1. **Modeling Neural Tube Defects in the Mouse**
   Irene E. Zohn and Anjali A. Sarkar

2. **The Etiopathogenesis of Cleft Lip and Cleft Palate: Usefulness and Caveats of Mouse Models**
   Amel Gritli-Linde

3. Murine Models of Holoprosencephaly
   Karen A. Schachter and Robert S. Krauss

4. Mouse Models of Congenital Cardiovascular Disease
   Anne Moon

5. Modeling Ciliopathies: Primary Cilia in Development and Disease
   Robyn J. Quinlan, Jonathan L. Tobin, and Philip L. Beales

6. Mouse Models of Polycystic Kidney Disease
   Patricia D. Wilson

7. Fraying at the Edge: Mouse Models of Diseases Resulting from Defects at the Nuclear Periphery
   Tatiana V. Cohen and Colin L. Stewart

8. Mouse Models for Human Hereditary Deafness
   Michel Leibovici, Saaid Safieddine, and Christine Petit

9. The Value of Mammalian Models for Duchenne Muscular Dystrophy in Developing Therapeutic Strategies
   Glen B. Banks and Jeffrey S. Chamberlain

## Volume 85

1. Basal Bodies: Platforms for Building Cilia
   Wallace F. Marshall

2. Intraflagellar Transport (IFT): Role in Ciliary Assembly, Resorption and Signalling
   Lotte B. Pedersen and Joel L. Rosenbaum

3. How Did the Cilium Evolve?
   Peter Satir, David R. Mitchell, and Gáspár Jékely

4. Ciliary Tubulin and Its Post-Translational Modifications
   Jacek Gaertig and Dorota Wloga

5. Targeting Proteins to the Ciliary Membrane
   Gregory J. Pazour and Robert A. Bloodgood

6. Cilia: Multifunctional Organelles at the Center of Vertebrate Left–Right Asymmetry
   Basudha Basu and Martina Brueckner

Contents of Previous Volumes

7. **Ciliary Function and Wnt Signal Modulation**
   Jantje M. Gerdes and Nicholas Katsanis

8. **Primary Cilia in Planar Cell Polarity Regulation of the Inner Ear**
   Chonnettia Jones and Ping Chen

9. **The Primary Cilium: At the Crossroads of Mammalian Hedgehog Signaling**
   Sunny Y. Wong and Jeremy F. Reiter

10. **The Primary Cilium Coordinates Signaling Pathways in Cell Cycle Control and Migration During Development and Tissue Repair**
    Søren T. Christensen, Stine F. Pedersen, Peter Satir, Iben R. Veland, and Linda Schneider

11. **Cilia Involvement in Patterning and Maintenance of the Skeleton**
    Courtney J. Haycraft and Rosa Serra

12. **Olfactory Cilia: Our Direct Neuronal Connection to the External World**
    Dyke P. McEwen, Paul M. Jenkins, and Jeffrey R. Martens

13. **Ciliary Dysfunction in Developmental Abnormalities and Diseases**
    Neeraj Sharma, Nicolas F. Berbari, and Bradley K. Yoder

## Volume 86

1. **Gene Regulatory Networks in Neural Crest Development and Evolution**
   Natalya Nikitina, Tatjana Sauka-Spengler, and Marianne Bronner-Fraser

2. **Evolution of Vertebrate Cartilage Development**
   GuangJun Zhang, B. Frank Eames, and Martin J. Cohn

3. ***Caenorhabditis* Nematodes as a Model for the Adaptive Evolution of Germ Cells**
   Eric S. Haag

4. **New Model Systems for the Study of Developmental Evolution in Plants**
   Elena M. Kramer

5. **Patterning the Spiralian Embryo: Insights from *Ilyanassa***
   J. David Lambert

6. The Origin and Diversification of Complex Traits Through Micro- and Macroevolution of Development: Insights from Horned Beetles
*Armin P. Moczek*

7. Axis Formation and the Rapid Evolutionary Transformation of Larval Form
*Rudolf A. Raff and Margaret Snoke Smith*

8. Evolution and Development in the Cavefish *Astyanax*
*William R. Jeffery*

## Volume 87

1. Theoretical Models of Neural Circuit Development
*Hugh D. Simpson, Duncan Mortimer, and Geoffrey J. Goodhill*

2. Synapse Formation in Developing Neural Circuits
*Daniel A. Colón-Ramos*

3. The Developmental Integration of Cortical Interneurons into a Functional Network
*Renata Batista-Brito and Gord Fishell*

4. Transcriptional Networks in the Early Development of Sensory–Motor Circuits
*Jeremy S. Dasen*

5. Development of Neural Circuits in the Adult Hippocampus
*Yan Li, Yangling Mu, and Fred H. Gage*

6. Looking Beyond Development: Maintaining Nervous System Architecture
*Claire Bénard and Oliver Hobert*

## Volume 88

1. The Bithorax Complex of *Drosophila*: An Exceptional *Hox* Cluster
*Robert K. Maeda and François Karch*

2. Evolution of the *Hox* Gene Complex from an Evolutionary Ground State
*Walter J. Gehring, Urs Kloter, and Hiroshi Suga*

3. Hox Specificity: Unique Roles for Cofactors and Collaborators
   Richard S. Mann, Katherine M. Lelli, and Rohit Joshi
4. *Hox* Genes and Segmentation of the Vertebrate Hindbrain
   Stefan Tümpel, Leanne M. Wiedemann, and Robb Krumlauf
5. *Hox* Genes in Neural Patterning and Circuit Formation in the Mouse Hindbrain
   Yuichi Narita and Filippo M. Rijli
6. Hox Networks and the Origins of Motor Neuron Diversity
   Jeremy S. Dasen and Thomas M. Jessell
7. Establishment of Hox Vertebral Identities in the Embryonic Spine Precursors
   Tadahiro Iimura, Nicolas Denans, and Olivier Pourquié
8. *Hox*, *Cdx*, and Anteroposterior Patterning in the Mouse Embryo
   Teddy Young and Jacqueline Deschamps
9. *Hox* Genes and Vertebrate Axial Pattern
   Deneen M. Wellik

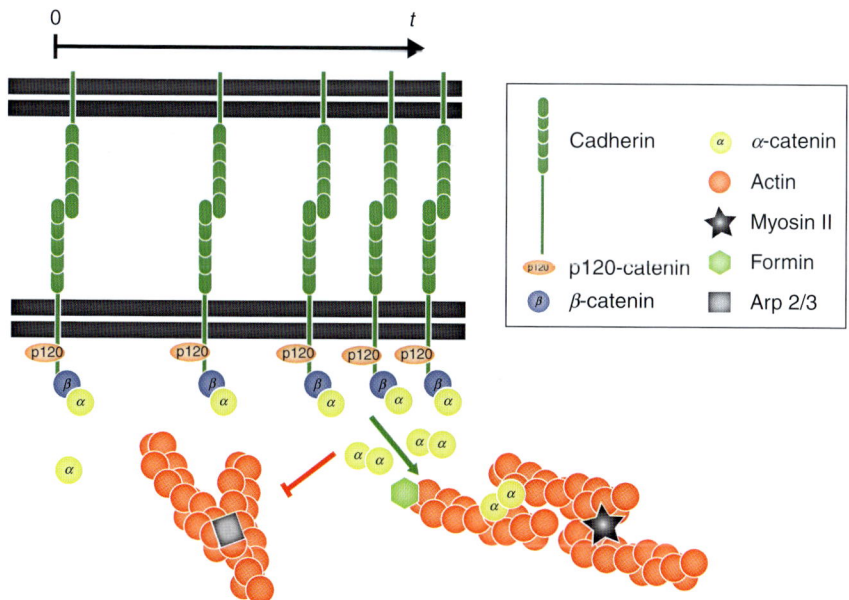

Nicolas Borghi and W. James Nelson, Figure 1.1   Please refer to the legend in the text

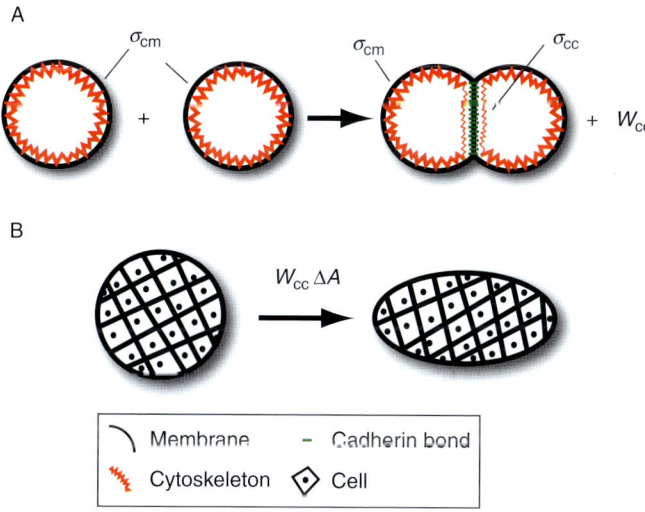

Nicolas Borghi and W. James Nelson, Figure 1.2   Please refer to the legend in the text.

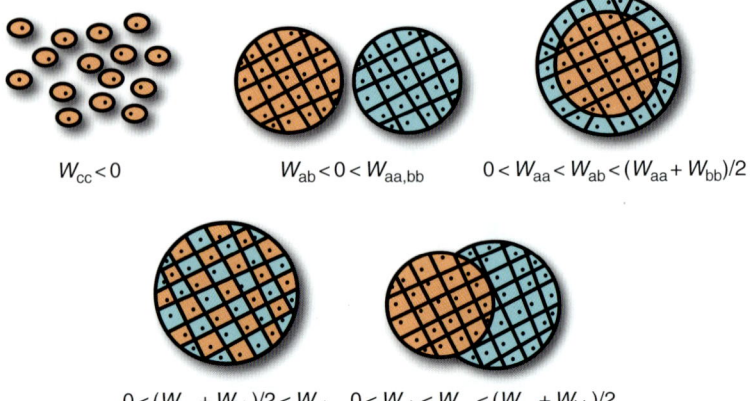

**Nicolas Borghi and W. James Nelson, Figure 1.4**  Please refer to the legend in the text.

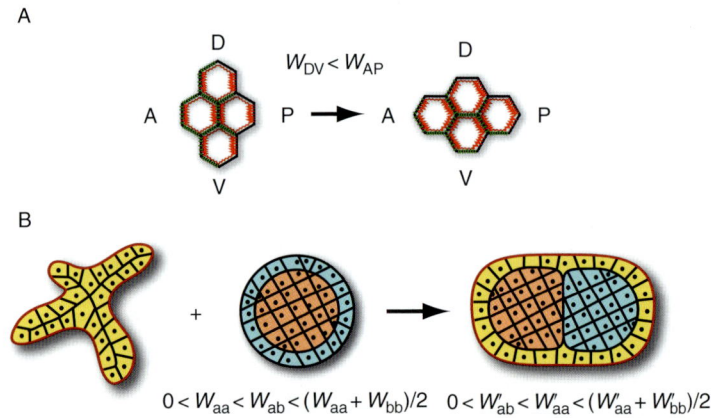

**Nicolas Borghi and W. James Nelson, Figure 1.5**  Please refer to the legend in the text.

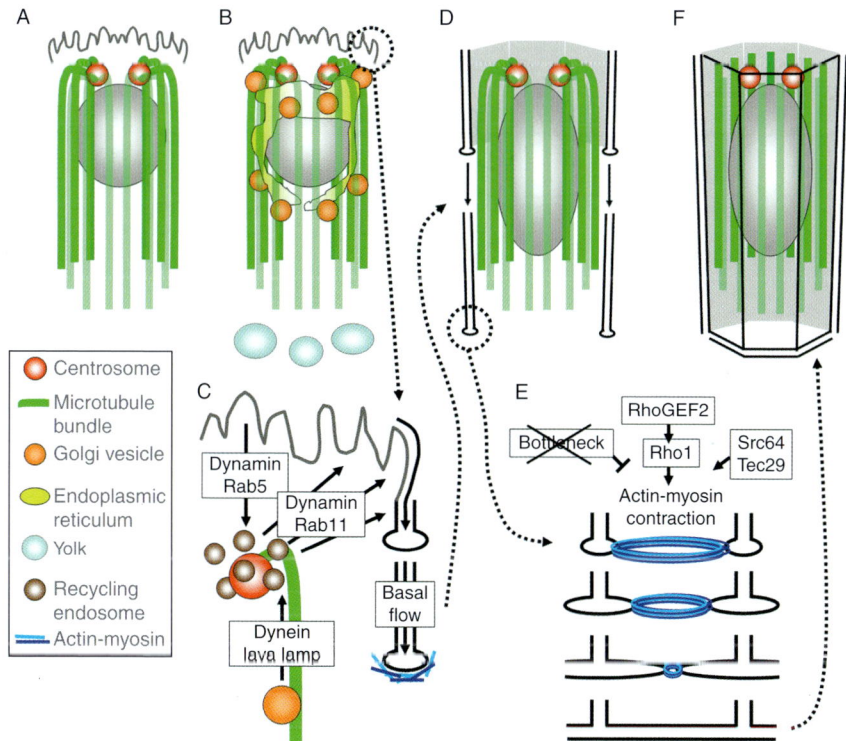

**Tony J. C. Harris et al., Figure 3.1**   Please refer to the legend in the text.

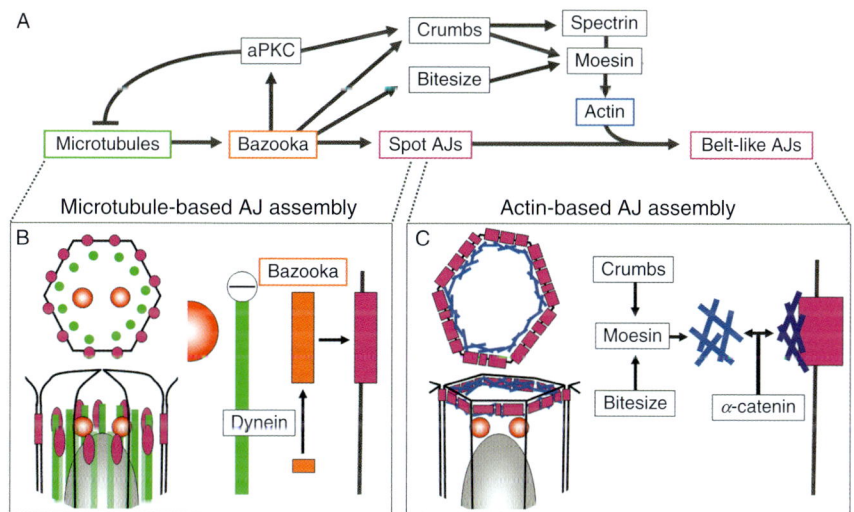

**Tony J. C. Harris et al., Figure 3.2**   Please refer to the legend in the text.

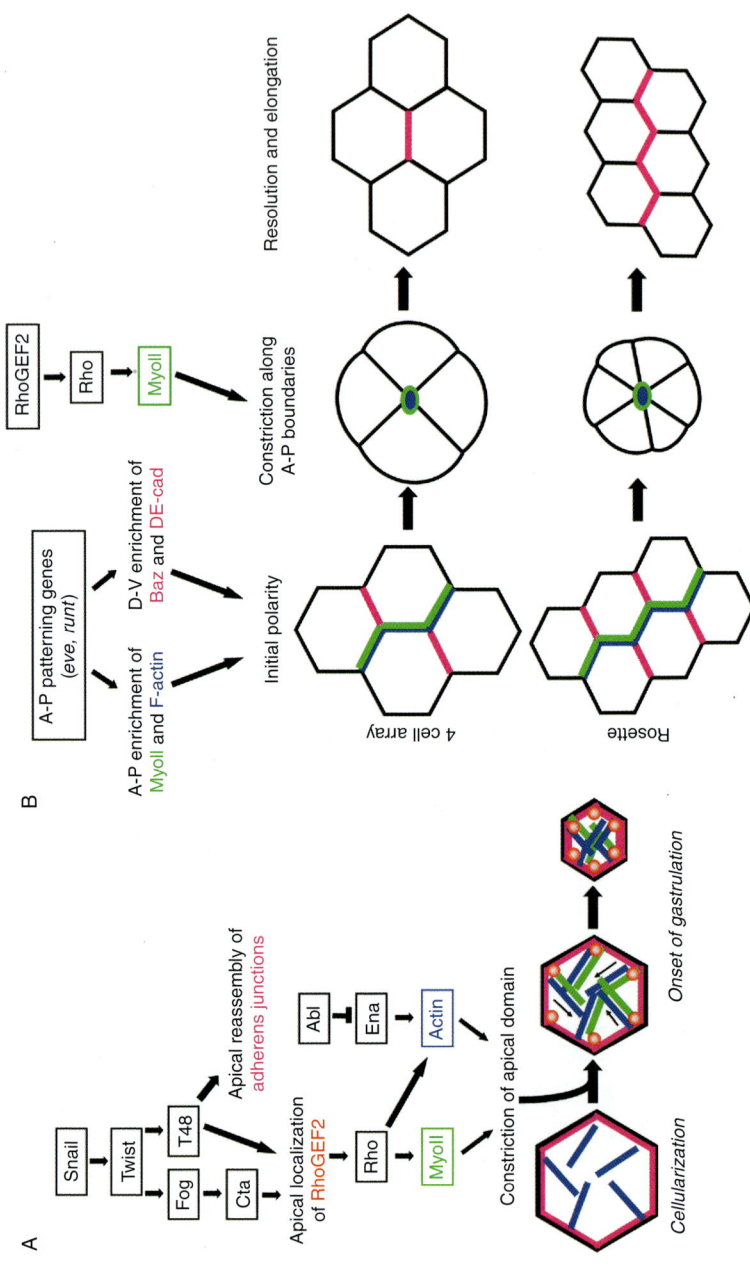

**Tony J. C. Harris *et al*., Figure 3.3** Please refer to the legend in the text.

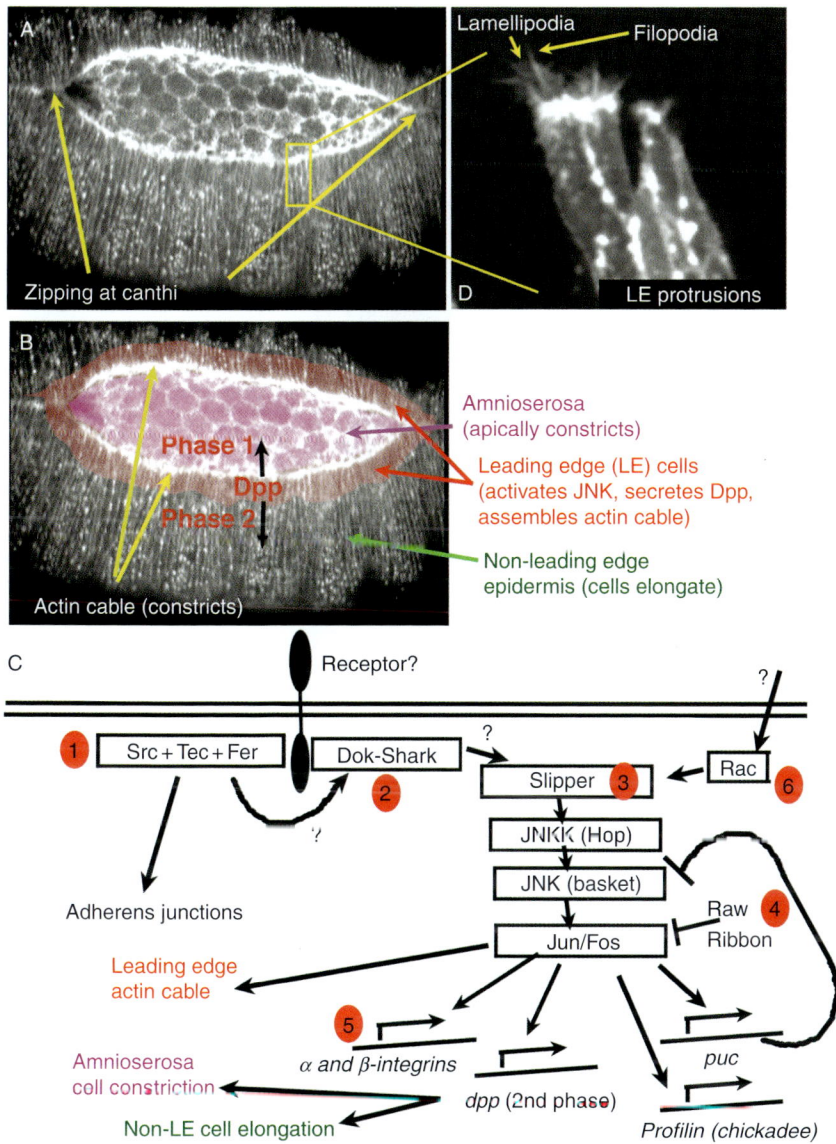

Tony J. C. Harris *et al.*, Figure 3.4  Please refer to the legend in the text.

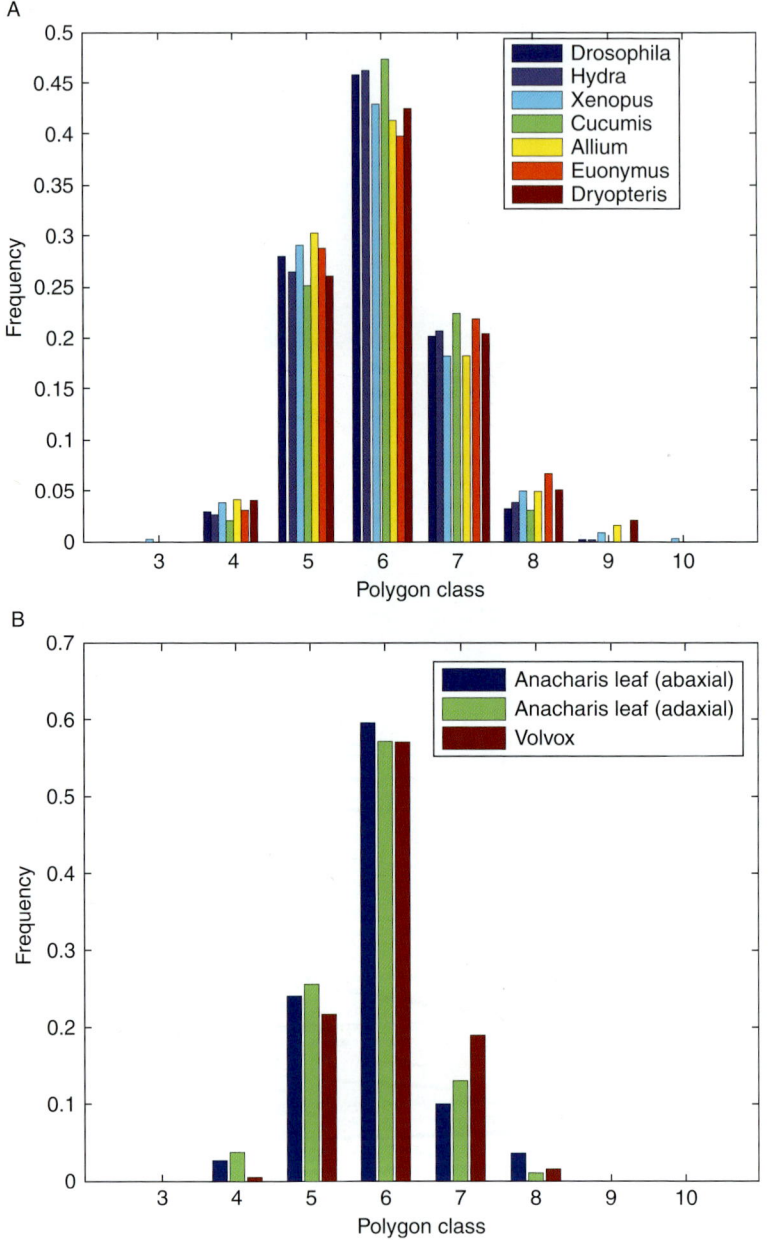

**William T. Gibson and Matthew C. Gibson, Figure 4.2** Please refer to the legend in the text.

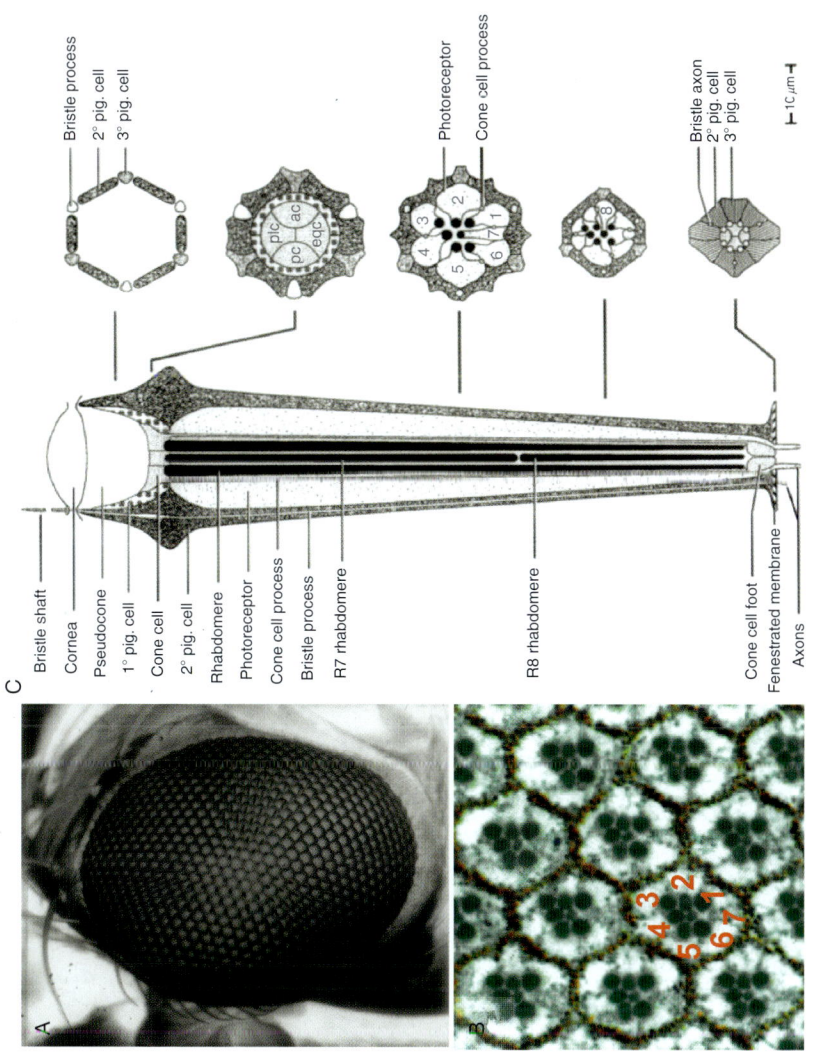

Ross Cagan, **Figure 5.1** Please refer to the legend in the text.

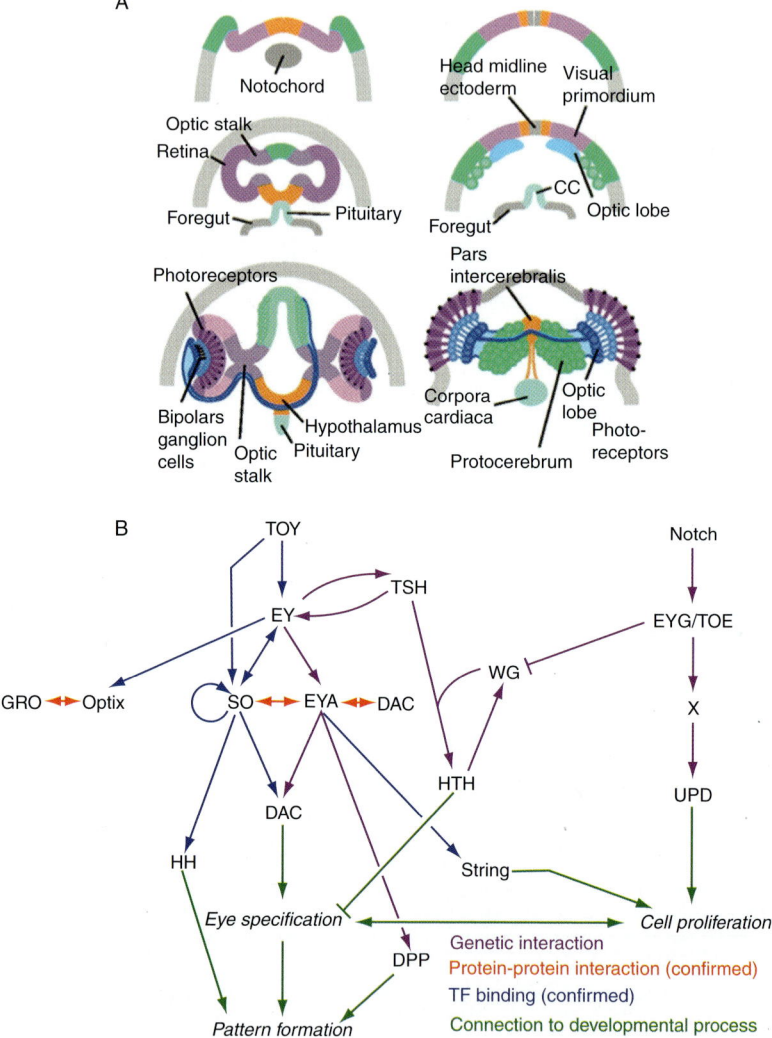

**Ross Cagan, Figure 5.2** Please refer to the legend in the text.

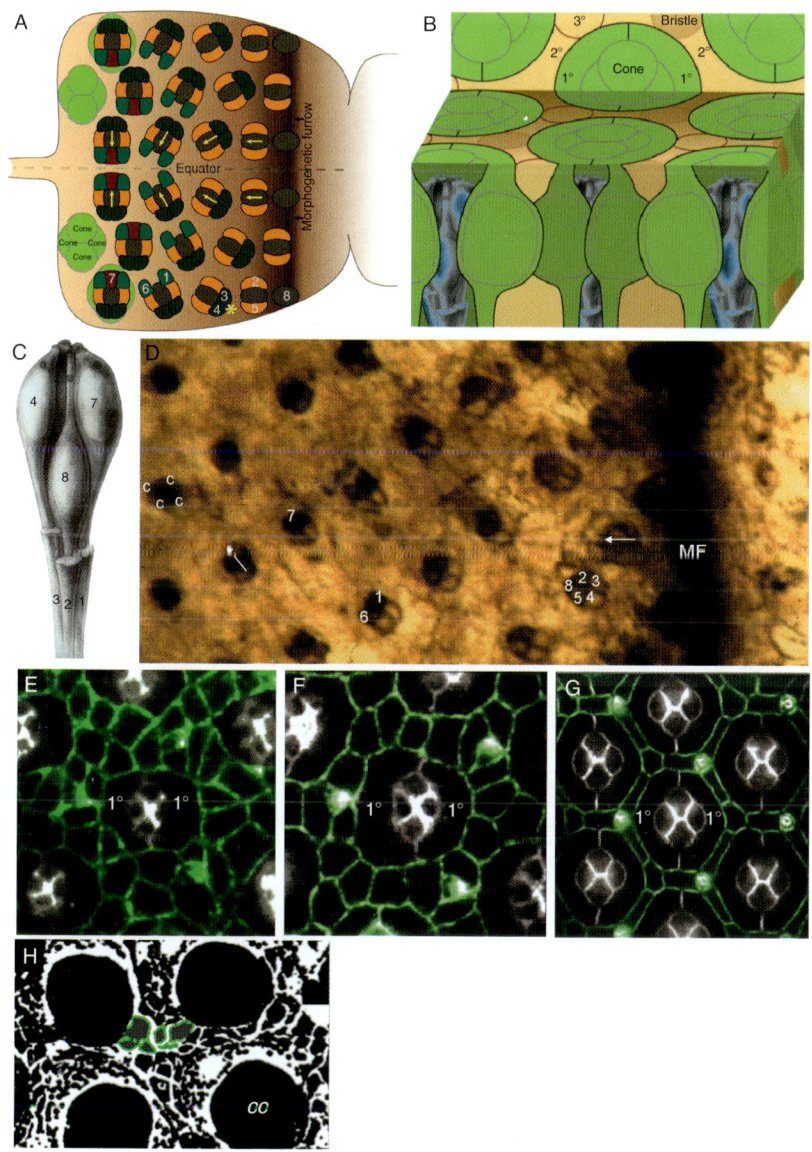

Ross Cagan, Figure 5.3   Please refer to the legend in the text.

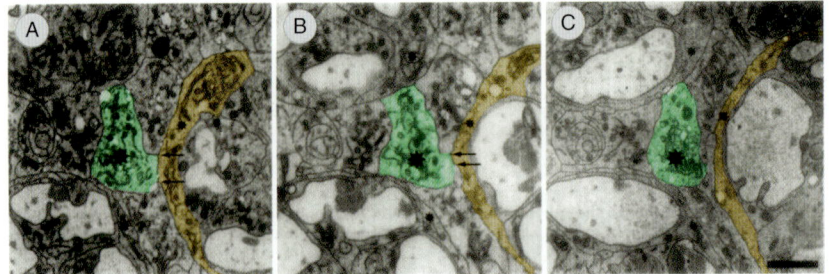

**Ross Cagan, Figure 5.4**  Please refer to the legend in the text.

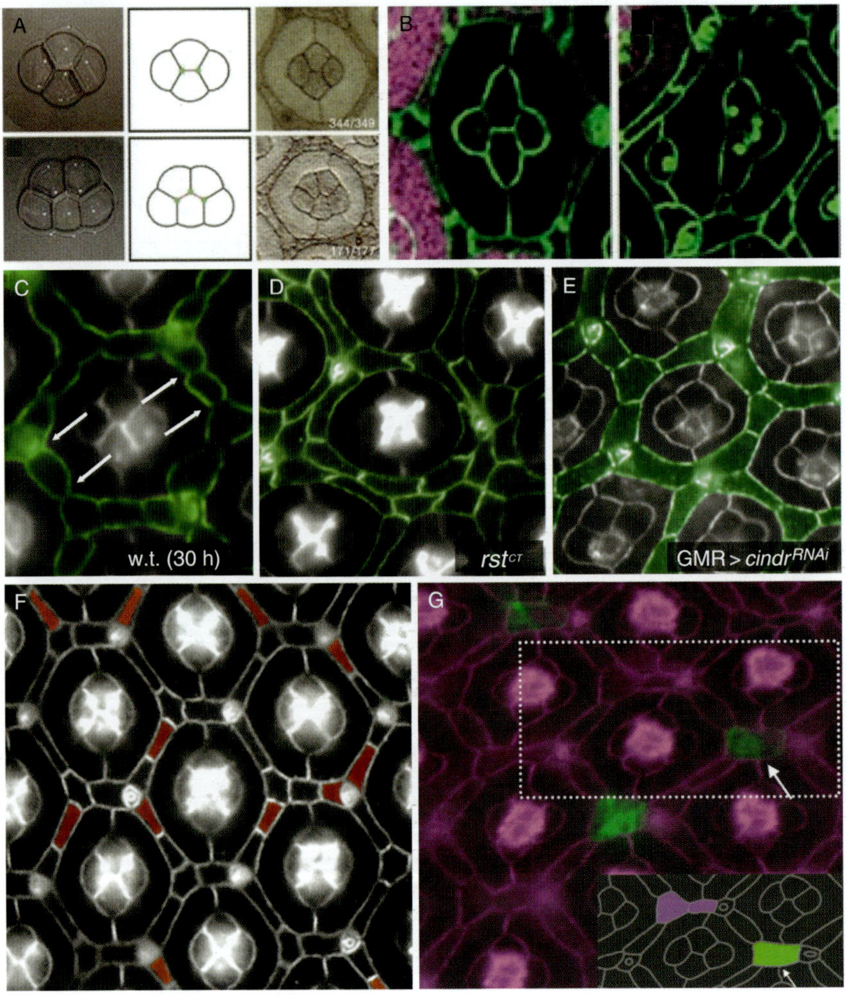

**Ross Cagan, Figure 5.5**  Please refer to the legend in the text.

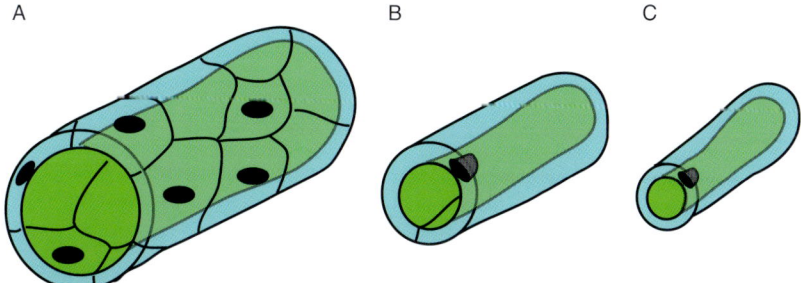

**Magdalena M. Baer *et al.*, Figure 6.1**  Please refer to the legend in the text.

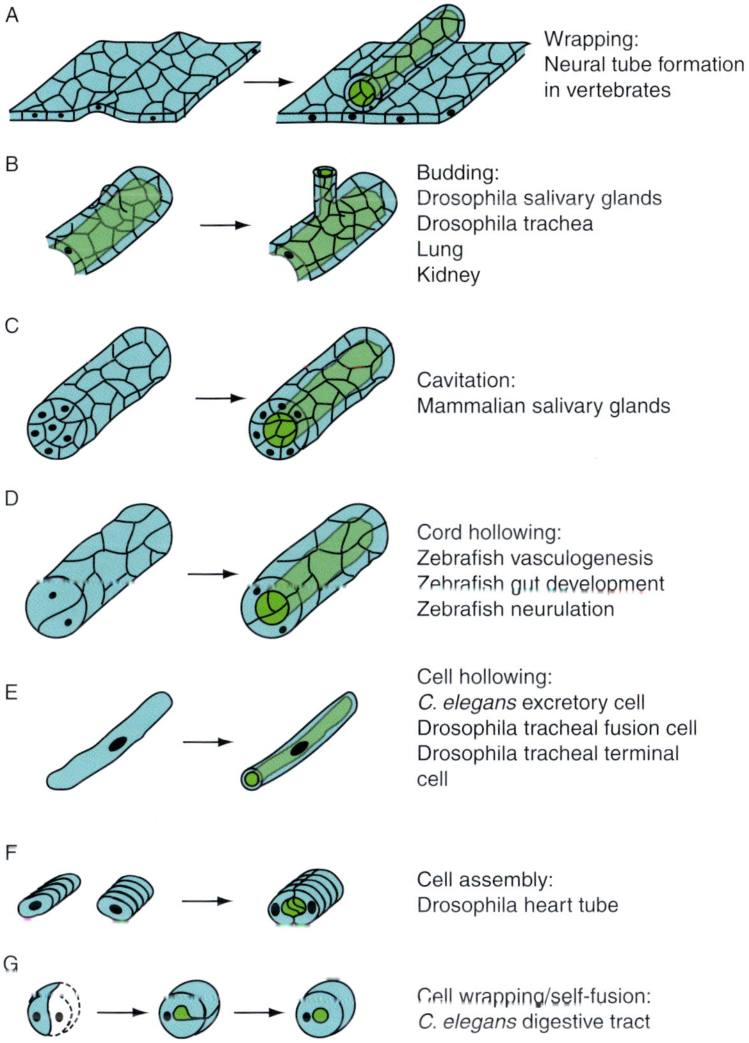

A — Wrapping: Neural tube formation in vertebrates

B — Budding: Drosophila salivary glands, Drosophila trachea, Lung, Kidney

C — Cavitation: Mammalian salivary glands

D — Cord hollowing: Zebrafish vasculogenesis, Zebrafish gut development, Zebrafish neurulation

E — Cell hollowing: *C. elegans* excretory cell, Drosophila tracheal fusion cell, Drosophila tracheal terminal cell

F — Cell assembly: Drosophila heart tube

G — Cell wrapping/self-fusion: *C. elegans* digestive tract

**Magdalena M. Baer *et al.*, Figure 6.2**  Please refer to the legend in the text.

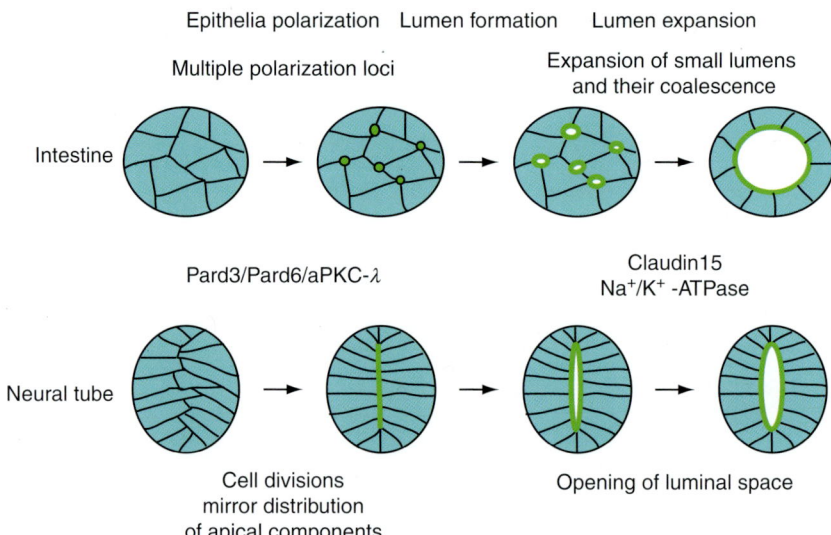

**Magdalena M. Baer *et al.*, Figure 6.3**  Please refer to the legend in the text.

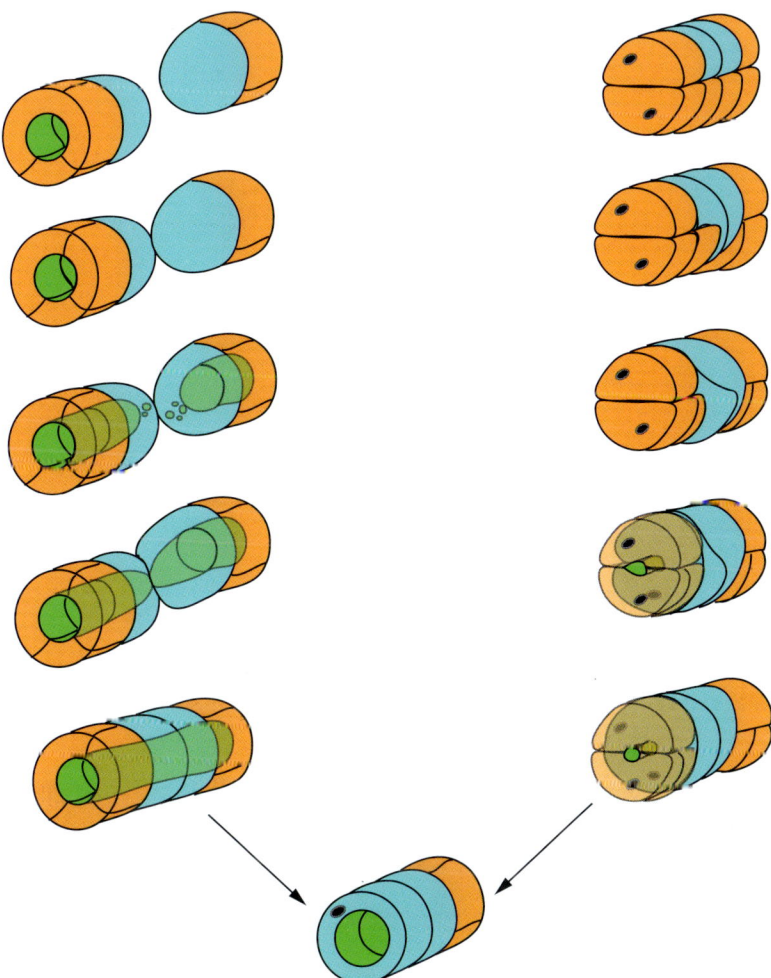

**Magdalena M. Baer *et al.*, Figure 6.6**  Please refer to the legend in the text.

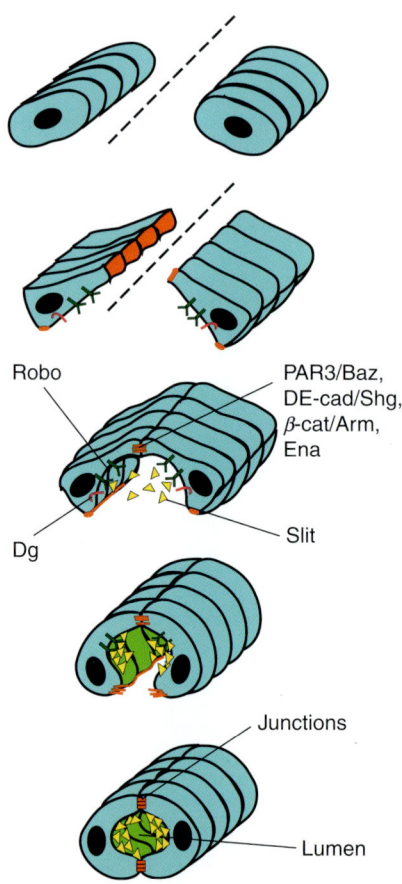

**Magdalena M. Baer *et al.*, Figure 6.7**   Please refer to the legend in the text.

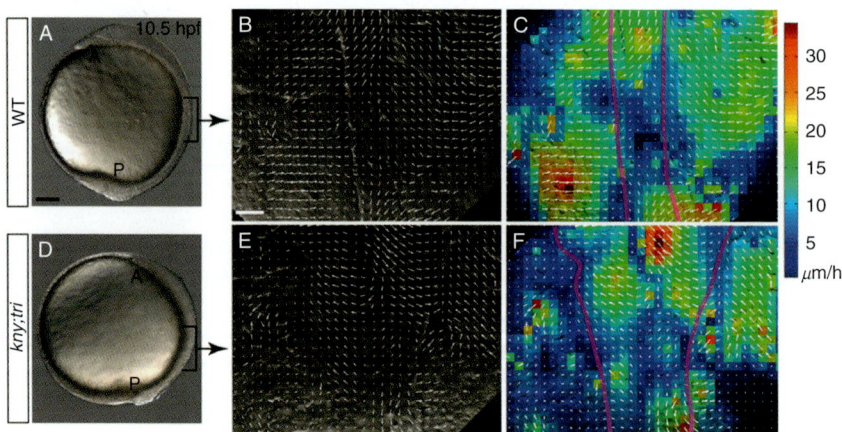

**Chunyue Yin *et al.*, Figure 7.1**   Please refer to the legend in the text.

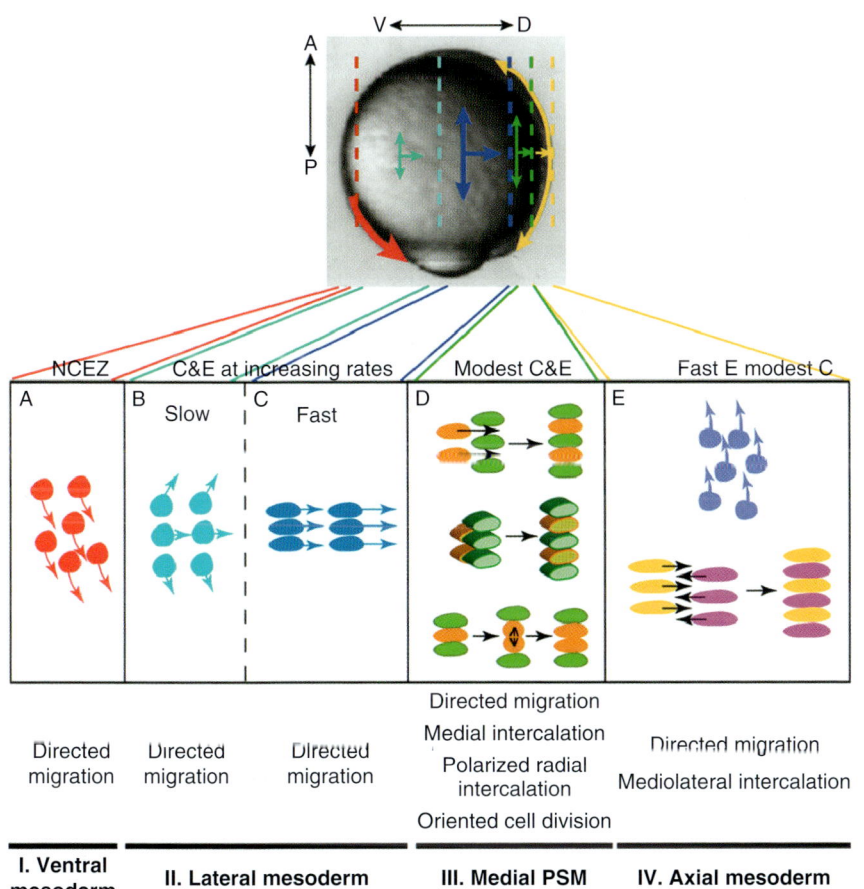

**Chunyue Yin *et al*., Figure 7.2**  Please refer to the legend in the text.

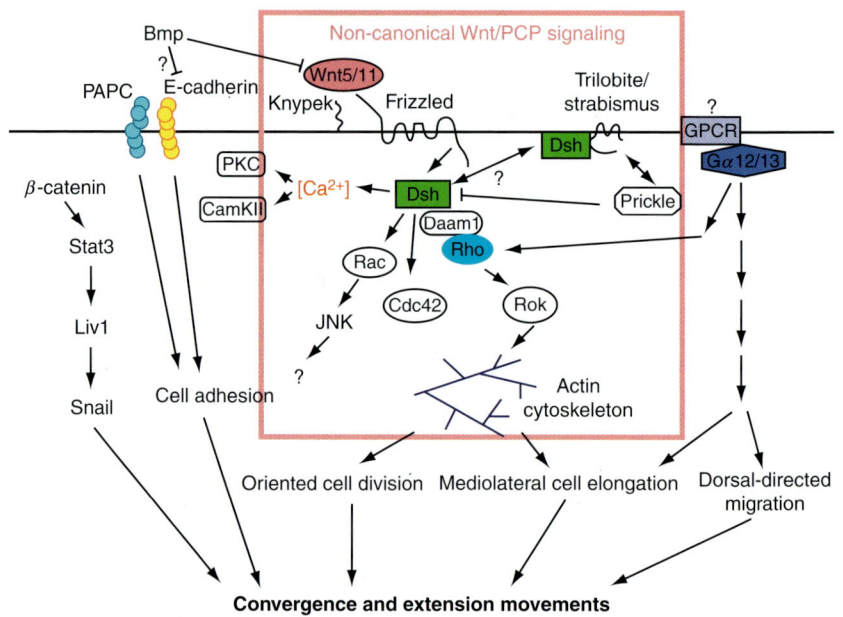

**Chunyue Yin *et al*., Figure 7.3**  Please refer to the legend in the text.